Pounder's
Marine Diesel Engines
Sixth edition

Pounder's Marine Diesel Engines

Sixth edition

C. T. Wilbur, CEng, MIMarE
Editor Marine Propulsion

D. A. Wight, BSc, CEng, MIMechE, FIMarE

Butterworth-Heinemann Ltd
Halley Court, Jordan Hill, Oxford OX2 8EJ

 PART OF REED INTERNATIONAL BOOKS

OXFORD LONDON GUILDFORD BOSTON
MUNICH NEW DELHI SINGAPORE SYDNEY
TOKYO TORONTO WELLINGTON

First published 1984
Reprinted 1991

© Butterworth-Heinemann Ltd 1984

All rights reserved. No part of this publication
may be reproduced in any material form (including
photocopying or storing in any medium by electronic
means and whether or not transiently or incidentally
to some other use of this publication) without the
written permission of the copyright holder except in
accordance with the provisions of the Copyright,
Designs and Patents Act 1988 or under the terms of a
licence issued by the Copyright Licensing Agency Ltd,
33–34 Alfred Place, London, England WC1E 7DP.
Applications for the copyright holder's written permission
to reproduce any part of this publication should be addressed
to the publishers.

British Library Cataloguing in Publication Data
Pounder, C. C.
  Marine diesel engines.—6th edition
  1. Marine diesel engines
  I. Title II. Wilbur, C. III. Wight, D.
  623.8′7236

ISBN 0 7506 0078 0

Library of Congress Cataloguing in Publication Data
Wilbur, C. (Christopher)
  Marine diesel engines
  Rev. ed. of: Marine diesel engines/C. C. Pounder.
  5th ed. 1972
  Bibliography: p.
  Includes index.
  1. Marine diesel motors. I. Wight, D. (Donald)
  II. Pounder, C. Coulson (Cuthbert Coulson) Marine
  diesel engines. II. Title
  VM770.W54 1984 623.8′7236 83-14459

ISBN 0 7506 0078 0

Printed and bound in Great Britain by
Courier International Ltd, Tiptree, Essex

Preface

Since its first publication in 1950 *Marine Diesel Engines* has been edited by the late Mr. C. C. Pounder, formerly of Harland and Wolff. Indeed it is often referred to simply as 'Pounders'.

The time has now come when C. C. Pounder can no longer continue to update the text, that is why the two present joint editors have agreed to undertake the preparation of the 6th Edition.

Perhaps it is a sign of the times that it has been found advisable to have two editors where CCP used to carry the burden alone. At all events the marine diesel engine and its environment have changed more in the last 10 years than in the previous three decades.

Medium-speed engines have, notwithstanding occasional setbacks, made sufficient progress in acceptance to warrant more detailed treatment of the leading makes. The cost of fuel, the nature of its constituents and its growing scarcity have become the dominant consideration of engine design; and the situation has yet to stabilise.

A revolution in the conditions in which sea-going engineers work has largely exchanged watchkeeping for Unmanned Machinery Space (UMS) and this has placed a very high premium on the ability to operate with utter reliability. Technological change has been rapid, to meet ever changing economic constraints, and this has sometimes brought with it fresh problems. By contrast, some of the problems which haunted two generations of earlier engineers are now much better understood and controlled, and some once popular solutions have given way to others.

While we have tried to follow the arrangement of previous editions we have felt it necessary to make some changes to reflect the current relative importance of the subjects.

It is appropriate to extend the treatment of medium-speed engines by giving separate chapters to the most popular makes, and, of course, more emphasis has been given to them in the more general chapters. In particular we have, apart from four chapters concerned with theory or major technical subjects, kept to a format based on practical problems.

The book is aimed at the same readership as formerly. That is to say, it is intended to be useful to the practising marine engineer, particularly in preparing for examinations for Certificate of Competence, but also, for reference, to those who are fully qualified.

As the subject is very large the book does not aim at detailed theoretical treatment (which can be found in textbooks specialising in each topic) nor at duplicating the manufacturers' instruction manuals. In any case it would in practice be more appropriate to refer to the Superintendent's Office or the maker concerned if fine detail is in question, since marine engine design, as well as technology, is constantly changing and it may well be crucial to have the latest information.

Rather it is our purpose to provide a fairly full background to help the marine engineer place the problems he is likely to meet in the correct perspective, and to point out and explain some of these, both the familiar and the less obvious, so that he can more effectively maintain the machinery in his care in a state of readiness or of efficient performance.

C. Wilbur
D. A. Wight

Acknowledgements

Acknowledgement must be made of the help of the manufacturers concerned with the text for those chapters dedicated to specific machinery, and for providing, where appropriate, material for Chapter 23.

Thanks are also due to Dr. A. D. Cameron of Bristol for his valuable advice in the preparation of the section headed Fatigue in Chapter 26, to Mr. A. J. Ellis of Gloucester for advice and much helpful comment on Chapter 5, and to Lucas Bryce Ltd. for permission to use several of the illustrations in Chapters 5 and 26, and to base the sections on medium-speed pumps and injectors in Chapter 25 on their service publications; to Mr. J. T. Hadfield and Mr. C. Hodgson, both of Newton-le-Willows, for help with the section on Diesel–Electric Propulsion in Chapter 13, and with compiling some of the material in Chapter 23; to Mr. D. Woodyard of *Motor Ship* for help in providing essential material for Chapters 13, 22 and 23 including many illustrations; to The General Electric Company for permission to use Figures 26.21 to 26.24 inclusive which were made available by GEC Traction Ltd., and to the GEC *Journal of Science and Technology* for permission to reproduce Figures 5.10, 13.5 and 25.8 which originally appeared in *English Electric Journal*; to *Marine Engineers Review* for permission to reproduce Figures 13.1 and 13.3 as well as three sections of text in Chapter 26 as indicated therein. Figures 23.28 and 23.29 originally appeared in the *Locomotive Cyclopaedia* published by McGraw-Hill and Figure 23.30 in a paper presented at CIMAC 71 in Stockholm. Figure 22.11 appeared in *Diesel Engine and Gas Turbine Catalogue*.

Thanks are also due to all those firms who have facilitated the references to machinery and equipment which appear in the text, as well as posthumously to Mr. C. C. Pounder for those parts of the 5th Edition, which, having stood the passage of time, we do not feel we can improve, principally in Chapters 25 and 26, and in Chapter 27 where we have adapted and expanded text originally provided by Mr. Niels Gram of Paul Bergsøe and Son.

Contents

1 Theory and general principles   1

2 Engine selection   27

3 Performance   36

4 Pressure charging   52

5 Fuel injection   72

6 Sulzer engines   86

7 Burmeister and Wain engines   108

8 MAN engines   138

9 Mitsubishi engines   153

10 GMT engines   164

11 Doxford engines   176

12 Götaverken engines   198

13 Medium-speed engines   214

14 SEMT Pielstick engines   225

15 MAN four-stroke engines   239

16 Stork Werkspoor engines   259

17 Sulzer four-stroke engines   273

18 GMT engines   292

19 MaK engines   310

20 Deutz engines   328

21 Mirrlees Blackstone engines   343

22 The rest of the field   359

23 High-speed engines and auxiliaries   374

24 Fuels and fuel chemistry   409

25 Operation, monitoring and maintenance   422

26 Significant operating problems   487

27 Engine alignment   543

28 Materials   556

   Index   567

1 Theory and general principles

THEORETICAL HEAT CYCLE

In the original patent by Rudolf Diesel, the diesel engine operated on the diesel cycle in which the heat was added at constant pressure. This was achieved by the blast injection principle. Nowadays the term is universally used to describe any reciprocating engine in which the heat induced by compressing air in the cylinders ignites a finely atomised spray of fuel.

This means that the theoretical cycle on which the modern diesel engine works is better represented by the dual or mixed cycle, diagrammatically illustrated in Figure 1.1. The area of the diagram, to a suitable scale, represents the work done on the piston during one cycle.

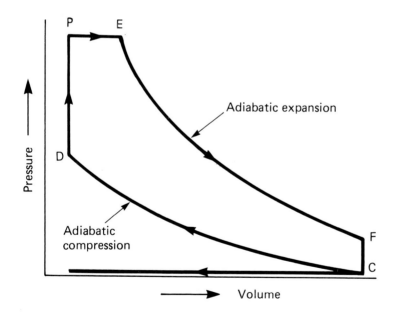

Figure 1.1 Theoretical heat cycle of true Diesel engine

1

Starting from point C, the air is compressed adiabatically to a point D. Fuel injection begins at D, and heat is added to the cycle partly at constant volume as shown by vertical line DP, and partly at constant pressure, as shown by horizontal line PE. At the point E expansion begins. This proceeds adiabatically to point F when the heat is rejected to exhaust at constant volume as shown by vertical line FC.

The ideal efficiency of this cycle (i.e. of the hypothetical indicator diagram) is about 55–60%: that is to say, about 40–45% of the heat supplied is lost to the exhaust. Since the compression and expansion strokes are assumed to be adiabatic, and friction is disregarded, there is no loss to coolant or ambient.

For a four-stroke engine the exhaust and suction strokes are shown by the horizontal line at C, and this has no effect on the cycle.

PRACTICAL CYCLES

While the theoretical cycle facilitates simple calculation, it does not exactly represent the true state of affairs. This is because:

1. The manner in which, and the rate at which, heat is added to the compressed air (the heat release rate) is a complex function of the hydraulics of the fuel injection equipment and the characteristic of its operating mechanism; of the way the spray is atomised and distributed in the combustion space; of the air movement at and after top dead centre (TDC), and to a degree also of the qualities of the fuel.
2. The compression and expansion strokes are not truly adiabatic. Heat is lost to the cylinder walls to an extent which is influenced by the coolant temperature and by the design of the heat paths to the coolant.
3. The exhaust and suction strokes on a four-stroke engine (and the appropriate phases of a two-stroke cycle) do create pressure differences which the crank shaft feels as 'pumping work'.

It is the designer's objective to minimise all these losses without prejudicing first cost or reliability, and also to minimise the cycle loss, that is, the heat rejected to exhaust. It is beyond the scope of this book to derive the formulae used in the theoretical cycle, and in practice designers have at their disposal sophisticated computer techniques which are capable of representing the actual events in the cylinder with a high degree of accuracy. But broadly speaking, the cycle efficiency is a function of the compression ratio (or more correctly the effective expansion ratio of the gas/air mixture after combustion).

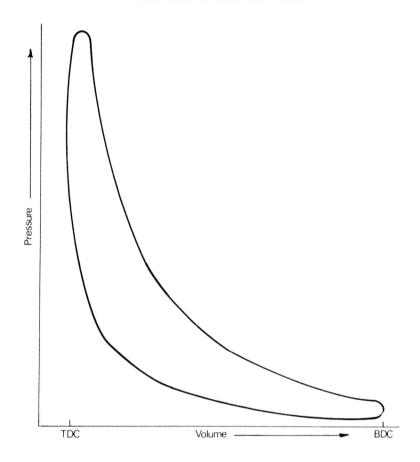

Figure 1.2 Typical indicator diagram (stroke based)

The theoretical cycle (Figure 1.1) may be compared with a typical actual diesel indicator diagram such as that shown in Figure 1.2.

Note that in higher speed engines combustion events are often represented on a crank angle, rather than a stroke basis, in order to achieve better accuracy in portraying events at the top dead centre, as in Figure 1.3. The actual indicator diagram is derived from it by transposition. This form of diagram is useful too when setting injection timing. If electronic indicators are used it is possible to choose either form of diagram.

An approximation to a crank angle based diagram can be made with mechanical indicators by disconnecting the phasing and taking a card quickly, pulling it by hand: this is termed a 'draw card'.

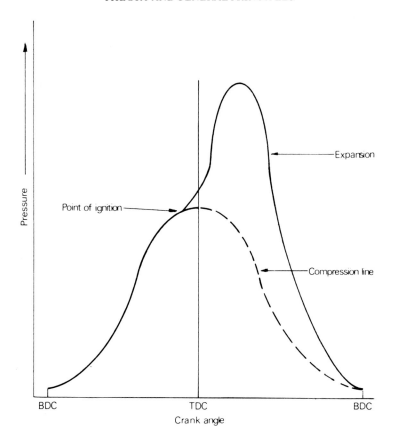

Figure 1.3 Typical indicator diagram (crank angle based)

EFFICIENCY

The only reason a practical engineer wants to run an engine at all is to achieve a desired output of useful work, which is, for our present purposes, to drive a ship at a prescribed speed, and/or to provide electricity at a prescribed kilowattage.

To determine this power he must, therefore, allow not only for the cycle losses mentioned above but for the friction losses in the cylinders, bearings and gearing (if any) together with the power consumed by engine-driven pumps, and other auxiliary machines. He must also allow for such things as windage. The reckoning is further complicated by the fact that the heat rejected from the cylinder to exhaust is not necessarily totally lost, as practically all modern engines use up to 25%

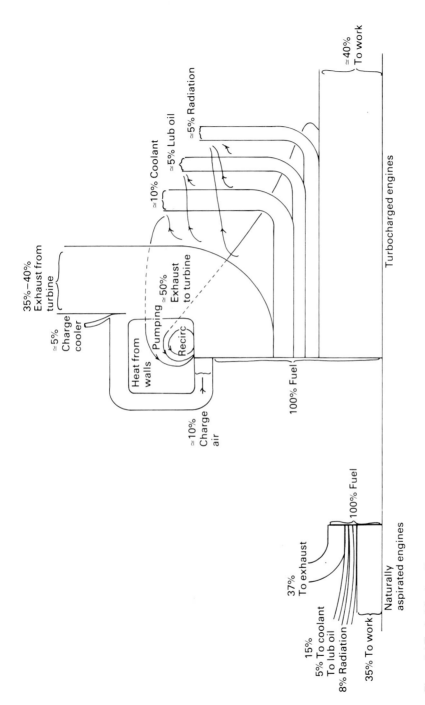

Figure 1.4 Typical Sankey diagrams

of that heat to drive a turbocharger. Many use much of the remaining high temperature heat to raise steam, and low temperature heat for other purposes.

The detail is beyond the scope of this book, but a typical diagram (usually known as a Sankey diagram) representing the various energy flows through a modern diesel engine, is reproduced in Figure 1.4. The right-hand side represents a turbocharged engine and an indication is given of the kind of interaction between the various heat paths as they leave the cylinders after combustion.

Note that the heat released from the fuel in the cylinder is augmented by the heat value of the work done by the turbocharger in compressing the intake air. This is apart from the turbocharger's function in introducing the extra air needed to burn an augmented quantity of fuel in a given cylinder, compared with what the naturally aspirated system could achieve, as in the left-hand side of the diagram.

It is the objective of the marine engineer to keep the injection settings, the air flow, coolant temperatures (not to mention the general mechanical condition) at those values which give the best fuel consumption for the power developed. How he does this is developed in later chapters, particularly in Chapter 25.

Note also that whereas the fuel consumption is not difficult to measure in tonnes per day, kilograms per hour or in other units, there are many difficulties in measuring work done in propelling a ship. This is because the propeller efficiency is influenced by the entry conditions created by the shape of the afterbody of the hull, by cavitation, etc., and also critically influenced by the pitch setting of a controllable pitch propeller. The resulting speed of the ship is much dependent, of course, on hull cleanliness, wind and sea conditions, draught, and so on.

Even when driving a generator it is necessary to allow for generator efficiency and instrument accuracy.

It is normal when talking of efficiency to base the work done on that transmitted to the driven machinery by the crankshaft. In a propulsion system this can be measured by torsionmeter; in a generator it can be measured electrically. Allowing for measurement error, these can be compared with figures measured on a brake in the test shop.

THERMAL EFFICIENCY

Thermal efficiency (Thη) is the overall measure of performance. In absolute terms it is equal to

$$\frac{\text{heat converted into useful work}}{\text{total heat supplied}} \qquad (1.1)$$

As long as the units used agree it does not matter whether the heat or work is expressed in pounds–feet, kilograms–metres, BTU, calories, kWh or joules. The recommended units to use are now those of the SI system.

Heat converted into work per hour $= N\,\text{kWh}$
$$= 3600\,N\,\text{kJ}$$
where $N =$ the power output in kW
Heat supplied $= M \times K$
where $M =$ mass of fuel used per hour in kg
and $K =$ calorific value of the fuel in kJ/kg

$$\text{therefore Th}\eta = \frac{3600\,N}{M \times K} \qquad (1.2)$$

It is now necessary to decide where the work is to be measured. If it is to be measured in the cylinders, as is usually done in slow-running machinery, by means of an indicator (though electronic techniques now make this possible directly and reliably even in high-speed engines), the work measured (and hence power) is that indicated within the cylinder, and the calculation leads to the *indicated* thermal efficiency.

If the work is measured at the crankshaft output flange, it is net of friction, auxiliary drives, etc., and is what would be measured by a brake, whence the term *brake* thermal efficiency. [Manufacturers in some countries do include as output the power absorbed by essential auxiliary drives but the present editors consider this to give a misleading impression of the power available.]

Additionally, the fuel is reckoned to have a higher (or gross) and a lower (or net) calorific value, according to whether one calculates the heat recoverable if the exhaust products are cooled back to standard atmospheric conditions, or assessed at the exhaust outlet. The essential difference is that in the latter case the water produced in combustion is released as steam and retains its latent heat of vaporisation. This is the more representative case — and more desirable as water in the exhaust flow is likely to be corrosive. (See Chapter 26.) Nowadays the net or lower calorific value (LCV) is more widely used.

Returning to our formula (Equation 1.2), if we take the case of an engine producing a (brake) output of 10 000 kW for an hour using 2000 kg of fuel per hour having a LCV of 42 000 kJ/kg

$$(\text{Brake})\,\text{Th}\eta = \frac{3600 \times 10\,000}{2000 \times 42\,000} \times 100\%$$

$$= 42.9\% \text{ (based on LCV)}$$

MECHANICAL EFFICIENCY

$$\text{Mechanical efficiency} = \frac{\text{output at crankshaft}}{\text{output at cylinders}} \qquad (1.3)$$

$$= \frac{\text{bhp}}{\text{ihp}} = \frac{\text{kW (brake)}}{\text{kW (indicated)}} \qquad (1.4)$$

The reasons for the difference are listed above. The brake power is normally measured with a high accuracy (98% or so) by coupling the engine to a dynamometer at the builders' works. If it is measured in the ship by torsionmeter it is difficult to match this accuracy and, if the torsionmeter cannot be installed between the output flange and the thrust block or the gearbox input, additional losses have to be reckoned due to the friction entailed by these components.

The indicated power can only be measured from diagrams where these are feasible and they are also subject to significant measurement errors.

Fortunately for our attempts to reckon the mechanical efficiency, test bed experience shows that the 'friction' torque (that is, in fact, *all* the losses reckoned to influence the difference between indicated and brake torque) is not very greatly affected by the engine's torque output, nor by the speed. This means that the friction power loss is roughly proportional to speed, and fairly constant at fixed speed over the output range. Mechanical efficiency, therefore falls more and more rapidly as brake output falls. It is one of the reasons why it is undesirable to let an engine run for prolonged periods at less than about 30% torque.

WORKING CYCLES

A diesel engine may be designed to work on the two-stroke or on the four-stroke cycle: both of these are explained below. They should not be confused with the terms 'single-acting' or 'double-acting', which relate to whether the working fluid (the combustion gases) acts on one or both sides of the piston. (Note, incidentally, that the opposed piston two-stroke engine is nowadays *single*-acting.)

The four-stroke cycle

Figure 1.5 shows diagrammatically the sequence of events throughout the typical four-stroke cycle of two revolutions. It is usual to draw such diagrams starting at TDC (firing), but the explanation will start at

THEORY AND GENERAL PRINCIPLES 9

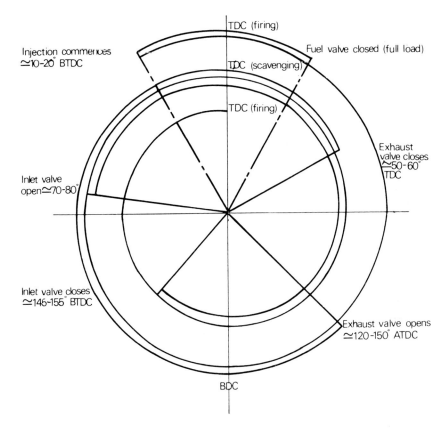

Figure 1.5 Four-stroke cycle

TDC (scavenge). Top dead centre is sometimes referred to as inner dead centre (IDC).

Proceeding clockwise round the diagram, both inlet (or suction) and exhaust valves are initially open. (All modern four-stroke engines have poppet valves.) If the engine is naturally aspirated, or is a small high-speed type with a centripetal turbocharger, the period of valve overlap, i.e. when both valves are open, will be short, and the exhaust valve will close some 10° after top dead centre (ATDC).

Propulsion engines and the vast majority of auxiliary generator engines running at speeds below 1000 rev/min will almost certainly be turbocharged and will be designed to allow a generous throughflow of scavenge air at this point in order to control the turbine blade temperature. (See also Chapter 4.) In this case the exhaust valve will remain open until exhaust valve closure (EVC) at 50–60° ATDC. As the piston descends to outer or bottom dead centre (BDC) on the suction stroke, it will inhale a fresh charge of air. To maximise this, balancing the reduced opening as the valve seats against the slight ram or inertia effect of the incoming charge, the inlet (suction) valve will normally be held open until about 25–35° ABDC (145–155° BTDC). This event is called inlet valve closure (IVC). The charge is then compressed by the rising piston until it has attained a temperature of some 550 °C. At about 10–20° BTDC (firing), depending on the type and speed of the engine, the injector admits finely atomised fuel which ignites within 2–7° (depending on type again) and the fuel burns over a period of 30–50° while the piston begins to descend on the expansion stroke, the piston movement usually helping to induce air movement to assist combustion.

At about 120–150° ATDC the exhaust valve opens (EVO), the timing being chosen to promote a very rapid blow-down of the cylinder gases to exhaust. This is done: (a) to preserve as much energy as is practicable to drive the turbocharger, and (b) to reduce the cylinder pressure to a minimum by BDC to reduce pumping work on the 'exhaust' stroke. The rising piston expels the remaining exhaust gas and at about 70–80° BTDC the inlet valve opens (IVO) so that the inertia of the outflowing gas, plus the positive pressure difference, which usually exists across the cylinder by now, produces a through flow of air to the exhaust to 'scavenge' the cylinder.

If the engine is naturally aspirated the IVO is about 10° BTDC. The cycle now repeats.

The two-stroke cycle

Figure 1.6 shows the sequence of events in a typical two-stroke cycle, which, as the name implies, is accomplished in one complete revolution of the crank. Two-stroke engines invariably have ports to admit air when uncovered by the descending piston (or air piston where the engine has two pistons per cylinder). The exhaust may be via ports adjacent to the air ports and controlled by the same piston (loop scavenge) or via piston controlled exhaust ports or poppet exhaust valves at the other end of the cylinder (uniflow scavenge). The principles of the cycle apply in all cases.

Starting at TDC combustion is already under way and the exhaust opens (EO) at 110–120° ATDC to promote a rapid blow-down before the inlet opens (IO) about 20–30° later (130–150° ATDC). In this way the inertia of the exhaust gases — moving at about the speed of sound — is contrived to encourage the incoming air to flow quickly through the cylinder with a minimum of mixing, because any unexpelled exhaust gas detracts from the weight of air entrained for the next stroke.

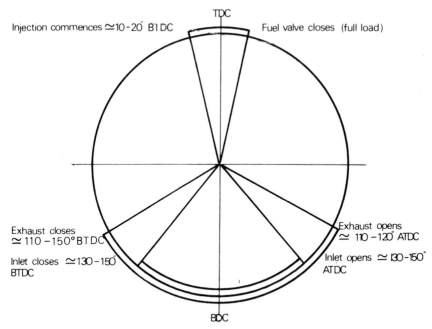

Figure 1.6 Two-stroke cycle

The exhaust should close before the inlet on the compression stroke to maximise the charge, but the geometry of the engine may prevent this if the two events are piston controlled. It can be done in an engine with exhaust valves, but otherwise the inlet and exhaust closure in a single-piston engine will mirror their opening. The inlet closure (IC) may be retarded relative to exhaust closure (EC) in an opposed piston engine to a degree which depends on the ability of the designer and users to accept greater out-of-balance forces.

At all events the inlet ports will be closed as many degrees ABDC as they opened before it (i.e. again 130–150° BTDC) and the exhaust in the same region. Where there are two cranks and they are not in phase, the timing is usually related to that coupled to the piston controlling the air ports. The dephasing is described as 'exhaust lead'.

Injection commences at about 10–20° BTDC depending on speed, and combustion lasts 30–50°, as with the four-stroke.

HORSEPOWER

Despite the introduction of the SI system, in which power is measured in kilowatts (kW), horsepower is too firmly entrenched to be discarded altogether just yet. Power is the rate of doing work. In linear measure it is the mean force acting on a piston multiplied by the distance it moves in a given time. The force here is the mean pressure acting on the piston. This is obtained by averaging the difference in pressure in the cylinder between corresponding points during the compression and expansion strokes. It can be derived by measuring the area of an indicator diagram and dividing it by its length. This gives naturally the indicated mean effective pressure (imep), also known as mean indicated pressure (mip). Let this be denoted by 'p'.

Mean effective pressure is mainly useful as a design shorthand for the severity of the loading imposed on the working parts by combustion. In that context it is usually derived from the horsepower. If the latter is 'brake' horsepower (bhp), the mep derived is the brake mean effective pressure (bmep); but it should be remembered that it then has no direct physical significance of its own.

To obtain the total force the mep must be multiplied by the area on which it acts. This in turn comprises the area of one piston, $a = \pi d^2/4$ multiplied by the number of cylinders in the engine denoted by N.

The distance moved per cycle by the force is the working stroke (l), and for the chosen time unit, the total distance moved is the product of $l \times n$, where n is the number of *working* strokes in one cylinder in the specified time. Gathering all these factors gives the well-known 'plan' formula:

$$\text{power} = \frac{p \times l \times a \times n \times N}{k} \qquad (1.5)$$

The value of the constant k depends on the units used. The units must be consistent as regards force, length and time.

If, for instance, SI units are used (newtons, metres and seconds) k will be 1000 and the power will be given in kW.

If imperial units are used (lb, feet and minutes) k will be 33 000 and the result is in imperial horsepower.

On board ship the marine engineer's interest in the above formula will usually be to relate mep and power for the engine with which he is directly concerned. In that case l, a, n become constants as well as k and

$$\text{power} = p \times C \times N \qquad (1.6)$$

where $C = \dfrac{l \times a \times n}{k}$

Note that in an opposed-piston engine '*l*' totals the sum of the strokes of the two pistons in each cylinder. To apply the formulae to double-acting engines is somewhat more complex since; for instance, allowance must be made for the piston rod diameter. Where double-acting engines are used it would be advisable to seek the builders' advice about the constants to be used.

TORQUE

The formula for power given in Equations (1.5) and (1.6) is based on the movement of the point of application of the force on the piston in a straight line. The inclusion of the engine speed in the formula arises in order to take into account the total distance moved by the force(s) along the cylinder(s), that is, the number of repetitions of the cycle in unit time.

Alternatively, power can be defined in terms of rotation.

If F = the effective resulting single force assumed to act tangentially at given radius from the axis of the shaft about which it acts
r = the nominated radius at which F is reckoned, and
n = revolutions per unit time of the shaft specified

then the circumferential distance moved by the tangential force in unit time is $2\pi rn$.

$$\text{Hence power} = \frac{F \times 2\pi rn}{K}$$

$$= \frac{Fr \times 2\pi n}{K} \qquad (1.7)$$

The value of K depends on the system of units used, which as before, must be consistent. In this expression $F \times r = T$, the torque acting on the shaft, and is measured (in SI units) in newton-metres.

Note that T is constant irrespective of the radius at which it acts. If n is in rev/min, and power is in kW, the constant $K = 1000$ so that

$$T = \frac{\text{power} \times 60 \times 1000}{2\pi n}$$

$$= \frac{30\,000 \times \text{power}}{\pi n} \quad \text{in newton-metres (NM)} \qquad (1.8)$$

If the drive to the propeller is taken through gearing, the torque acts at the pitch circle diameter of each of the meshing gears. If the pitch circle diameters of the input and output gears are respectively d_1 and d_2 and the speeds of the two shafts are n_1 and n_2, the circumferential distance travelled by a tooth on either of these gears must be $\pi d_1 \times n_1$ and $\pi d_2 \times n_2$ respectively. But since they are meshed $\pi d_1 n_1 = \pi d_2 n_2$.

Therefore $\dfrac{n_1}{n_2} = \dfrac{d_2}{d_1}$

The tangential force F on two meshing teeth must also be equal. Therefore the torque on the input wheel

$$T_1 = \frac{Fd_1}{2}$$

and the torque on the output wheel

$$T_2 = \frac{Fd_2}{2}$$

or

$$\frac{T_1}{T_2} = \frac{d_1}{d_2}$$

If there is more than one gear train the same considerations apply at each. In practice there is a small loss of torque at each train due to friction, etc. This is usually of the order of 1.5–2% at each train.

MEAN PISTON SPEED

This parameter is sometimes used as an indication of how highly rated an engine is. However, although in principle a higher piston speed can imply a greater degree of stress, as well as wear, etc., in modern practice the lubrication of piston rings and liner, as well as of other rubbing surfaces, has become much more scientific. It does not any longer follow that a 'high' piston speed is of itself more detrimental than a lower one in a well-designed engine.

Mean piston speed is simply $\dfrac{l \times n}{30}$ \hfill (1.9)

This is given in metres/sec if l = stroke in metres and n = revolutions per minute.

THEORY AND GENERAL PRINCIPLES 15

FUEL CONSUMPTION IN 24 HOURS

In SI units:

$$W = \frac{w \times kW \times 24}{1000} \qquad (1.10)$$

$$w = \frac{1000\,W}{kW \times 24} \qquad (1.11)$$

where w = fuel consumption rate, kg/kW h
$\quad\quad\quad W$ = total fuel consumed per day in tonnes.
 1 tonne = 1000 kg

VIBRATION

Many problems have their roots in, or manifest themselves as, vibration. Vibration may be in any linear direction, and it may be rotational (torsional). Vibration may be resonant, at one of its natural frequencies, or forced. It may affect any group of components, or any one. It can occur at any frequency up to those which are more properly called noise.

That vibration failures are less dramatic now than formerly is due to the advances in our understanding of it during the last 70 years. It can be controlled, once it is recognised, by design and by correct maintenance, by minimising it at source, damping, and by arranging to avoid exciting resonance. Vibration is a very complex subject and all that will be attempted here is a brief outline.

Any elastically coupled shaft or other system will have one or more natural frequencies which, if excited, can build up to an amplitude which is perfectly capable of breaking crankshafts. 'Elastic' in this sense means that a displacement or a twist from rest creates a force or torque tending to return the system to its position of rest, and which is proportional to the displacement. An elastic system, once set in motion in this way, will go on swinging, or vibrating, about its equilibrium position forever, in the theoretical absence of any damping influence. The resulting time/amplitude curve is exactly represented by a sine wave, i.e. it is sinusoidal.

In general, therefore, the frequency of torsional vibration of a single mass will be:

$$f = \frac{1}{2\pi}\sqrt{\frac{q}{I}} \text{ cycles per second} \qquad (1.12)$$

where q is the stiffness in newton metres per radian

and I is the moment of inertia of the attached mass in kg metres^2.

for a transverse or axial vibration

$$f = \frac{1}{2\pi}\sqrt{\frac{s}{m}} \text{ cycles per second} \tag{1.13}$$

where s is the stiffness in newtons per metre of deflection

and m is the mass attached in kg

The essence of control is to adjust these two, q and I, (or s and m) to achieve a frequency which does not coincide with any of the forcing frequencies.

Potentially the most damaging form of vibration is the torsional mode, affecting the crankshaft and propeller shafting (or generator shafting). Consider a typical diesel propulsion system, say a six cylinder two-stroke engine with a flywheel direct coupled to a fixed pitch propeller. There will be as many 'modes' in which the shaft can be induced to vibrate naturally as there are shaft elements: seven in this case. For the sake of simplicity, let us consider the two lowest, the one-node mode and the two-node mode (Figure 1.7 (a) and (b)).

In the one-node case, when the masses forward of the node swing clockwise, those aft of it swing anti-clockwise and vice versa. In the two-node case when those masses forward of the first node swing clockwise, so do those aft of the second node, while those between the two nodes swing anti-clockwise: and vice versa.

The diagrams in Figure 1.7 show (exaggerated) at left: the angular displacements of the masses at maximum amplitude in one direction. At right: they plot the corresponding circumferential deflections from the mean or unstressed condition of the shaft when vibrating in that mode. The line in the right-hand diagrams connecting the maximum amplitudes reached simultaneously by each mass on the shaft system is called the 'elastic curve'.

A node is found where the deflection is zero and the amplitude changes sign. The more nodes that are present the higher the corresponding natural frequency.

The problem arises when the forcing frequencies of the externally applied, or input, vibration coincide with, or approach closely, one of these natural frequencies. A lowish frequency risks exciting the one-node mode; a higher frequency will possibly excite the two-node mode, and so forth. Unfortunately the input frequencies or — to give them their correct name the 'forcing frequencies' — are not simple.

THEORY AND GENERAL PRINCIPLES

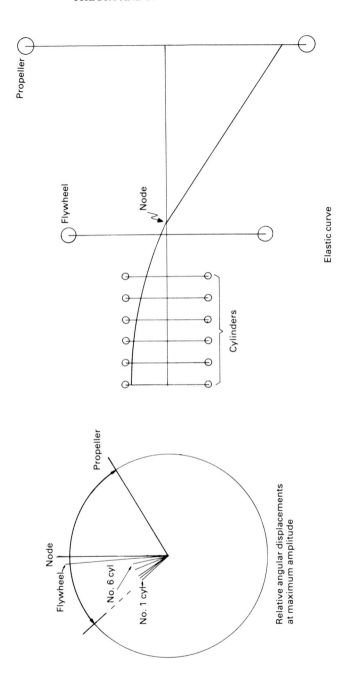

Figure 1.7(a) One-node mode

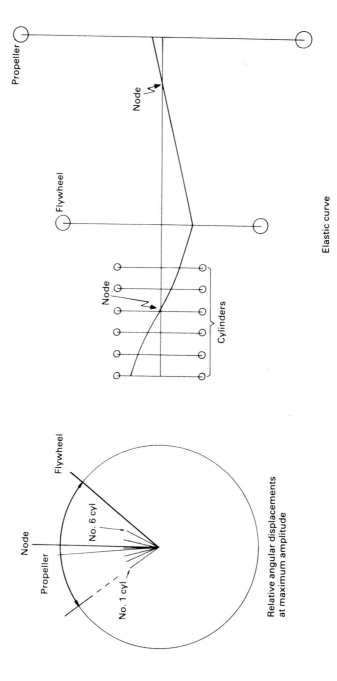

Figure 1.7(b) Two-node mode

THEORY AND GENERAL PRINCIPLES

As far as the crankshaft is concerned, the forcing frequencies are caused by the firing impulses in the cylinders. But the firing impulse put into the crankshaft at any loading by one cylinder firing is not a single sinusoidal frequency at once per cycle. It is a complex waveform which has to be represented for calculation purposes by a component at $1 \times$ cycle frequency; another, usually lower in amplitude, at $2 \times$ cycle frequency; another at $3 \times$ and so on up to at least 10 before the components become small enough to ignore. These components are called the 1st, 2nd, 3rd up to the 10th orders or harmonics of the firing impulse. For four-stroke engines whose cycle speed is half the running speed the convention has been adopted of basing the calculation on running speed. There will therefore be 'half' orders as well, for example, 0.5, 1.5, 2.5, etc.

Unfortunately from the point of view of complexity, but fortunately from the point of view of control, these corresponding impulses have to be combined from all the cylinders according to the firing order. For the 1st order the interval between successive impulses is the same as the crank angle between successive firing impulses. For most engines, therefore, and for our six cylinder engine in particular, the one-node 1st order would tend to cancel out, as shown in the vector summation in the centre of Figure 1.8. The length of each vector shown in the diagram is scaled from the corresponding deflection for that cylinder shown on the elastic curve such as is in Figure 1.7.

On the other hand, in the case of our six cylinder engine, for the 6th order, where the frequency is six times that of the 1st order (or fundamental order) to draw the vector diagram (right of Figure 1.8) all the 1st order phase angles have to be multiplied by 6. Therefore all the cylinder vectors will combine linearly and become much more damaging.

If, say, the natural frequency in the one-node mode is 300 vibrations per minute (vpm) and our six cylinder engine is run at 50 rev/min, the 6th harmonic ($6 \times 50 = 300$) would coincide with the one-node frequency, and the engine would probably suffer major damage. 50 rev/min would be termed the '6th order critical speed' and the 6th order in this case is termed a 'major critical'.

Not only the engine could achieve this. The resistance felt by a propeller blade varies periodically with depth while it rotates in the water, and with the periodic passage of the blade tip past the stern post, or the point of closest proximity to the hull in the case of a multi-screw vessel. If a three-bladed propeller were used and shaft run at 100 rev/min, a 3rd order of propeller excited vibration could also risk damage to the crankshaft (or whichever part of the shaft system was most vulnerable in the one-node mode).

THEORY AND GENERAL PRINCIPLES

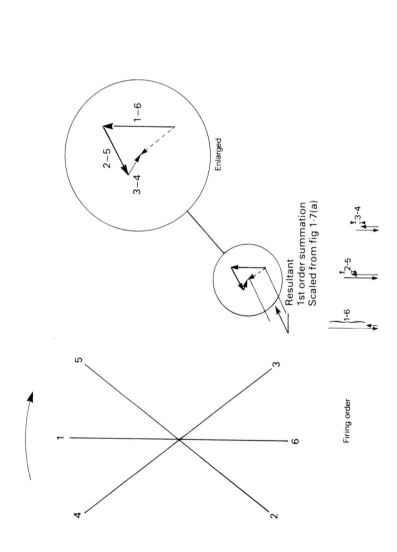

Figure 1.8 Vector summations based on identical behaviour in all cylinders

The most significant masses in any mode of vibration are those with the greatest amplitude on the corresponding elastic curve. That is to say, changing them would have the greatest effect on frequency. The most vulnerable shaft sections are those whose combination of torque and diameter induce in them the greatest stress. The most significant shaft sections are those with the steepest change of amplitude on the elastic curve and therefore the highest torque. These are usually near the nodes but this depends on the relative shaft diameter. Changing the diameter of such a section of shaft will also have a greater effect on the frequency.

The two-node mode is usually of a much higher frequency than the one-node mode in propulsion systems, and in fact usually only the first two or three modes are significant. That is to say that beyond the three-node mode the frequency components of the firing impulse that could resonate in the running speed range will be small enough to ignore.

The Classification Society chosen by the owners will invariably make their own assessment of the conditions presented by the vessel's machinery, and will judge by criteria based on their experience.

Designers can nowadays adjust the frequency of resonance, the forcing impulses and the resultant stresses by adjusting shaft sizes, number of propeller blades, crankshaft balance weights and firing orders, as well as by using viscous or other dampers, detuning couplings and so on. Gearing, of course, creates further complications — and possibilities.

Branched systems, involving twin input or multiple output gearboxes, introduce complications in solving them; but the principles remain the same.

The marine engineer needs to be aware however, that designers tend to rely on reasonably correct balance among cylinders. It is important to realise that an engine with one cylinder cut out for any reason, or one with a serious imbalance between cylinder loads or timings, may inadvertently be aggravating a summation of vectors which the designer, expecting it to be small, had allowed to remain near the running speed range.

If an engine were run at or near a major critical speed it would sound rough, because, at mid-stroke, the torsional oscillation of the cranks with the biggest amplitude would cause a longitudinal vibration of the connecting rod. This would set up in turn a lateral vibration of the piston and hence of the entablature. Gearing, if on a shaft section with a high amplitude, would also probably be distinctly noisy.

The remedy, if the engine appears to be running on a torsional critical speed, would be to run at a different and quieter speed while an investigation is made. Unfortunately noise is not always distinct enough to be relied upon as a warning.

It is usually difficult, and sometimes impossible, to control all the possible criticals, so that in a variable speed propulsion engine it is sometimes necessary to 'bar' a range of speeds where vibration is considered too dangerous for continuous operation.

Torsional vibrations can sometimes affect camshafts also. Linear vibrations usually have simpler modes, except for those which are known as axial vibrations of the crankshaft. These arise because firing causes the crank pin to deflect and this causes the crankwebs to bend. This in turn leads to the setting up a complex pattern of axial vibration of the journals in the main bearings.

Vibration of smaller items, such as brackets holding components, or pipework, can often be controlled either by using a very soft mounting whose natural frequency is *below* that of the lowest exciting frequency, or by stiffening. These matters will be discussed further in Chapter 26.

BALANCING

The reciprocating motion of the piston in an engine cylinder creates out-of-balance forces acting along the cylinder, while the centrifugal force associated with the crankpin rotating about the main bearing centres creates a rotating out-of-balance force. These forces, if not in themselves necessarily damaging, create objectionable vibration and noise in the engine foundations, and through them to the ship (or building) in which the engine is operating.

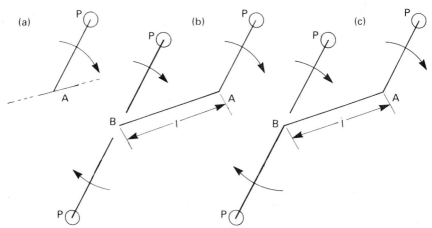

Figure 1.9 Principle of balancing
    (a) Force (P) to be balanced acting at A
    (b) Equal and opposite forces, of magnitude (P) assumed to act at B, a distance.
        (1) arbitrarily chosen to suit calculation
    (c) System is equivalent to couple (P × 1) plus force (P) now acting at B.

Balancing is a way of controlling vibrations by arranging that the overall summation of the out-of-balance forces and couples cancels out, or is reduced to a more acceptable amount.

The disturbing elements are in each case forces, each of which acts in its own plane, usually including the cylinder axis. The essence of balancing is that a force can be exactly replaced by a parallel force acting in a reference plane (chosen to suit the calculation) and a couple whose arm is the distance perpendicularly between the planes in which these two forces act (Figure 1.9). Inasmuch as balance is usually considered at and about convenient reference planes, all balancing involves a consideration of forces and of couples.

In multi-cylinder engines couples are present because the cylinder(s) coupled to each crankthrow act(s) in a different plane.

There are two groups of forces and couples. These relate to the revolving and to the reciprocating masses. (This section of the book is not concerned with balancing the power outputs of the cylinders. In fact, cylinder output balance, while it affects vibration levels, has no effect on the balance we are about to discuss.)

Revolving masses are concentrated at a radius from the crankshaft, usually at the crankpin, but are presumed to include a proportion of the connecting rod shank mass adjacent to the crankpin. This is done to simplify the calculations. Designers can usually obtain rotating balance quite easily by choice of crank sequence and balance weights.

Reciprocating masses are concentrated at the piston/crosshead, and similarly are usually assumed to include the rest of the connecting rod shank mass.

Revolving masses turning at a uniform speed give rise to a single system of revolving forces and couples, each system comprising the related effects of each crank and of any balance weights in their correct phase relationship.

Reciprocating masses are more complicated, because of the effect of the obliquity of the connecting rod. This factor, particularly as the connecting rod length is reduced in relation to crank radius, causes the motion of the piston to depart from the simple sinusoidal time/displacement pattern along the cylinder axis. It has instead to be represented by a primary (sinusoidal) component at a frequency corresponding to engine speed in rev/min, a secondary component at 2 × engine speed, a tertiary at 3 × engine speed, and so on. (It is seldom necessary to go beyond secondary.)

In practical terms, the reversal of direction involves greater acceleration at TDC than at BDC. This can be thought of as an average reversal force (the primary) at both dead centres, on which is superimposed an extra *inward* force at twice the frequency which adds to the primary at

TDC and detracts from it at BDC. Reciprocating forces and couples always act in the plane of the cylinder. For Vee engines they are combined vectorially. (The effect of side-by-side design for Vee engine big ends is usually ignored.)

Most cylinder combinations of two- and of four-stroke engines lend themselves to complete balance of either primary forces, primary couples, secondary forces or secondary couples; some to any two, some to any three, and some to all of them.

Where one or more systems do not balance, the designer can contain the resulting vibration by low frequency anti-vibration mountings (as in the four-stroke four cylinder automotive engine which normally has severe unbalanced secondary forces) or he can ameliorate it by adding extra rotational balance weights on a different shaft or shafts running at crankshaft speed (for primary unbalance) or at twice it (for secondary unbalance).

Note that a reciprocating unbalance force can be cancelled out if two shafts, each carrying a balance weight equal to half the offending unbalanced weight, and phased suitably, are mounted in the frame, running in opposite directions. The lateral effect of these two contra-rotating weights cancel out. This is the principle of Lanchester balancing used to cope with the heavy unbalanced secondary forces of four in-line and eight Vee four-stroke engines with uniform crank sequence.

In general it is nowadays dangerous to consider only the crankshaft system as a whole, because of the major effect that line-by-line balance has on oil film thickness in bearings. It is often necessary to balance internal couples much more closely, i.e. for each half of the engine or even line-by-line. These factors will be discussed further in Chapter 26.

NOISE

Noise is unwanted sound. It is also a pervading nuisance and a hazard to hearing, if not to health itself. Noise is basically a form of vibration, and it is a complex subject; but a few brief notes will be given here.

To the scientist noise is reckoned according to its sound pressure level, which can be measured fairly easily on a microphone. Since the topic is essentially subjective, that is to say, it is of interest because of its effect on people via the human ear, the mere measurement of sound pressure level is of very limited use.

Noises are calibrated with reference to a sound pressure level of 0.0002 dynes per cm^2 for a pure 1000 Hz sine wave. A dyne is the force that gives 1 g an acceleration of 1 cm/s^2. It is 10^{-5} newtons. A linear scale gives misleading comparisons, so noise levels are measured on a logarithmic scale of bels. (1 bel means 10 times the reference level, 3 bel means 1000 times the reference level, i.e. 10^3.) For convenience the bel is divided into 10 parts, hence the decibel or db. This means that 80 db (8 bel) is 10^8 times the reference pressure level.

The human ear hears high and lower frequencies (over say 1 kHz or below 50 Hz) as louder than the middle frequencies, even when sound pressure levels are the same. For practical purposes, sounds are measured according to a frequency scale weighted to correspond to the response of the human ear. This is the 'A' scale and the readings are quoted in dBA.

The logarithmic-based scale is sometimes found confusing, so it is worth remembering that:

30 db corresponds to a gentle breeze in a meadow
70 db to an open office
100 db to a generator room

Any prolonged exposure to levels of 85 db or above is likely to lead to hearing loss in the absence of ear protection. 140 db or above is likely to be physically painful.

The scope for reducing at source noise emitted by a diesel engine is limited, without fundamental and very expensive changes of design principle. However, the measures which lead to efficiency and economy — higher pressures, faster pressure rise and higher speeds — all, sadly, lead also to greater noise, and to noises emitted at higher frequencies and therefore more objectionably.

The noise from several point sources is also added in the logarithmic scale. Two point sources (e.g. cylinders) are twice as noisy as one. $\log_{10} 2 = 0.3010$, i.e. 0.301 bel or 3.01 db louder than one. Six cylinders, since $\log_{10} 6 = 0.788$, are 7.88 db louder than one.

If there is no echo, sound diminishes according to the square of the distance from the source (usually measured at a reference distance of 0.7 metres in the case of diesel engines). So moving ten times further away reduces the noise to 1/100th, or by 20 db.

Unfortunately a ship's engine room has plenty of reflecting surfaces, so that there is little benefit to be had from moving away. The whole space tends to fill with noise only a few db less than the source. Lining as much as possible with sound-absorbing materials does two things:

1. It reduces the echo, so that moving away from the engine gives a greater reduction in perceived noise.
2. It tends to reduce sympathetic (resonant) vibration of parts of the ship's structure, which, by drumming, add to the vibration and noise which is transmitted into the rest of the ship.

Anti-vibration mounts help in the latter case also but are not always practical. The only other measure which can successfully reduce noise is to put weight, particularly if a suitable cavity can be incorporated (or a vacuum), between the source and the observer.

A screen, of almost any material, weighing 5 kg/m^2 will effect a reduction of 10 dB in perceived noise, and pro rata. Where weight is increased by increasing thickness it is more effective to do so in porous/flexible material than in rigid material. For the latter the dB reduction in transmitted noise is proportional to the $\log_{10}$ of the weight, but for the former it is proportional to the thickness. A tenfold increase in the thickness of the rigid material of 5 kg/m^2 would double the attenuation; in porous material it would be tenfold. The screen must be totally effective. A relatively small aperture will destroy much of the benefit. For example, most ships' engine rooms now incorporate a control room to provide some noise protection. The last point will be appreciated every time the door is opened.

D.A.W.

2 Engine selection

The choice of a main propulsion engine for a motorship is by no means an easy one. A few years ago a shipowner had the straight choice of a direct coupled slow-speed engine driving a fixed pitch propeller or a geared four-stroke, medium-speed engine driving either a fixed or controllable pitch propeller. Today, vessels are entering service with geared and direct coupled two-stroke engines driving either fixed or controllable pitch propellers and geared four-strokes; while for certain ships, particularly those involved with offshore oil exploration and production, diesel–electric power plants are becoming increasingly popular. The choice of either direct or indirect drive of a ship is governed more by the operating profile of the ship and economic factors concerning the power plant as a whole, rather than the characteristics of a particular make or type of diesel main propulsion engine.

There are, of course, many shipowners who may prefer to remain with a particular make or type of engine for various reasons such as reliable past operating experience, crew familiarity, spares control, good service back-up, and so on; but new designs of engine are rapidly brought out these days because of intense competition between builders. In many cases an owner who has not built a new vessel for several years will have to choose between a number of engines and possible arrangements with which he is unfamiliar. Much decision making about the choice of engine is governed by cost considerations regarding type of fuel burnt, maintenance costs, manning levels, availability of spares and — not least — the initial purchase price of the engine. There is, however, more of a tendency to consider total life cycle costs rather than simply the purchase price of the main engine. These total purchase and operating costs over, say, 20 years, can be greatly influenced by the initial choice of main engine.

The main factors which influence the choice of engine are:

1. Ability to burn heavy fuel of poor quality without detrimental effect on the engine components and maintenance cost.
2. The maintenance work load, i.e., the number of cylinders, valves, etc., requiring periodic attention related to the number of crew

carried, which is far less nowadays than, say, 10 years ago.
3. Suitability for unattended operation through application of automatic controls and monitoring systems.
4. Propulsive efficiency, i.e. an engine or propeller shaft turning at a low enough speed to drive the largest diameter and hence most efficient propeller.
5. The size and weight of the propelling machinery.
6. The cost of the engine itself.

As the size of the machinery space is largely governed by the size of the main engine, too large an engine room may penalise the cargo-carrying capacity of the vessel. Headroom available is also important in such ships as ferries with vehicle decks, and insufficient headroom can make the overhaul of some engines difficult if not impossible.

DIRECT DRIVE

The direct drive of a larger ship's propeller by a slow-speed two-stroke engine and a small ship's by a four-stroke, medium-speed engine still remain the most popular methods of marine propulsion. At one time a slight loss of propulsive efficiency was accepted for the sake of simplicity, but the introduction of 'long stroke' and very recently 'super long stroke' crosshead engines has done much to reduce propulsive losses. For a large ship a direct coupled speed of, say, 110 rev/min is not necessarily the most suitable as it is proven that a larger propeller turning at speeds even as low as 60 rev/min is more efficient than one of smaller diameter absorbing the same horsepower at around 110 rev/min. The long and super long stroke engines now on the market develop their rated outputs at speeds ranging from as low as 65 rev/min up to around 180 rev/min for the smallest bore (around 350 mm) models. It is now possible to install a direct drive diesel engine which will achieve very nearly the optimum propulsive efficiency.

Direct coupled engines develop high outputs per cylinder, particularly large bore models, and it is easy to obtain the power required from an engine of a small number of cylinders. An owner will prefer an engine with as few cylinders as possible, provided that problems of vibration, balance, etc. do not ensue, because this directly affects the maintenance workload, spare parts carried and held in stock and the overall size of the engine and hence the machinery space. In most vessels height is less of a problem than length, so a larger bore engine with fewer cylinders will inevitably result in a shorter machinery space and more space for cargo. It is also proven in practice that larger bore engines have a better specific fuel consumption than smaller engines and there seems to be a better tolerance to burning heavy fuels of poor quality.

A direct coupled main propulsion engine cannot operate unaided as it requires service pumps for cooling, lubrication, and fuel and lubricating oil handling and treatment. These items of auxiliary machinery need a power source, which is usually provided by generators driven by four-stroke, medium-speed diesel engines which normally burn fuel of a lighter quality than that used by the main engine. Manufacturers of small auxiliary engines have stepped up their efforts to produce machines capable of burning not only the same heavy fuel as main engines, but marine diesel fuel or blended fuel (heavy fuel and distillate mixed in various proportions, usually 70:30) either supplied as an intermediate fuel or blended on board.

The cost of auxiliary power generation can weigh heavily on the choice of main machinery, so developments recently have tended to maximise the use of waste heat recovery for auxiliary power generation, the use of alternators directly driven from the main engine through speed increasing gears and the drive of certain auxiliary machinery items from the main engine. The designers of slow-speed direct coupled power plants are moving towards the 'one fuel ship', whereby low quality heavy fuel is used aboard ship for all purposes using the methods previously mentioned. Examples of such vessels are now in service.

INDIRECT DRIVE

The most satisfactory and economical form of indirect drive of a ship's propeller is one or more four-stroke, medium-speed diesel engines coupled through clutches and couplings to a reduction gearbox to drive either a fixed or controllable pitch propeller, Figure 2.1. The latter obviates the need for a direct reversing main engine while the gear allows the best propeller speed to be obtained with certainty. There is, nevertheless, a loss of efficiency in the transmission, but this in most cases would be cancelled out by the improvement in propulsive efficiency when making a comparison of direct and indirect drive engines of the same horsepower. The additional cost of the transmission can also be compensated for by the lower cost of main engine, as two-strokes, being large and heavier, inevitably cost more.

In general terms, the following advantages are claimed for geared medium-speed engines:

1. For vessels having more than one main engine reliability is improved as the vessel can still operate with one engine if a breakdown at sea occurs.

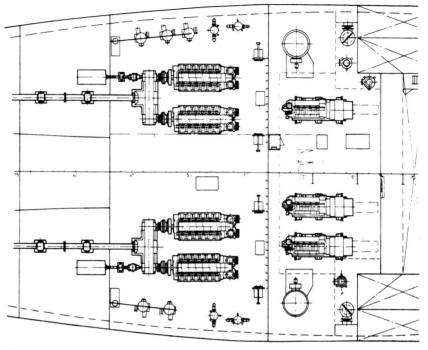

Figure 2.1 A typical multi-engine geared installation

2. When vessels are running light, partially loaded or slow steaming, one engine can operate at its normal rating and others be shut down rather than a direct coupled engine which may be called upon to operate for long periods at reduced output, at which it is inefficient. This has indeed been the case in the last few years with many large oil tankers.
3. Maintenance is easier because the engine components are a more manageable size, and generally major components such as cylinder covers, pistons, liners, etc. are much cheaper than corresponding two-stroke engine components.
4. Engines can be overhauled at sea by steaming on other engine(s) rather than is the case for the direct coupled engine which can only be overhauled in port. Ships today spend little time in port, so this facility of repair can be an important factor.
5. By modifying the number of engines per ship and cylinder numbers per engine to suit individual horsepower requirements, the propelling machinery for a fleet of ships can be standardised on a single cylinder bore size with subsequent savings in spares costs, availability, etc. With certain vessels with high auxiliary

loads, such as ferries, passenger ships, offshore service craft and so on, so called 'uniform machinery' can be employed with engines of the same bore and stroke used for both main and auxiliary duties.
6. The weight of the machinery and space required (particularly headroom) is much reduced — an important feature for vehicle deck vessels such as car ferries.

The above are true advantages for the geared medium-speed engine and are much exploited by manufacturers in promoting their sales. However, there are many shipowners who are reluctant to burn heavy fuels (say 3500 seconds Redwood 1 viscosity) in many types of four-stroke engines, while none of the direct coupled two-stroke engines appears to have great problems, though in some cases excessive wear has been caused by contaminants in fuels. (See Chapter 24 on fuel quality.)

Lubricating oil consumption is another factor to be considered, as the specific oil consumption of, for example, a trunk piston engine is inevitably higher. A direct coupled engine requires cylinder lubrication, but barring accidents the system oil consumption is negligible. There is a fine balance between the higher cost of cylinder oil and the higher consumption of cheaper system oil used in trunk piston engines.

Diesel–electric drive

The use of an indirect drive employing multiple trunk piston engines driving alternators (or dc generators) to power electric propulsion motors has a number of advantages for certain ship types. Notwithstanding the losses from conversion of mechanical to electrical power and back to mechanical drive of the propeller, a number of advantages are claimed for electric drive:

1. An increase in reliability by having several prime movers instead of one.
2. Easier maintenance because the main engines tend to be smaller in order to avoid large electrical machines; lower weight and greater interchangeability of spare parts.
3. One or more engines can be shut down at sea and the service speed of the ship can be varied by the number of generator sets in the circuit. This is of significance for a ferry operating to a fixed timetable, and any loss of speed through adverse weather could be made up by using more prime movers than normal.
4. When the ship is running at slower speed, or in ballast, main engines can be more effectively loaded by shutting down other

engines. There is a case in point with offshore support vessels. Many of these diesel–electric vessels have high electrical consumers such as bow and stern thrusters, winches, etc., and when on station with the main propellers not in use the running generator set can supply the power requirements.
5. The total installation weight is less and the engine room height is much reduced.
6. It is only necessary to employ engines of one bore and stroke size throughout the vessel, while there is the further option of using the prime movers for both main and auxiliary purposes, or at least using motor-generator sets if auxiliary power supplies need to be at a different voltage or frequency.
7. Flexibility of power distribution, as any desired number of engines per propeller shaft can be installed.
8. The main engines can be located anywhere and the motors can be sited right aft, thus eliminating propeller shafting.
9. The prime movers are non-reversible, so inevitably cheaper on a horsepower comparison basis.
10. Control of the propeller shaft speed and direction from the bridge is simplified through the use of direct electrical controls, not requiring the usual interlocks and safeguards for direct drive slow-speed engines or reversible four-stroke engines driving fixed pitch propellers.
11. Port turnround times are much quicker as all maintenance work can be undertaken at sea; a very important feature for a ferry on relatively short sea voyages.

The major disadvantages of a diesel–electric installation are the higher operating costs resulting from the increased fuel consumption brought about by the loss of transmission efficiency and the higher initial cost of the machinery. Though reduction gears are not necessarily required (they may be used between the propulsion motor(s) and propeller(s)) electrical machinery is not cheap and therefore the overall plant will cost much more.

While a direct or geared diesel installation is usually rated to meet a given service speed, in the case of diesel–electric plant the installed horsepower is usually much higher. A prudent operator of, say, a passenger ferry will install much more horsepower than is needed for the service speed to meet contingencies. For example, a ferry equipped with, say, six generator sets of around 2000 bhp each may maintain its service speed on three in fine weather, will perhaps need four in heavy weather, and could even on occasions need five in order to maintain the schedule. In all probability the sixth engine would be undergoing periodic maintenance. Hence, as much as double the required propulsive power could be installed to give sufficient flexibility with the correspondingly much higher initial cost for the ship as a whole.

There are some vessel types for which diesel–electric drive can be the best of all solutions, particularly ships with a high electrical load capacity. Vessels in the offshore industry with many electrically driven thrusters today use dynamic positioning systems to keep on station. These motors (or propeller pitches) are operated automatically by computer and full power on any or all thruster motors could be called up at any instance. It is impracticable to have these machines driven by diesel engines which could at one moment be fully loaded, the next idling, so the diesel–electric 'power station' concept is the best method of ensuring that power is available when it is required.

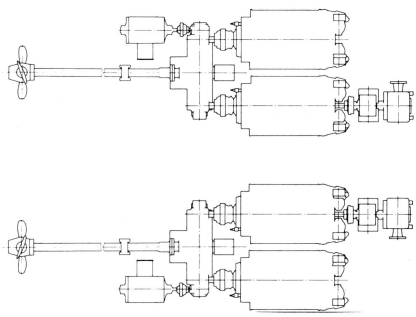

Figure 2.2 A combined diesel and diesel-electric plant

An interesting variant now being installed in offshore support vessels is a combination of geared diesel and diesel–electric drive (Figure 2.2). Supply vessels have two roles: to travel from the shore support base with materials to the offshore rigs and structures, and to stand by the rigs during cargo handling, rescue operations, handling anchors of rigs and so on. In the first case the propulsive power required is that necessary to maintain the free running service speed. In the second, little power is required for main propulsion but sufficient electrical power is needed for winches, thrusters, cargo handling gear, etc. The machinery installation solution for these dual roles is to have the main propellers

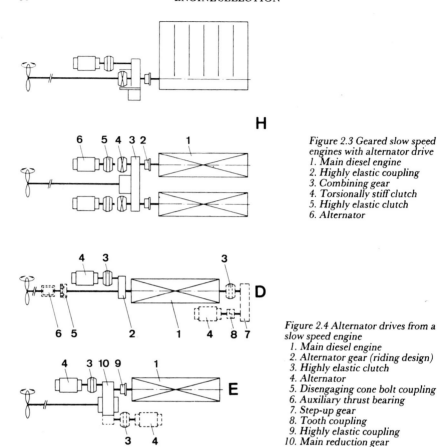

Figure 2.3 Geared slow speed engines with alternator drive
1. Main diesel engine
2. Highly elastic coupling
3. Combining gear
4. Torsionally stiff clutch
5. Highly elastic clutch
6. Alternator

Figure 2.4 Alternator drives from a slow speed engine
1. Main diesel engine
2. Alternator gear (riding design)
3. Highly elastic clutch
4. Alternator
5. Disengaging cone bolt coupling
6. Auxiliary thrust bearing
7. Step-up gear
8. Tooth coupling
9. Highly elastic coupling
10. Main reduction gear

driven from a medium-speed engine through a reduction gear while from these gear units are driven large shaft alternators. By using one shaft alternator as a motor and the other to generate electric power, it is possible to have a diesel–electric main propulsion system of low horsepower and at the same time provide sufficient power for the thrusters. Such a twin screw plant could simultaneously have one main engine driving the alternator and the propeller, while the other main engine would be shut down but its propeller driven by the shaft alternator acting as a motor.

A further option is to fit dedicated propulsion motors as well as the shaft alternators, which though much more complicated and costly, gives a greater degree of flexibility and potential for fuel savings. For such power plants to operate successfully controllable pitch main and thruster propellers are essential.

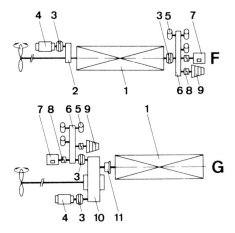

Figure 2.5 Alternator and auxiliary drives from a slow speed engine
1. Main diesel engine
2. Alternator gear (riding design)
3. Highly elastic clutch
4. Alternator
5. Pump
6. Distribution gear
7. Electric motor
8. Tooth coupling
9. Compressor
10. Main reduction gear
11. Highly elastic coupling

To summarise, the choice of main machinery for a given ship will depend on many factors, each of which needs to be carefully examined by a shipowner or builder. The ship's overall size, tonnage and range to perform its required duties are first established and the choice of direct drive, geared drive, or electric drive will depend on these requirements but much more on degree of reliability, initial cost of the machinery, cost of maintenance over the life cycle, quality and cost of fuel that can be burnt in both main and auxiliary engines, the cost of lubricating oil and, to some degree, passenger or crew comfort through excessive noise and/or vibrations. Multi-medium-speed engines could place an intolerably high work load burden on the crews of vessels with minimum manning levels, which could make the finding of suitably qualified crews difficult. When choosing a main propelling engine a shipowner today has much more to consider than the list price quoted by the main engine manufacturer.

C.T.W.

3 Performance

A most important parameter for a marine diesel engine is the rating figure, usually stated as bhp or kW per cylinder at a given rev/min.

Although engine builders talk of continuous service rating (csr) and maximum continuous rating (mcr) as well as overload ratings, the rating which concerns a shipowner most is the maximum output guaranteed by the engine builder at which the engine will operate continuously day in and day out. It is most important that an engine be sold for operation at its true maximum rating and that a correctly sized engine be installed in the ship in the first place, as an under-rated main engine, or more particularly an auxiliary, will inevitably be operated at its limits most of the time. It is wrong for a ship to be at the mercy of two or three undersized and thus over-rated auxiliary engines, or a main engine that needs to operate at its maximum continuous output to maintain the desired service speed.

Prudent shipowners usually insist that the engines be capable of maintaining the desired service speed fully loaded, when developing not more than 80% (or some other percentage) of their rated brake horsepower. However, such a stipulation can leave the full-rated power undefined and therefore does not necessarily ensure a satisfactory moderate continuous rating, hence the appearance of continuous service rating and maximum continuous rating. The former is the moderate in-service figure, the latter is the engine builder's set point of mean pressures and revolutions which the engines can carry continuously. Normally a ship will run sea trials to meet the contract trials speed (at a sufficient margin above the required service speed) and the continuous service rating should be applied when the vessel is in service. It is not unknown for shipowners to then stipulate that the upper power level of the engines in service should be somewhere between 85–100% of the service speed output, which could be as much as 20% less than the engine maker's guaranteed maximum continuous rating.

PERFORMANCE

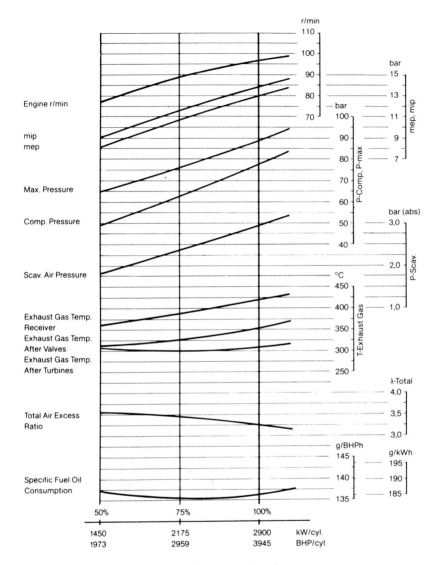

Figure 3.1 Typical performance curves for a two-stroke engine

MAXIMUM RATING

The practical maximum output of a diesel engine may be said to have been reached when one or more of the following factors operate:

1. The maximum percentage of fuel possible is being burnt effectively in the cylinder volume available (combustion must be com-

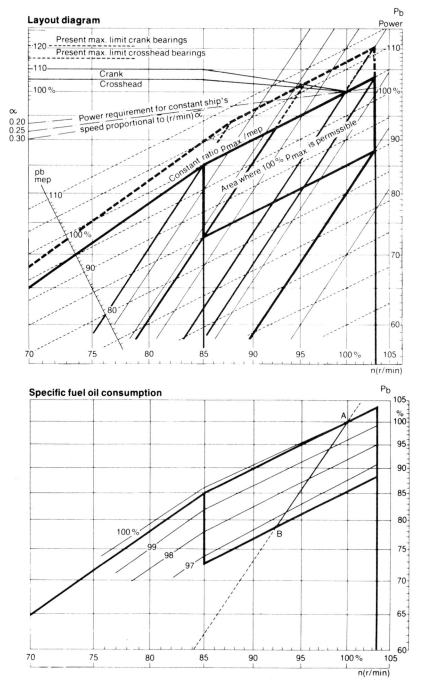

Figure 3.2 Layout diagrams showing maximum and economy ratings and corresponding fuel consumptions

pleted fully at the earliest possible moment during the working stroke).
2. The stresses in the component parts of the engine generally, for the mechanical and thermal conditions prevailing, have attained the highest safe level for continuous working.
3. The piston speed and thus revolutions per minute cannot safely be increased.

For a given cylinder volume, it is possible for one design of engine effectively to burn considerably more fuel than one of another design. This may be the result of more effective scavenging, higher pressure turbocharging, by a more suitable combustion chamber space and design, and by a more satisfactory method of fuel injection. Similarly, the endurance limit of the materials of cylinders, pistons and other parts may be much higher for one engine than for another; this may be achieved by the adoption of more suitable materials, by better design of shapes, thicknesses, etc., more satisfactory cooling and so on. A good example of the latter is the bore cooling arrangements now adopted for piston crowns, cylinder liner collars and cylinder covers in way of the combustion chamber.

The piston speed is limited by the acceleration stresses in the materials, the speed of combustion and the scavenging efficiency, that is, the ability of the cylinder to become completely free of its exhaust gases in the short time of one part cycle. Within limits, so far as combustion is concerned, it is possible sometimes to increase the speed of an engine if the mean pressure is reduced. This may be of importance for auxiliary engines built to match a given alternator speed.

For each type of engine, therefore, there is a top limit beyond which the engine should not be run continuously. It is not easy to determine this maximum continuous rating; in fact, it can only be satisfactorily established by exhaustive tests for each size and type of engine, depending on the state of development of the engine at the time.

If a cylinder is overloaded by attempting to burn too much fuel, combustion may continue to the end of the working stroke and perhaps also until after exhaust has begun. Besides suffering an efficiency loss, the engine will become overheated and piston seizures or cracking of engine parts may result; or, at least, sticking piston rings will be experienced, also dirty and sticking fuel valves.

EXHAUST TEMPERATURES

The temperature of the engine exhaust gases can be a limiting factor for the maximum output of an engine. An exhaust-temperature graph plotted with mean indicated pressures as abscissae and exhaust temperatures as ordinates will generally indicate when the economical combustion limit, and sometimes when the safe working limit, of an engine has been attained. The economical limit is reached shortly after the exhaust temperature begins to curve upwards from what was, previously, almost a straight line. Very often the safe continuous working load is also reached at the same time, as the designer naturally strives to make all the parts of an engine equally suitable for withstanding the respective thermal and mechanical stresses to which they are subjected. When comparing different engine types, however, exhaust temperature cannot be taken as proportionate to mean indicated pressure.

Sometimes it is said and generally thought that engine power is limited by exhaust temperature. What is really meant is that torque is so limited and exhaust temperature is a function of torque and not of power. The exhaust temperature is influenced by the lead and dimensions of the exhaust piping. The more easily the exhaust gases can flow away, the lower their temperature, and vice versa.

DERATING

A recent development to reduce the specific fuel consumption of diesel engines is the availability from engine builders of derated or so-called 'economy' ratings for engines. This means operation of an engine at its normal maximum cylinder pressure for the design continuous service rating, but at lower mean effective pressure and shaft speed.

By altering the fuel injection timing to adjust the mean pressure/ maximum pressure relationship the result is a worthwhile saving in fuel consumption. The horsepower required for a particular speed by a given ship is calculated by the naval architect and once the chosen engine is coupled to a fixed pitch propeller, the relationship between engine horsepower, propeller revolutions and ship speed is set according to the fixed propeller curve. A move from one point on the curve to another is simply a matter of giving more or less fuel to the engine. Derating is the setting of engine performance to maximum cylinder pressures at lower than normal shaft speeds, at a point lower down the propeller curve. For an existing ship and without changing the propeller this will result in a lower ship speed, but in practice when it is applied to new buildings, the derated engine horsepower is that which will drive the ship at a given speed with the propeller optimised to absorb this horsepower at a lower than normal shaft speed.

Savings in specific fuel consumption by fitting a derated engine can be as much as 5 g/bhph. However, should it be required at some later date to operate the engine at its full output potential (normally about 15–20% above the derated value) the ship would require a new propeller to suit both higher revolutions per minute and greater absorbed horsepower. The injection timing would also have to be reset.

MEAN EFFECTIVE PRESSURES

The term brake mean effective pressure (bmep) is widely quoted by engine builders, and is useful for industrial and marine auxiliary diesel engines that are not fitted with a mechanical indicator gear. However, the term has no useful meaning for shipboard propelling engines. It is artificial and superfluous as it is derived from measurements taken by a dynamometer (or brake), which are then used in the calculation of mechanical efficiency. Aboard ship where formerly the indicator, and now pressure transducers producing PV diagrams on an oscilloscope, are the means of recording cylinder pressures, mean indicated pressure (mip) is the term used, particularly in the calculation of indicated horsepower.

Many ships now have permanently mounted torsionmeters. By using the indicator to calculate mean indicated pressure and thus indicated horsepower, and the torsionmeter to calculate shaft horsepower from torque readings and shaft revolutions, the performance of the engine both mechanically and thermally in the cylinders can be readily determined.

Instruments such as pressure transducers, indicators, tachometers and pressure gauges (many of which are of the electronic digital or analogue type of high reliability) allow the ship's engineer to assess accurately the performance of the engine at any time.

The values of brake horsepower, mean indicated pressure and revolutions per minute are, of course, capable of mutual variation within reasonable limits, the horsepower developed per cylinder being the product of mean indicated (or effective) pressure, the revolutions per minute and the cylinder constant (based on bore and stroke). The actual maximum values for horsepower, and revolutions to be used in practice, are those quoted by the engine builder for the given continuous service rating.

PROPELLER SLIP

The slip of the propeller is normally recorded aboard ship as a useful pointer to overall results. While it may be correct to state that the amount of apparent slip is no indication of propulsive efficiency in a new ship design, particularly as a good design may have a relatively high propeller slip, the daily variation in slip (based on ship distance travelled compared with the product of propeller pitch and revolutions turned by the engine over a given period of time) can be symptomatic of changes in the relationship of propulsive power and ship speed; and slip, therefore, as an entity, is a useful parameter. The effects on ship speed 'over the ground' by ocean currents is sometimes considerable. For example, a following current may be as much as 2.5% and heavy weather ahead may have an effect of more than twice this amount.

PROPELLER LAW

An engine builder is at liberty to make the engine mean pressure and revolutions what he will, within the practical and experimental limits of the engine design. It is only after the maximum horsepower and revolutions are decided and the engine has been coupled to a propeller that the propeller law operates in its effect upon horsepower, mean pressure and revolutions.

shp varies as V^3
shp varies as R^3
T varies as R^2
P varies as R^2

where shp = aggregate shaft horsepower of engine, metric or imperial;
V = speed of ship in knots
R = revolutions per minute
T = torque, in kg metres or lbft = Pr
P = brake mean pressure kgf/cm^2 or lbf/in^2
r = radius of crank, metres or feet

If propeller slip is assumed to be constant:

$$shp = KV^3$$

where K = constant from shp and R for a set of conditions.

But R is proportional to V, for constant slip,

$$\therefore shp = K_1 R^3$$

where K_1 = constant from shp and R for a set of conditions.

But $$shp = \frac{p \times A \times c \times r \times 2\pi \times R}{33\,000}$$

$$= K_1 R^3 \text{ (Imperial)}$$

when A = aggregate area of pistons, cm² or in²
c = 0.5 for two-stroke, 0.25 for four-stroke engines:

or $T = PAcr = \dfrac{33\,000}{2\pi \times R} \times K_1 R^3 = K_2 R_2$ (Imperial)

or $T = PAcr = \dfrac{4500}{2\pi \times R} \times K_1 R^3 = K_2 R_2$ (metric)

i.e. $T = K_2 R^2$

where K_2 = constant, determinable from T and R for a set of conditions.

$PAcr = T$ or $\dfrac{T}{Acr} = \dfrac{K_2}{Acr} \times R^2$

i.e. $P = K_3 R^2$

where K_3 = constant determinable from P and R for a set of conditions.

The propeller law index is not always 3, nor is it always constant over the full range of speeds for a ship. It could be as much as 4 for short high-speed vessels but 3 is normally satisfactory for all ordinary calculations. The index for R, when related to the mean pressure P, is one number less than that of the index for V.

Propeller law is most useful for engine builders at the test beds where engine loads can be applied with the dynamometer according to the load and revolutions calculated from the law, thus matching conditions to be found on board the ship when actually driving a propeller.

FUEL COEFFICIENT

An easy yardstick to apply when measuring machinery performance is the fuel coefficient:

$$C = \dfrac{D^{2/3} \times V^3}{F}$$

where C = fuel coefficient
D = displacement of ship in tons
V = speed in knots
F = fuel burnt per 24 hours in tons

This method of comparison is applicable only if ships are similar, are run at approximately corresponding speeds, operate under the same conditions, and burn the same quality of fuel. The ship's displacement in relation to draught is obtained from a scale provided by the shipbuilders.

ADMIRALTY CONSTANT

$$C = \frac{D^{2/3} \times V^3}{shp}$$

where C = Admiralty constant, dependent on ship form, hull finish and other factors.

If C is known for a ship the approximate *shp* can be calculated for given ship conditions of speed and displacement.

APPARENT PROPELLER SLIP

$$\text{Apparent slip, \%} = \left(\frac{P \times R - 101.33 \times V}{P \times R}\right) \times 100$$

where P = propeller pitch in ft
R = speed of ship in knots
101.33 is one knot in ft/min

The true propeller slip is the slip relative to the wake stream, which is something very different. The engineer, however, is normally interested in the apparent slip.

PROPELLER PERFORMANCE

Many variables affect the performance of a ship's machinery at sea so the only practical basis for a contract to build to a specification and acceptance by the owner is a sea trial where everything is under the builder's control. The margin between the trial trip power and sea service requirements of speed and loading must ensure that the machinery is of ample capacity. One important variable on the ship's performance is that of the propeller efficiency.

Propellers are designed for the best combinations of blade area, diameter, pitch, number of blades, etc. and are matched to a given horsepower and speed of propulsion engine; and in fact each propeller is specifically designed for the particular ship. It is important that the engine should be able to provide heavy torque when required, which implies an ample number of cylinders with ability to carry high mean pressures. However, when a propeller reaches its limit of thrust capacity under head winds, an increase in revolutions can be to no avail.

In tank tests with models for powering experiments the following particulars are given:

Quasi-propulsive coefficient (Q)

$$Q = \frac{\text{model resistance} \times \text{speed}}{2\pi \times \text{torque} \times \text{rev/min}} = \frac{\text{work got out per min}}{\text{work put in per min}}$$

Total shaft horsepower at propeller (EHP)

$$\text{EHP} = \frac{ehp \times p}{Q}$$

where ehp = effective horsepower for model as determined by tank testing;
P = increase for appendages and air resistance equal to 10–12% of the naked model EHP, for smooth water conditions.

The shaft horsepower at the propeller for smooth sea trials is about 10% more than in tank tests. The additional power, compared with sea trials, for sea service is about 11–12% more for the South Atlantic and 20–25% more for the North Atlantic. This is due to the normal weather conditions in these areas. The size of the ship affects these allowances: a small ship needs a greater margin. By way of example: 15% margin over trial conditions equals 26.5% over tank tests.

The shaft horsepower measured by torsionmeter abaft the thrust block exceeds the EHP by the power lost in friction at the stern tube and plummer blocks and can be as much as 5–6%.

The brake horsepower (bhp) exceeds the torsionmeter measured shp by the frictional power lost at the thrust block. Bhp can only be calculated on board ship by multiplying the recorded indicated horsepower by the mechanical efficiency stated by the engine builder.

Required bhp for engine = EHP + sea margin + hp lost at stern tube and plummer blocks + hp lost at thrust block.

Typical values for the quasi-propulsive coefficient (Q) are: tanker 0.67–0.72; slow cargo vessel 0.72–0.75; fast cargo liner 0.70–0.73; ferry 0.58–0.62; passenger ship 0.65–0.70.

POWER BUILD-UP

Figures 3.3 and 3.4 are typical diagrams showing the propulsion power data for a twin screw vessel. In Figure 3.3 curve A is the EHP at the trial draught; B is the EHP corrected to the contract draught; C is the shp at trial draught on the Firth of Clyde; D is the shp corrected to the contract draught; E is the power service curve from voyage results; F shows the

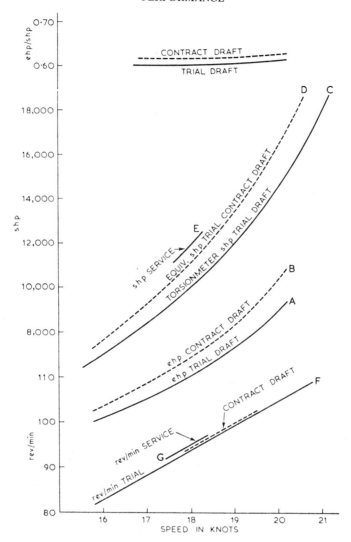

Figure 3.3 Propulsion data

relation between speed of the ship and engine revolutions on trials; G is the service result. The full rated power of the propelling engines is 18 000 bhp; the continuous service rating 15 000 bhp.

In Figure 3.3 curves A to D show shp as ordinates and the speed of the ship as abscissae. In Figure 3.4 powers are shown as ordinates, revolutions as abscissae.

Figure 3.5 shows the relationship between revolutions, power and brake mean pressure for the conditions summarised in Figures 3.3 and

PERFORMANCE

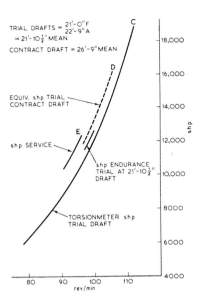

Figure 3.4 Propulsion data

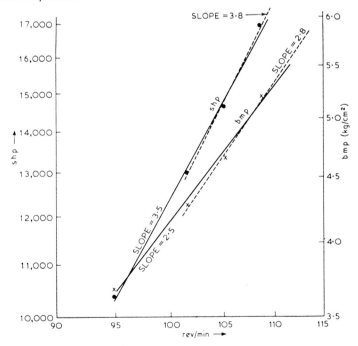

Figure 3.5 Engine trials: power, revolutions and mean pressure

3.4. A fair line drawn through the observed points for the whole range shows the shp to increase approximately as the cube of the revolutions and the square of the bmep. For the range 95–109 rev/min, the index increases to 3.5 for the power and 2.5 for the bmep. Between 120 and 109 rev/min a more closely drawn curve shows the index to rise to 3.8 and the bmep to 2.8. In Figure 3.5 ordinates and abscissae are plotted to a logarithmic base, thus reducing the power/revolution and the pressure/revolution curves to straight lines, for simplicity.

TRAILING AND LOCKING OF PROPELLER

In Figure 3.6 there are shown the normal speed/power curves for a twin screw motor vessel on the measured mile and in service.

The effect upon the speed and power of the ship when one of the propellers is trailed, by 'free-wheeling', is indicated in the diagram. The effect of one of the propellers being locked is also shown.

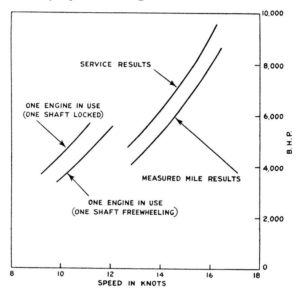

Figure 3.6 Speed/power curves, twin screw vessel

Figure 3.7 shows the specd/power curves for a four screw motorship:

1. When all propellers are working.
2. When the vessel is propelled only by the two centre screws, the outer screws being locked.
3. When the ship is being propelled only by the two wing screws, the two inner screws being locked.

PERFORMANCE

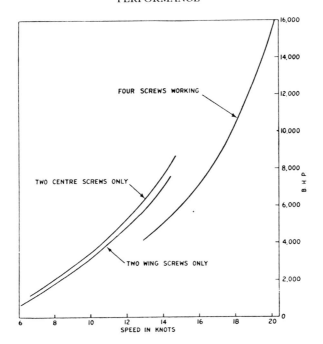

Figure 3.7 Speed/power curves, quadruple screw ship

ASTERN RUNNING

Figure 3.8 summarises a series of tests made on the trials of a twin screw passenger vessel, 716 ft long, 83 ft 6 in beam, trail draught 21 ft forward, 26 ft aft, 26 000 tons displacement.

As plotted in Figure 3.8., tests I to VI show distances and times, the speed of approach being as stated at column 2 in Table 3.1. The dotted curves show reductions of speed and times.

The dotted curves A to F respectively correspond to curves I to VI. In test I, after the ship had travelled, over the ground, a distance of two nautical miles (1 mm = 6080 ft) the test was terminated and the next test begun.

In Table 3.2 a typical assortment of observed facts related to engine stopping and astern running is given. Where two or three sets of readings are given, these are for different vessels and/or different engine sizes.

Trials made with a cargo liner showed that the ship was brought to rest from 20 knots in 65 seconds. Another cargo liner, travelling at 16 knots was brought to a stop in a similar period. The engine, running full power ahead, was brought to 80 rev/min astern in 32 seconds, and had settled down steadily at full astern revolutions in 50 seconds.

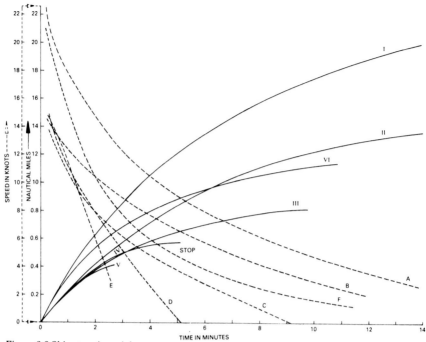

Figure 3.8 Ship stopping trials

Table 3.1 Ship stopping trials

1	2	3	4		5	6
	Ahead speed of approach knots		Propellers stopped (min)		Ship stopped	Distance travelled (nautical
Test No.	(rev/min)	Propellers	P.	S.	(min)	miles)
I	23.0 (119)	Trailing; unlocked	16.4	14.9	—	2.0
II	14.5 (75)	Trailing; unlocked	12.5	12.1	17.0	1.4
III	13.5 (75)	Trailing; locked	1.5	1.5	13.0	0.8
IV	15.0 (75)	Ahead running checked; no additional astern power	1.3	1.3	5.2	0.5
V	14.8 (75)	Engine stopped; astern as quickly as possible	0.7	0.8	3.1	0.4
VI	22.4 (116)	Trailing; unlocked	1.5	1.6	15.0	1.1

Table 3.2 Engine reversing and ship stopping

Ship	Engine type	Ahead (rev/min)	Time for engine stopping (sec)	Engine moving astern (sec)	Astern running rev/min	Astern running sec	Ship stopped min	Ship stopped sec
Large passenger	D.A. 2C. (twin)	65	—	30	—	—	4	2
Small fast passenger	S.A. 2C. tr. (twin)	217	53	63.5	160	72.5	2	15
		195	35	45	160	60	1	54
		220	45	59	170	81	2	5
Passenger	Diesel-electric (twin)	92	110	—	80	225	5	0
Cargo	S.A. 2C. (twin)	112	127.5	136	100	141	—	
Cargo	D.A. 2C. (single)	90	24	26	90	115	2	26
		88	12	14	82	35	3	25
		116	35	40	95	50	—	
Cargo	S.A. 2C. (single)	95	31	33	75	45	—	
		110	12	21	110	53	3	21
		116	10	55	110	110	4	6

A shipbuilder will think of ship speed in terms of the trial performance in fair weather but to the shipowner ship speed is inevitably related to scheduled performance on a particular trade route. Sea trials are invariably run with the deadweight limited to fuel, fresh water and ballast. Because of the difference between loaded and trials draught, the hull resistance may be 25–30% greater for the same speed. This has a consequential effect upon the relation between engine torque and power, and in the reaction on propeller efficiency. Adverse weather, marine growth and machinery deterioration necessitate a further power allowance, if the service speed is to be maintained. The mean wear and tear of the engine may result in a reduction of output by 10–15% or a loss of speed by up to one knot may be experienced.

When selecting a propulsion engine for a given ship a suitable power allowance for all factors such as weather, fouling, wear and tear, and other factors, as well as the need to maintain the service speed at around 85% of the maximum continuous rating, should all be taken into consideration.

C.T.W.

4 Pressure charging

A naturally aspirated engine draws air of the same density as the ambient atmosphere and as this air density determines the maximum weight of fuel that can be effectively burned per working stroke in the cylinder, it also determines the maximum power that can be developed by the engine. By increasing the density of the charge air by the use of a suitable compressor between the air intake and the cylinder, the weight of air induced per working stroke is increased, and thereby a greater weight of fuel can be burnt, with a consequent increase in the cylinder power output. This increase in charge air density is accomplished on most modern diesel engine types by use of exhaust gas turbocharging, in which a turbine wheel driven by exhaust gases from the engine is rigidly coupled to a centrifugal type air compressor. This is a self-governing process which does not require an external governor.

The power expended in driving the compressor has an important influence on the operating efficiency of the engine. It is relatively uneconomical to drive the compressor direct from the engine by chain or gear drive because some of the additional power is thereby absorbed and there is an increase in specific fuel consumption for the extra power obtained. About 35% of the total heat energy in the fuel is wasted to the exhaust gases, so by using the energy in these gases to drive the compressor an increase in power is obtained in proportion to the increase in the charge air density.

The turbocharger comprises a gas turbine driven by the engine exhaust gases, mounted on the same spindle as a blower with the power generated in the turbine equal to that required by the compressor. There are a number of advantages of pressure charging by means of an exhaust gas turboblower system:

1. A substantial increase in the engine power output for any stated size and piston speed, or conversely, a substantial reduction in engine dimensions and weight for any stated horsepower.
2. An appreciable reduction in the specific fuel consumption rate at all engine loads.
3. A reduction in initial engine cost.

PRESSURE CHARGING

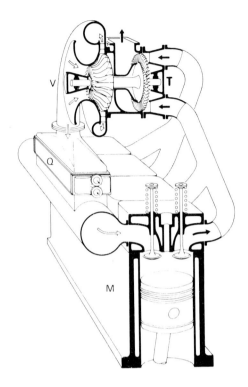

Figure 4.1 Basic arrangement of exhaust turbocharging. M. Diesel engine; O. Charge air cooler; T. Turbocharger exhaust turbine; V. Turbocharger compressor.

4. Increased reliability and reduced maintenance costs, resulting from less-exacting conditions at the cylinders.

FOUR-STROKE ENGINES

Exhaust gas turbocharged single-acting four-stroke cycle marine engines will deliver up to three times as much power as naturally aspirated engines of the same speed and dimensions. Even higher power output ratios are achieved on some more recent engine types employing two-stage turbocharging where turbochargers are arranged in series. At one time almost all four-stroke engines operated on the pulse system, though constant pressure turbocharging is now becoming more common as it provides greater fuel economy while considerably simplifying the arrangement of exhaust piping.

In matching the turboblower to the engine, a free air quantity in excess of the swept volume is required to allow for the increased density of the charge air and to provide sufficient air for through scavenging of the cylinders after combustion. For example, an engine with a full load bmep of 10.4 bar would need about 100% of excess free air, about 60% of which is retained in the cylinders, with the remaining 40% being used for scavenging. Modern engines carry bmep's up to 22 bar in some cases, requiring greater proportions of excess air which is made possible by the latest designs of turbocharger with pressure ratios as high as 4:1.

To ensure adequate scavenging and cooling of the cylinders, a valve overlap of approximately 140° is normal with, in a typical case, the air inlet valve opening at 80° before top dead centre and closing at 40° after bottom dead centre. The exhaust valve opens at 50° before bottom dead centre and closes at 60° after top dead centre.

For a low-rated engine with a bmep of 10.4 bar compared with 5.5 bar for a naturally aspirated engine, a boost pressure of some 0.5 bar is required, corresponding to a compressor pressure to ratio of 1.5:1. As average pressure ratios today are between 2.5 and 4, considerable boost pressures are being carried on modern high bmep medium-speed four-stroke engines.

Optimum values of power output and specific fuel consumption can be achieved only by utilisation of the high energy engine exhaust pulses. The engine exhaust system should be so designed that it is impossible for gases from one cylinder to contaminate the charge air in another cylinder, either by blowing back through the exhaust valve or by interfering with the discharge of gases from the cylinder. During the period of valve overlap it follows that the exhaust pressure must be less than air charging pressure to ensure effective scavenging of the cylinder to remove residual gases and cooling purposes. It has been found in practice that if the period between discharge of successive cylinders into a common manifold is less than about 240°, then interference will take place between the scavenging of one cylinder and the exhaust of the next. This means that engines with more than three cylinders must have more than one turbocharger or, as is more common, separate exhaust gas passages leading to the turbine nozzles.

The exhaust manifold system should be as small as possible in terms of pipe length and bores. The shorter the pipe length, the less likelihood there is of pulse reflections occurring during the scavenge period. The smaller the pipe bore, the greater the preservation of exhaust pulse energy, though too small a bore may increase the frictional losses of the high velocity exhaust gas to more than offset the increased pulse energy. Sharp bends or sudden changes in pipe cross sectional area should be avoided wherever possible.

TWO-STROKE ENGINES

Compared with four-stroke engines, the application of pressure charging to two-stroke engines is more complicated because until a certain level of speed and power is reached, the turboblower is not self-supporting. At low engine loads there is insufficient energy in the exhaust gases to drive the turboblower at the speed required for the necessary air-mass flow. In addition, the small piston movement during the through scavenge period does nothing to assist the flow of air, as in the four-stroke engine. Accordingly, starting is made very difficult and off-load running can be very inefficient. Below certain loads it may even be impossible. A solution is found by having mechanical scavenge pumps driven from the engine arranged to operate in series with the turboblowers or, as is more common on the latest engines, electrically driven auxiliary blowers.

Two-stroke engine turbocharging is achieved by two distinct methods, respectively termed the 'constant pressure' and 'pulse' systems. It is the constant pressure system that is now used by all slow-speed two-stroke engines.

For constant pressure operation, all cylinders exhaust into a common receiver, which tends to dampen-out all the gas pulses to maintain an almost constant pressure. The advantage of this system is that it eliminates complicated multiple exhaust pipe arrangements and leads to higher turbine efficiencies and hence lower specific fuel consumptions. An additional advantage is that the lack of restriction, within reasonable limits, on exhaust pipe length permits greater flexibility in positioning of the turboblower relative to the engine. Typical positions are at either or both ends of the engine, at one side above the air manifold or on a flat adjacent to the engine. The main disadvantage of the constant pressure system is the poor performance at part load conditions and, owing to the relatively large exhaust manifold, the system is insensitive to changes in engine operating conditions. The resultant delay in turboblower acceleration, or deceleration, results in poor combustion during transition periods.

For operation under the pulse system the acceptable minimum firing order separation for cylinders exhausting to a common manifold is about 120°. The sudden drop in manifold pressure, which follows each successive exhaust pulse, results in a greater pressure differential across the cylinder during the scavenge period than is obtained with the constant pressure system. This is a factor which makes for better scavenging.

Figure 4.2 shows the variation in rotational speed of a turboblower during each working cycle of the engine, i.e. one revolution. This

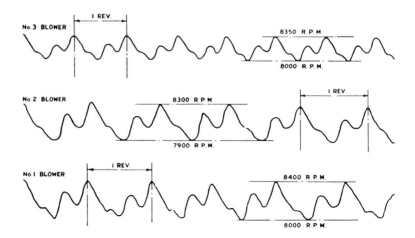

Figure 4.2 Cyclic variations in turbocharger revolutions

diagram clearly illustrates the reality of the impulses given to the turbine wheel. The fluctuation in speed is about 5%. Each blower is coupled to two cylinders, with crank spacing 135°, 225°. It is now standard practice to fit charge air coolers to turbocharged two-stroke engines, the coolers being located between the turbochargers and the cylinders.

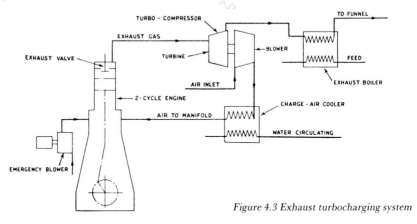

Figure 4.3 Exhaust turbocharging system

CHARGE AIR COOLING

The increased weight or density of air introduced into the cylinder by pressure charging enables a greater weight of fuel to be burnt, and this in turn brings about an increase in power output. The increase in air density is, however, fractionally offset by the increase of air temperature resulting from adiabatic compression in the turboblower, the amount of which is dependent on compressor efficiency. This reduction of air density due to increased temperature implies a loss of potential power for a stated amount of pressure charging. For example, at a charge air pressure of, say, 0.35 bar, the temperature rise is of the order of 33 °C — equivalent to a 10% reduction in charge air density. As the amount of pressure charging is increased, the effect of turboblower temperature rise becomes more pronounced. Thus, for a charge air pressure of 0.7 bar, the temperature rise is some 60 °C, which is equivalent to a reduction of 17% in the charge air density.

Much of this potential loss can be recovered by the use of charge air coolers. For moderate amounts of pressure charging, cooling of the charge air is not worthwhile, but for two-stroke engines especially, it is an advantage to fit charge air coolers, which are standard on all makes of two-strokes and most medium-speed four-stroke engines.

Charge air cooling has a double effect on engine performance. By increasing the charge air density, it thereby increases the weight of air flowing into the cylinders, and by lowering the air temperature it reduces the maximum cylinder pressure, the exhaust temperature and the engine thermal loading. The increased power is obtained without loss — and, in fact, with an improvement in fuel economy. It is important that charge air coolers should be designed for low pressure drop on the air side; otherwise, to obtain the required air pressure the turboblower speed must be increased.

The most common type of cooler is the water cooled design with finned tubes in a casing carrying seawater over which the air passes. To ensure satisfactory effectiveness and a minimum pressure drop on the charge air side and on the water side, the coolers are designed for air speeds of around 11 m/sec and water speeds in the tubes of 0.75 m/sec. Charge air cooler effectiveness is defined as the ratio of charge air temperature drop to available temperature drop between air inlet temperature and cooling water inlet temperature. This ratio is approximately 0.8.

SCAVENGING

It is essential that each cylinder should be adequately scavenged of gas before a fresh charge of air is compressed, otherwise this fresh air charge is contaminated by residual exhaust gases from the previous cycle. Further, the cycle temperature will be unnecessarily high if the air charge is heated by mixing with residual gases and by contact with hot cylinders and pistons.

In the exhaust turbocharged engine the necessary scavenging is obtained by providing a satisfactory pressure difference between the air manifold and the exhaust manifold. The air flow through the cylinder during the overlap period has a valuable cooling effect; it helps to increase the volumetric efficiency and to ensure a low cycle temperature. Also, the relatively cooler exhaust allows a higher engine output to be obtained before the exhaust temperature imposes a limitation on the satisfactory operation of the turbine blades.

In two-stroke engines the exhaust/scavenge overlap is necessarily limited by the engine design characteristics. In Figure 4.4 a comparison of the exhaust and scavenge events for poppet valve engines and opposed piston engines is given. In the poppet valve engine the camshaft lost motion coupling enables the exhaust pre-opening angle to be 52° ahead and astern. In the opposed piston engine the exhaust pre-opening angle is only 34° ahead and 20° astern. Against this, however, the rate of port opening in opposed piston engines is quicker than in poppet valve engines. It should be emphasised, however, that opposed piston slow-speed engines are no longer in production and poppet valves are used in the majority of new designs.

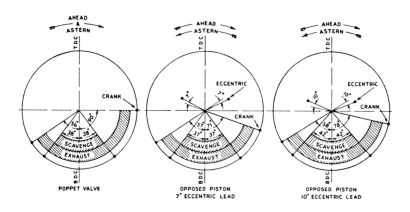

Figure 4.4 *Single-acting two-stroke engine timing*

In the four-stroke engine the substantial increase in power per cylinder, obtained from turbocharging, is achieved without increase of cylinder temperature. In the two-stroke engine the augmented cylinder loading — if any — is without significance.

A strange fact is that the engine exhaust gas raises the blower air to a pressure level greater than the mean pressure of the exhaust gas itself.

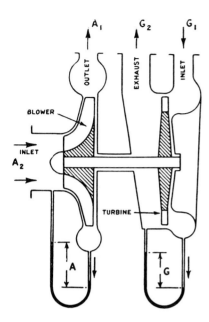

Figure 4.5 Scavenge gradient

This is because of the utilisation of the kinetic energy of the exhaust gas leaving the cylinder, and the energy of the heat drop as the gas passes through the turbine. In Figure 4.5 the blower pressure gradient A exceeds the turbine pressure gradient G by the amount of the scavenge gradient. The design of the engine exhaust pipe system can have an important influence on the performance of turboblowers.

The test results of a turbocharged engine of both two- and four-stroke type will show that there is an increase in temperature of the exhaust gas between the cylinder exhaust branch and the turbine inlet branch, the rise being sometimes as much as 95 °C. The reason for this apparent anomaly is that the kinetic energy of the hot gas leaving the cylinder is converted, in part, into additional heat energy as it adiabatically compresses the column of gas ahead of it until, at the turbine inlet, the temperature exceeds that at the cylinder branch. At the turbine some of the heat energy is converted into horsepower, lowering the gas temperature somewhat, with the gas passing out of the turbine to atmosphere or to a waste heat recovery boiler for further conversion to energy. Though the last thing to emerge from the exhaust parts or valves is a slug of cold scavenge air which can have a cooling effect on the recording thermometer, adiabatic compression can still be accepted as the chief cause of the temperature rise.

MATCHING OF TURBOBLOWERS

The correct matching of a turboblower to an engine is extremely important. With correct matching, the engine operating point should be close to optimum efficiency, as shown by the blower characteristic curve (Figure 4.6). In this diagram the ordinates indicate pressure ratio; the abscissae show blower capacity.

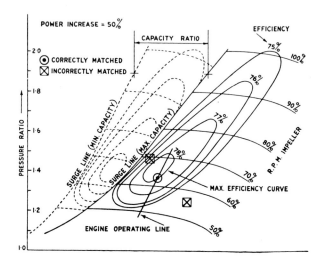

Figure 4.6 Turboblower characteristics

During the initial shop trials of a new engine, with the turbocharger recommended by the manufacturer, test data are recorded and analysed. If the turboblower is correctly matched, nothing more needs to be done. Should the matching be incorrect, however, the turboblower will supply charge air at either too low or too high a pressure, or surging may occur at the blower. Mis-matching can usually be corrected by a change of turbine capacity and/or blower diffuser.

Compressors can be designed which maintain a high efficiency at a constant pressure ratio over a wide range of air mass flows by providing alternate forms of diffuser for any one design of impeller. This range can be further extended by the use of different impeller designs, each with its own set of diffusers, within a given frame size of turboblower. By this means, each frame size can provide a mass flow capacity range having a maximum to minimum ratio of about 3:1. By some overlap of capacity for each frame size, each size is increased in capacity from the next smaller size by a factor of about 1.6:1.

BLOWER SURGE

Too low an air-mass flow at a given speed, or pressure ratio, will cause the blower to surge, while too high a mass flow causes the blower to choke, resulting in loss of pressure ratio and efficiency at a given speed.

The blower impeller, as it rotates, accelerates the air flow through the impeller, and the air leaves the blower with a velocity that is convertible into a pressure at the diffuser. If, for any reason, the rate of air flow decreases, then its velocity at the blower discharge will also decrease; thus there will come a time when the air pressure that has been generated in the turboblower will fall below the delivery pressure. There will then occur a sudden breakdown of air delivery, followed immediately by a backward wave of air through the blower which will continue until the delivery resistance has decreased sufficiently for air discharge to be resumed. 'Surging' is the periodic breakdown of air delivery.

In the lower speed ranges surging is manifested variously as humming, snorting and howling. If its incidence is limited to spells of short duration it may be harmless and bearable. In the higher speed ranges, however, prolonged surging may cause damage to the blower, as well as being most annoying to engine room personnel. Close attention to the surging limit is always necessary in the design and arrangement of blowers.

If one cylinder of a two-stroke engine should stop firing, or is cut out for mechanical reasons, when the engine is running above say 40–50% of engine load, it is possible that the turboblower affected may begin to surge. This is easily recognised from the repeated changing in the pitch of the blower noise. In these circumstances the engine revolutions should be reduced until the surging stops or until firing can be resumed in all cylinders.

TURBOCHARGER TYPES

Today the designs of exhaust gas driven turbochargers have kept pace with the products of the engine manufacturers. Higher bmeps have called for greater efficiency and pressure ratios from the turbochargers. Similarly, the introduction of more efficient turbochargers has allowed engine designers to improve the performance of their engines, and to achieve significant reductions in specific fuel consumptions as a result of high cylinder maximum pressure to mean pressure ratios.

For marine engines, three manufacturers — Brown Boveri, Napier and Mitsubishi — supply practically all the turbochargers used, though the German builders MAN also produce quite a number of their own design turboblowers. Cooling of turbochargers on the gas side has been almost exclusively by water jackets, but the most recent trend is towards non-cooled (or air cooled) models, as a means of promoting higher gas exit temperatures for waste heat recovery systems.

Four basic turbocharging arrangements are now used on marine diesel engines employing turbochargers:

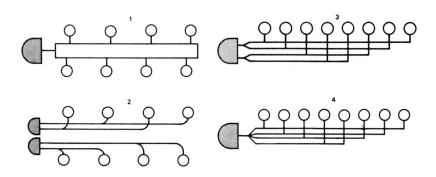

Figure 4.7 Exhaust turbocharging methods: 1. Constant pressure; 2. Impulse; 3. Pulse converter; 4. Multi-pulse system

CONSTANT PRESSURE METHOD. All the exhaust pipes of the cylinders of an engine end in a large common gas manifold to reduce the pressure pulses to a minimum with a given loading. The turbine can be built with all the gas being admitted through one inlet and therefore a high degree of efficiency is reached. For efficient operation the pressure generated by the compressor should always be slightly higher than that of the exhaust gas after the cylinder.

IMPULSE TURBOCHARGING METHOD. The exhaust gas flows in pulsating form into the pipes leading to the turbine. From there it flows out in a continuous stream. The gas pulses from the separate cylinders are each fed to a corresponding nozzle ring segment of the expansion turbine. By overlapping the opening times of inlet and exhaust valves after the pulse decay, efficient scavenging of the cylinders is possible.

PULSE CONVERTER METHOD. Interactive interference limits the ways in which the normal impulse system can be connected to groups of cylinder exhaust pipes. However, with a pulse converter such cylinder groups can be connected to a common ejector. This prevents return flows and has the effect of smoothing out the separate impulses. It also improves turbine admission, increases efficiency and does not mechanically load the blading as much as normal impulse turbocharging.

MULTI-PULSE METHOD. This is a further development of the pulse converter. In this case, a number of exhaust pipes feed into a common pulse converter together with a number of nozzles and a mixing pipe. With this form of construction the pressure waves are fed through with practically no reflection because the turbine nozzle area is larger. The multi-pulse method makes a noticeable performance increase possible, in comparison with the normal impulse turbocharging method.

Brown Boveri turbochargers

The development of Brown Boveri exhaust gas turbochargers dates back to 1924 and the present VTR range of turboblowers was conceived as far back as 1943. Since that date the external dimensions of nine frame sizes of units have remained the same. With the VTR design one turbocharger can be assembled in a number of different ways to give an optimum match between it and a given engine of either four- or two-stroke type with various ratings, numbers of cylinders, etc.

Turbocharger construction

The basic design of VTR turbochargers has remained unchanged for many years and a typical example is shown in the cross section Figure 4.8.

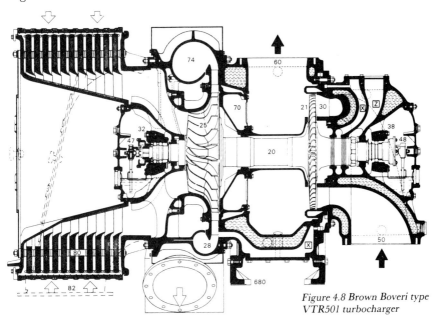

Figure 4.8 Brown Boveri type VTR501 turbocharger

The diesel engine exhaust gases enter through the water cooled gas inlet casing (50), expand in the nozzle ring (30) and supply energy to the shaft (20) by flowing through the blading (21). The gases exhaust to the open air through the gas outlet casing (60) which is also water cooled, and the exhaust piping. The charge air enters the compressor through an inlet stub (82) or through the silencer filter (80). It is then compressed in the inducer and the impeller (25), flows through the diffuser (28) and is fed to the engine via the pressure stubs on the compressor casing (74). Air and gas spaces are separated by the heat insulating bulkhead (70). In order to prevent exhaust gases from flowing into the balance channel (2) and the turbine side reservoir, barrier air is fed from the compressor to the turbine rotor labyrinth seal via channel (X).

The rotor (20) has easily accessible bearings (32, 38) at both ends, which are supported in the casing with vibration damping spring elements. Either roller or plain bearings are used but for the most common construction using roller bearings, a closed loop lubrication system with an oil pump directly driven from the rotor is used (47, 48). These pumps are fitted with oil centrifuges to separate out the dirt in the lubricating oil. The bearing covers are each fitted with an oil filter, an oil drain opening and an oil gauge glass. On models with plain bearings where the quantity of oil required is large, these are fed from the main engine lubricating oil system.

A major feature of the VTR range of turbochargers is the modular construction to match a wide variety of diesel engine types. The separate modules of the turbocharger as seen from Figure 4.8 are: the silencer filter (80); the air inlet casing (82); the compressor housing (74); the gas outlet casing (60); outlet casing feet (680); the gas inlet casing (50). The fixing screws are placed so that the radial position of all the separate casings can be arranged in any position relative to the other.

Precise thermal matching to the diesel engine can be achieved by the small differences between the individual sizes of turbochargers. Each model has been provided with a large number of variants of nozzle rings, turbine blading, inducer and compressor impellers with diffusers. Gas inlet casings (50) with 1 to 4 gas inlet stubs in various configurations and also compressor housings with 1 or 2 outlet stubs are produced.

Performance

Figure 4.9 shows a typical pressure–volume curve of a turbocharger, the operating characteristics of a two-stroke diesel engine fitted solely

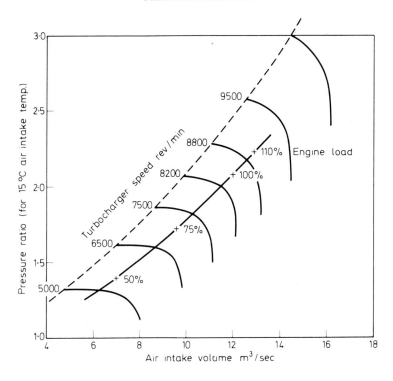

Figure 4.9 Typical pressure-volume curve

with turbochargers being indicated. As will be seen, the operating curve runs almost parallel to the surge line. Every value of the engine output corresponds to a point on the curve, and this point, in turn, corresponds to a particular turbocharger speed which is obtained automatically. There is consequently no need for any system of turbocharger speed control. If, for instance, at a definite turbocharger speed the charge air pressure is lower than normal, the conclusion may be drawn that the compressor is contaminated. By spraying a certain amount of water into the compressor the deposit can be removed, provided it has not already become too hard. A special duct for water injection is provided on VTR turbochargers.

The operating curve and the behaviour of the turbocharger may vary, depending upon whether constant pressure or pulse turbocharging is employed, and depending also upon how volumetric (i.e. mechanically driven) compressors and turbochargers are connected together. By observing changes in behaviour it is usually possible to deduce causes and to prescribe the measures necessary to rectify the difficulties.

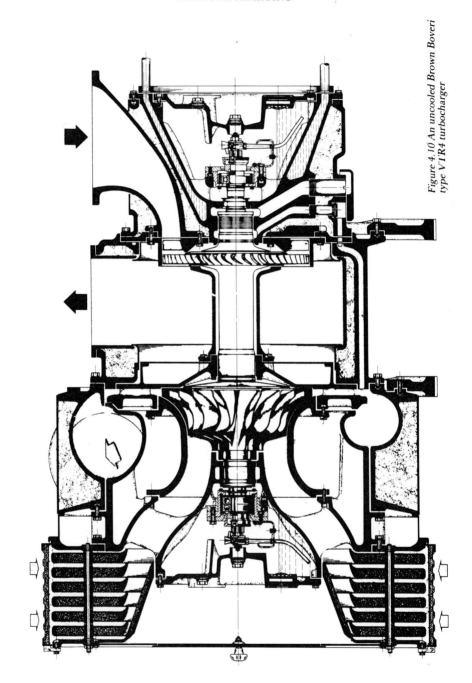

Figure 4.10 An uncooled Brown Boveri type VTR4 turbocharger

Latest uncooled types

As a result of enormous rises in fuel prices, shipowners are specifying diesel engine plants that utilise waste heat to generate steam for electricity production.

Based on a gas temperature at the turbine outlet of about 300 °C for two-stroke engines and about 400 °C for four-stroke engines, it is possible to utilise a temperature difference of 120° and 200 °C, respectively. If the gas casing of the turbocharger is uncooled, the temperature of the exhaust gases leaving the turbocharger can be raised by 10° to 15 °C compared with water cooled turbochargers. That is to say, the usable heat is increased by approximately 10%.

Brown Boveri have introduced a non-cooled unit as a variant of their most recent series '4' VTR models. A sketch of an uncooled VTR turbocharger is given in Figure 4.10. Well-established design features have been retained but the supporting outer walls of the newly designed gas casings have been totally separated from the ducts carrying hot gases. The resulting cavities are thereby filled with a special insulating material to back up efforts to keep the surface temperatures down to a desirable level. In order that the temperature of the oil on the turbine side remains within tolerable limits, the bearing housing on the turbine side is water cooled, but this cooling has no effect whatsoever on the temperature of the exhaust gases.

Napier turbochargers

The NA range of turbochargers (Figure 4.11) built by the British company Napier Turbochargers Ltd., follows well-established design principles some of which are exclusive to this company's product. The latest variant of the NA range is the series '5' with pressure ratios of up to 4:1. This latter type has been introduced to meet engine builders' requirements for greater efficiency.

Four separate casings make up the assembly of the Napier turbochargers types NA150, NA250 and NA350. The turbine outlet casing forms the main structure to which the other sub-assemblies are secured, incorporating the mounting foot which can be located in any one of three different positions around the casing. Bolted to the turbine casing is the inlet casing with alternative configurations of entry casings in axial or right angle form. A central casing carries the rotating assembly and is inserted in cartridge form into the turbine outlet casing. This special Napier design feature allows maintenance of the rotating assembly without disturbing the exhaust inlet and outlet and water connections to the engine; the cartridge being withdrawn through the compressor end of the outlet casing.

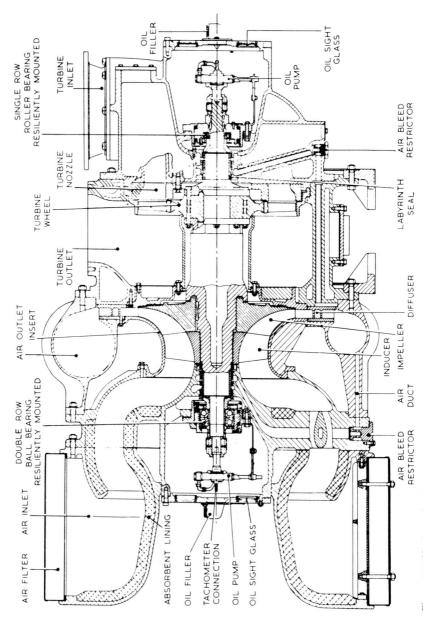

Figure 4.11 Napier turbocharger

The rotating assembly consist of a single-stage centrifugal compressor with separate inducer and a single-stage axial turbine. The short steel rotor shaft runs in split steel-backed bronze-lined plain bearings located inboard of the compressor impeller and turbine wheel. Thrust is taken on the bronze-lined faces of the bearings at the compressor end of the shaft. Lubrication is from the main engine oil system. The turbine disc and blades are made from Nimonic 90 with the blades fixed to the disc by the fir tree root method, secured by locking tabs and wire laced to damp vibration. The aerodynamically designed compressor impeller is machined from an aluminium forging and the inducer from a steel casting or aluminium forging. An air filter is secured to the compressor casing or, alternatively, a side entry casing can be fitted for ducted air to the inlet of the compressor.

The latest series '5' turbochargers have the same overall dimensions as the previous NA '0' series to allow retrofit to existing installations. Major components, such as the main casings, rotor shaft and bearings, are common with the old series, but the improved performance of the NA455, NA555, and NA655 units is the result of detailed development of the aerodynamic components, such as a new backward swept compressor impeller and the turbine itself. Outwardly both turbochargers are the same with similar servicing procedures and with built-in water washing facilities to cope with the effects of dirty heavy fuels.

Mitsubishi turbochargers

The Japanese manufacturer, Mitsubishi Heavy Industries Ltd., was the first to introduce the non-water cooled turbocharger with the first MET unit introduced in 1965. The MET turbocharger (Figure 4.12) has recently been supplemented by the much improved Super MET model. The Super MET turbocharger (Figure 4.13) has a pressure ratio of 3.5 at its maximum allowable revolutions and is built in five basic types, from the MET 35S for engines between 2300 and 3900 bhp up to the MET 71S for engines of 9100 to 15 300 bhp. They are all of the non-water cooled casing type and can be matched to either two- or four-stroke diesel engines. Three variants of the basic model are built with different configurations of gas inlet and outlet, the type 'B' having an axial flow gas inlet and another type 'B' and type 'T' having radial gas inlet and exhaust in opposite relative positions.

A number of features are built into the Super MET turbocharger. The turbine side casing needs no water cooling and is hence free from possible sulphuric acid corrosion; and the turbine nozzles and blades can be cleaned without removing the rotating assembly and disturbing the pipe connections. The bearings are housed in a tapered casing

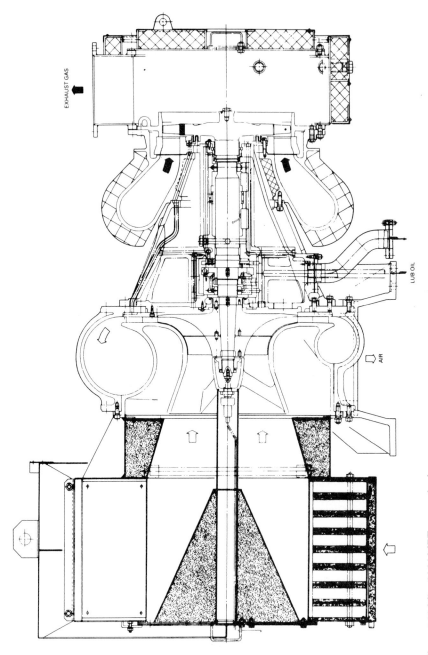

Figure 4.12 Mitsubishi MET type turbocharger

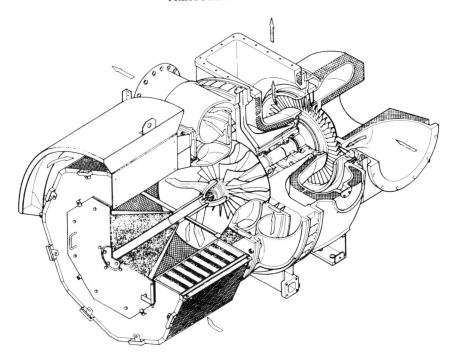

Figure 4.13 Mitsubishi Super MET B type turbocharger

which is aligned with the bearing pedestal; the bearings are white-metal lined with lubrication from an external supply.

Apart from the advantage of a higher gas exit temperature giving more useful heat for exhaust gas boiler systems, non-water cooled turbochargers are said to last longer and require less maintenance than traditional jacketed casing types because high temperatures above the acid dew point prevent corrosion, they are quieter in operation and corrosion from cooling water is eliminated, allowing lighter construction. For these reasons the main manufacturers of turbochargers are now offering non-water cooled models of either own design or built under licence.

C.T.W.

5 Fuel Injection

The emphasis in this chapter is on medium-speed engines which all have camshaft-actuated individual jerk pumps for each cylinder. Higher speed engines are also covered. including those which use camshaft pumps (or block pumps); that is to say, those in which all the jerk pump elements are grouped into one or more complete units, each equipped with a common camshaft.

This section does not apply, (except in a general way) to the direct drive two-stroke engines which favour the common rail system, or other systems described later under individual makes.

INJECTION AND COMBUSTION

The essence of a diesel engine is the introduction of finely atomised fuel into the air compressed in the cylinder during the piston's inward stroke. It is, of course, the heat generated by this compression, which is normally nearly adiabatic, that is crucial in achieving ignition. Although the pressure in the cylinder at this point is likely to be anything up to 60–70 bar, the fuel pressure at the atomiser will be of the order of 10 times as great.

There is a body of evidence to suggest that high injection pressure at full load confers advantages in terms of fuel economy, and also in the ability to digest inferior fuel. Most modern medium-speed engines, i.e. up to 1000 rev/min, attain 800–1000 bar in the injection high-pressure pipe. Some recent engine designs achieve as much as 1200 bar when pumping heavy fuel. For reasons of available technology, the earliest diesel engines had to use compressed air to achieve atomisation of the fuel as it entered the cylinder (air blast injection), and while airless, or solid, injection conferred a significant reduction in parasitic loads, it also presented considerable problems in the needs for high-precision manufacture, and the containment of very high and complex stresses.

The very high standard of reliability and life now attained by modern fuel injection systems, notwithstanding their basic simplicity, belies a considerable achievement in painstaking research by fuel injection equipment (FIE) manufacturers.

FUEL INJECTION

In the early days of airless injection, many ingenious varieties of combustion chamber were used, sometimes mainly to reduce noise, sometimes mainly to reduce smoke, sometimes to ease starting; but often in part to reduce, or to use modest, injection and combustion pressures.

With some significant exceptions growing emphasis on economy and specific output (which also encourages economy in order to reduce heat flux), coupled with material and calculation advances which allowed greater loads to be carried safely, have conspired to leave the direct injection principle almost unchallenged in modern medium-speed and many high-speed engines.

Direct injection is what it says it is, the fuel is delivered directly into a single combustion chamber formed in the cylinder space (Figure 5.1), atomisation being achieved as the fuel issues from small drillings in the nozzle tip.

For complete combustion of the fuel to take place, every droplet of fuel must be exposed to the correct proportion of air to achieve complete oxidation, or to an excess of air. In the direct injection engine the fuel/air mixing is achieved by the energy in the fuel spray propelling the droplets into the hot, dense air. Additional mixing may be achieved by the orderly movement of the air in the combustion chamber, which is called 'air swirl'. Naturally aspirated engines usually have a degree of swirl and an injection pressure of around 800 bar. Highly turbocharged engines with four valve heads have virtually no swirl, but generally have an injection pressure of 1200 bar to provide the mixing energy. Where indirect injection is concerned some high-speed engines retain a pre-chamber in the cylinder head into which fuel is injected as a relatively coarse spray at low pressure, sometimes using a single hole. Combustion is initiated in the pre-chamber, the burning gases issuing through the throat of the chamber to act on the piston (Figure 5.2).

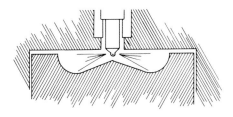
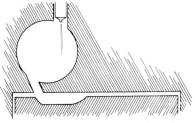

Figure 5.1 Cross section of direct injection combustion chamber

Figure 5.2 Indirect injection: the Ricardo pre-chamber

Fuel/air mixing is achieved by a very high air velocity in the chamber. The air movement scours the walls of the chamber promoting good heat transfer. Thus the wall can be very hot — requiring heat resistant materials — but it can also absorb too much heat from the air in the initial compression strokes during starting, and prevent ignition. It is these heat losses that lead to poor starting and inferior economy. Further forms of assistance, such as glowplugs, are therefore sometimes necessary in order to achieve starting when ambient pressures are low. The throttling loss entailed by the restricting throat also imposes an additional fuel consumption penalty.

One manufacturer (SEMT) has, however, achieved an ingenious combination of the two systems by dividing the pre-chamber between cylinder head and piston crown. At TDC a stud on the piston enters the pre-chamber to provide a restricted outlet. On the expansion stroke the restriction is automatically removed and fuel economy comparable with normal direct injection engines is attainable (Figure 5.3).

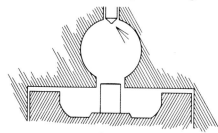

Figure 5.3 The variable geometry combustion chamber of SEMT Pielstick

Direct injection too has variants, which reflect the fact that despite a vast expenditure on research into its mechanism, the detail of how combustion develops after ignition is achieved is still almost entirely empirical.

In the author's view, the essentials are:

1. That some at least of the fuel injected is atomised sufficiently finely to initiate combustion. Ignition cannot take place until a droplet of fuel has reached the temperature for spontaneous ignition. Since heat is taken up as a function of surface area (proportional to the square of the diameter) and the quantity of heat needed to achieve a temperature rise is a function of volume (varies as the cube of the diameter) only a small number of fine droplets are needed to initiate combustion. High-speed photography of combustion does indeed show that ignition takes place in a random manner near the injector tip and usually outside the main core of the spray.

2. That the fuel should mix with the air in order to burn. Since most of the air in a roughly cylindrical space is, for geometrical reasons, near the periphery, most of the fuel must penetrate there, and this is easier with large droplets. Hence also the use of wide corse angles and multiple spray holes.

FUEL INJECTION

3. That under no circumstances must fuel reach the liner walls or it will contaminate the lubricating oil. An advantage of combustion spaces formed in piston crowns is that the walls of the chamber form a safe target at which spray may be directed. This type of combustion chamber in the piston has the further advantage that during the piston descent, air above the piston periphery is drawn into the combustion process in a progressive manner.
4. The injection period should be reasonably short, and must end sharply. Dribble and secondary injection are frequent causes of smoke, and also of lubricating oil becoming diluted with fuel or loaded with insoluble residues.

Dribble is the condition where fuel continues to emerge from the nozzle at pressures too low to atomise properly. It is caused by bad seating faces or slow closure.

Secondary injection is what happens when the pressure wave caused by the end of the main injection is reflected back to the pump and then again to the injector, reaching it with sufficient pressure to reopen the injector at a relatively late stage in combustion. Any unburnt or partly burnt fuel may find its way on to the cylinder walls and be drawn down by the piston rings to the sump.

INJECTOR

Working backwards from the desired result to the means to achieve it, the injector has to snap open when the timed high-pressure wave from the pump travelling along the high-pressure pipe has reached the injector needle valve. Needle lift is limited by the gap between its upper shoulder and the main body of the holder. Needle lift is opposed by a spring, set to keep the needle seated until the 'blow-off pressure' or 'release pressure' of the injector is reached by the fuel as the pressure wave arrives from the pump. This pressure is chosen by the engine builder to ensure that there is no tendency for the needle to reopen as the closing pressure waves are reflected back and forth along the high-pressure pipe from the pump. The setting also has some effect on injection delay and the quality of injection. It is usually chosen to be between 200–300 bar.

The needle valve is invariably provided with an outer diameter on which the fuel pressure acts to overcome the spring pressure, and cause the initial lift. This brings into play the central diameter of the needle so that it snaps open to the lift permitted.

The needle and seat cones are usually ground to a differential angle so that contact is made at the larger diameter. When the needle is only slightly open the greater restriction to flow is at the outer rather than the inner diameter of the seat. This ensures that as the needle lifts, there is a sudden change of pressurised area and the needle force against the spring changes correspondingly, giving a very rapid lift to fully open; and conversely when closing. (It is for this reason that it is bad practice to lap the needle to its seat.)

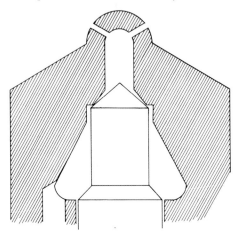

Figure 5.4 Detail of needle and seat showing differential angle (exaggerated)

Control of the temperature of the injector, particularly of the sensitive region round the needle seat and the sac, is very important. This is especially so when heavy fuels are used, both because the fuels themselves have to be heated, and because they tend to burn with a more luminous flame that increases the heat input to all the metal surfaces. If the tip temperature rises too high, the fuel remaining in the sac after injection boils and spits from the spray holes: it is not properly burnt and forms a carbon deposit around each spray hole. Such carbon formation, which builds up fairly rapidly around the boundary of the sprays, causes a rapid deterioration in the quality of combustion.

With smaller engines up to about 300 mm bore, it is usually sufficient to rely on the passage and recirculation of the pumped fuel itself to achieve the necessary cooling. The injector clearances may, in fact, be eased slightly to promote such circulation. This back leakage, normally about 1% of the pumped fuel, is led away to waste, or via a monitoring tank to the fuel tank. For larger engines it is usually necessary to provide separate circuits specifically for a coolant flow, either of treated water, or of light fuel or of lubricating oil.

All of these considerations lead to a relatively bulky injector, and thus to tricky compromises with the available space within the cylinder head on single piston engines, where space is also needed to permit the largest possible valve areas, cages and cooling. Nonetheless there are performance advantages in highly rated engines in reducing the inertia of the components directly attached to the needle of the injector, and this leads FIE manufacturers to try to find space for the injector spring at the bottom rather than at the top of the injector, as in Figure 5.5.

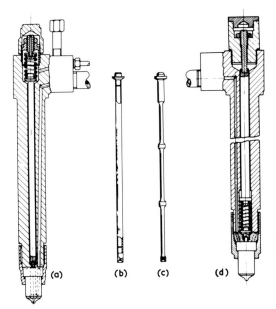

Figure 5.5 Comparison of traditional and low inertia injectors:
(a) conventional injector
(b) and (c) lightweight spindles
(d) low inertia injector
(Lucas Bryce Ltd.)

FUEL LINE

Figure 5.6 shows a typical fuel line pressure diagram and also the needle lift diagram for an engine having a camshaft speed of 300 rev/min both at overload (Figure 5.6(a)); and at half load (Figure 5.6(b)). Figure 5.6(a) shows a very brief partial reopening of the nozzle, or secondary injection, after the main opening period. In Figure 5.6 the breaks in each trace are made to indicate intervals of 3 cam degrees.

The fuel line has to preserve as much as possible of the rate of pressure rise, the maximum pressure and the brevity of injection which the

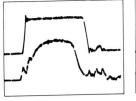 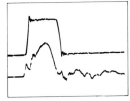

Figure 5.6 Fuel line pressure (lower line) and needle lift diagrams
(a) left. At high load
(b) right. At low load
(Lucas Bryce Ltd)

fuel pump has been designed to achieve. To do this the fuel line must have the lowest practical volume and the maximum possible stiffness against dilation. There is also a limit to the minimum bore because of pressure loss considerations.

The only real degree of freedom left to the designer is to make the fuel line as short as possible. Even so, a given molecule of fuel may experience two or three injection pulses on its way from the pump chamber to the engine cylinder in a medium-speed engine running at full load. This rises to 30 or more at idling.

If the fuel pump is mounted at a greater distance from the injector, the elastic deformation of the longer pipe and compression of the larger volume of fuel will dissipate more and more of the pressure developed at the fuel pump and prolong the period of injector needle lift compared with the pump delivery. In all medium-speed engines this compromise on injection characteristics, and its adverse effect on fuel economy (and heat flux in combustion chamber components) is considered unacceptable, and each cylinder has an individual fuel pump with a high-pressure fuel line as short as possible.

The smaller the engine, the more manufacturers (in fact all manufacturers of automotive sizes) opt for the production convenience of a camshaft pump (i.e. the pump elements serving all or several cylinders are mounted in one housing with their own camshaft) and accept any compromise in performance. In such engines a given molecule of fuel may endure 20 or more injection pulses while migrating along the high-pressure pipe at full load. In such cases the elastic characteristics of the HP pipe (and of the trapped fuel) come to equal the pumping element as a major influence on the qualities and duration of injection.

Naturally all injection pipes must have identical lengths to equalise cylinder behaviour. On a 150 kW auxiliary engine of approximately 150 mm bore running at 1200 rev/min, the effect of a 600 mm difference in pipe length, due to running pipes each of different length directly from pump to the cylinder it served, was 8° in effective injection timing, and consequently about 25 bar difference in firing pressure.

Camshaft or block pumps are, of course much cheaper, and one maker (Deutz) has adopted for his higher speed engine the ingenious solution of grouping up to three or four pump elements in one housing, each such housing being mounted close to the cylinders concerned.

The HP fuel lines are made of very high quality precision drawn seamless tube almost totally free from internal or external blemishes. They must, moreover, be free of installation stress, or they risk fracture at the end connections. They must be adequately clipped, preferably with a damping sleeve, if there is any risk of vibration. Fuel line failures due to pressure are not as frequent as they once were, because of improvements in the production of seamless tube. Nevertheless current regulations insist that in marine use fuel lines are sleeved or enclosed to ensure that in the event of a fracture, escaping fuel is channelled to a tank where an alarm may be fitted. This is to avoid the fire hazard of fuel finding a surface hot enough to ignite it, or of accumulating somewhere where it could be otherwise accidentally ignited. Most designers now also try to ensure that leaking fuel cannot reach the lubricating oil circuit via the rocker gear return drainage or by other routes.

Several manufacturers shorten and segregate the HP line by leading it to the injector through, rather than over, the cylinder head. One or two incorporate the pump and the injector in a single component, avoiding the high-pressure pipe altogether.

PUMP

The final item for consideration is the fuel injection pump, which is the heart of the fuel injection system. Essentially the pumping element is a robust sleeve or barrel which envelops a close fitting plunger. Each is produced to a very fine surface finish of 1–2 microinches and to give a clearance of approximately 7–10 µm (microns) depending on the plunger diameter.

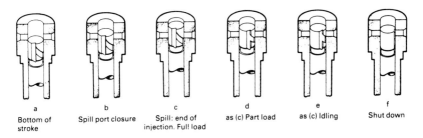

Figure 5.7 Principle of operation of the fuel injection pump   (Lucas Bryce Ltd)

In its simplest form (Figure 5.7) the barrel has a supply port at one side and a spill port at the other. (In early times these were combined.) The plunger is actuated by the fuel cam through a roller follower. Before, and as the plunger starts to rise (Figure 5.7(a)), the chamber is free to fill with fuel, and to spill back into the gallery in the pump housing outside, until the rising top edge of the plunger closes the supply and spill ports (Figure 5.7(b)). Thereafter the fuel is pressurised and displaced through the delivery valve towards the injectors. The initial travel ensures that the plunger is rising fast when displacement commences, so that the pressure rises sharply to the desired injection pressure.

When the plunger has risen far enough, a relieved area on it uncovers the spill port, the pressure collapses, and injection ceases (Figure 5.7(c)). The relief on the plunger has a helical top (control) edge so that rotation of the plunger by means of the control rod varies the lift of the plunger during which the spill port is closed, and therefore the fuel quantity injected and the load carried by the engine, as shown in Figures 5.7 (c, d, e). At minimum setting the helical edge joins the top of the plunger and if the plunger is rotated so that this point, or the groove beyond it, coincides with the spill port, the latter is never closed. In that case no fuel is pumped and that cylinder cuts out (Figure 5.7(f)). The area of the plunger between the top and the control edge is termed 'the land'.

There are, of course, many complications of this simple principle. The first is that a delivery valve, which is essentially a non-return valve, is needed at the pump outlet to keep the fuel line full, as shown in Figure 5.8. The delivery valve usually has an unloading collar to allow for the elasticity of the pipe and fuel, and to help eliminate the pressure pulsations which cause secondary injection. Some types of delivery valve, however, sense and unload the line pressure directly.

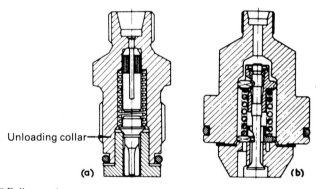

Figure 5.8 Delivery valves
  (a) volume unloading (b) direct pressure unloading   *(Lucas Bryce Ltd)*

The second cause of complication is that, when the spill port opens and a pressure of anything up to 1200 bar is released, cavitation and/or erosion is likely to occur, affecting both the housing directly opposite the port, and the plunger land which is still exposed to the port at the instant of release. Modern designs seek to eliminate these effects by using respectively a sacrificial (or in some cases a specially hardened) plug in the housing wall, and by special shaping of the ports.

Another complication is the filling process. When the pump spills, further plunger lift displaces fuel through the spill port. As it falls on the descending flank of the cam, the opposite effect applies until the falling control edge closes the spill port. Further fall, under the action of the plunger return spring, has to expand the fuel trapped above the plunger until the top edge uncovers the supply port again. This it can only do by expanding trapped air and fuel vapour, and by further vaporisation, so that a distinct vacuum exists, and this is exploited to refill the pump.

There is also the effect of dilation due to pressure, the need for lubrication, and the need to prevent fuel from migrating into spaces where it could mingle with crankcase oils.

Early pumps could get by without undue complications: the fuel itself provided adequate plunger lubrication, and fuel which had leaked away in order to do this was simply rejected to a tank which was emptied by a junior engineer to waste.

Nowadays grooves are provided on the plunger or in the barrel wall. The first collects leaking fuel. Where there are two grooves the second will spread lubrication, but there is sometimes an additional groove between them to collect any further migration either of fuel or lubricant. Figure 5.9 shows a section of a modern heavy duty fuel pump incorporating most of the features described above.

The pump is actuated by a follower which derives its motion from the cam. Block pumps incorporate all these components in a common housing for several cylinders, while individual pumps used on some larger engines incorporate the roller. This not only makes servicing more straightforward, it also ensures unified design and manufacturing responsibility.

The roller is a very highly stressed component, as is the cam, and close attention to the case hardening of the cam and to the roller transverse profile is essential to control Hertzian (contact) stresses. The roller is sometimes barrelled very slightly, and it must bear on the cam track parallel to the roller axis under all conditions. On the flanks of the cam this is a function of slight freedom for the follower assembly to rotate, while on the base circle and the top dwell of the cam it depends on precise geometry in the assembly.

The roller pin requires positive lubrication, and the cam track must at least receive a definite spray. Lubrication channels have to be provided from the engine system through the pump housing to the roller surface, its bearings, and to the lower part of the plunger.

Timing

The desired timing takes into account all the delays and dynamic effects between the cam and the moment of ignition. The desired timing is that which ensures that the combustion starts and continues while the cylinder pressure generated can press on the piston and crank with the

FUEL INJECTION

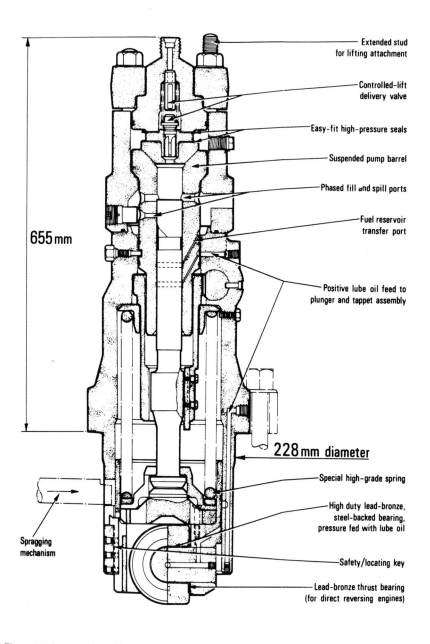

Figure 5.9 Cross section of Bryce FFOAR type flange-mounted unit pump suitable for four-stroke engines with crankshaft speeds up to 450 rev/min

(Lucas Bryce Ltd)

greatest mechanical advantage, and is complete before the exhaust valve opens at 110–130° ATDC. The engine builder always specifies this from his development work, but it usually means that the only settable criterion, the moment of spill port closure (SPC), is about 20–25° BTDC.

In earlier days this was done by barring (or manually turning over) the engine with the delivery valve on the appropriate cylinder removed, and with the fuel circulating pump on (or more accurately with a separate small gravity head), and the plunger rotated to a working position. When fuel ceased to flow, the spill port had closed. Sometimes this criterion was used to mark a line on a window in the pump housing through which the plunger guide could be seen, to coincide with a reference line on the guide.

The accuracy of this method is, however, disappointing. There are differences between static and dynamic timing which vary from pump to pump and many makers now rely on the accuracy of manufacture, and specify a jig setting based on the geometry of the pump and the SPC point to set the cam rise, or follower rise at the required flywheel (crank) position.

Some makers allow users to make further adjustments of the timing according to the cylinder pressure readings, but this places a premium on the accuracy not only of the calibration of instruments, but of the method itself which in the author's view is not very reliable. It should be borne in mind that neither the instruments nor the method are as accurate as the pump manufacture, by an order of magnitude.

If the cam is fixed, it is usually possible, within stated limits, to adjust the height of the tappet in relation to the cam. Raising it has the effect that the plunger cuts off the supply spill port (and starts injection) sooner for a given cam position, and vice versa (Figure 5.10). The limit is set at one extreme by the need for there to be 'top clearance' at the top of the follower lift (i.e. of the plunger stroke) between the top of the plunger and the underside of the delivery valve, in order to avoid damage. Too small a clearance also means that injection is commencing too near the base of the cam, and therefore at too low a velocity to give the required injection characteristics. On the other hand, too large a clearance (to compensate for an over-advanced cam by lowering the plunger) risks interfering with the proper spill function which terminates injection.

The cam with its follower may be designed to give a constant velocity characteristic throughout the injection stroke. Recent designs with a high initial velocity aim to give a constant pressure characteristic, but many manufacturers find that the performance advantage or stress advantage over the constant velocity design is too small to warrant the greater production cost.

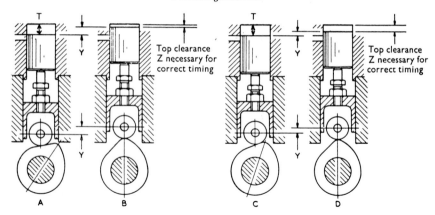

Figure 5.10 Adjusting timing by top clearance. Configuration of cam, tappet and pump at commencement of injection (A and C); and maximum cam and follower lift (B and D). In each case the pump is set correctly for the required injection point. A shows the case for an over-retarded cam, C that for an over-advanced cam within permitted limits of top clearances. In either case Y (the residual lift) + Z = the timing dimension of the pump, T (GEC Ltd)

Uniformity

Bearing in mind the random way in which ignition occurs during each injection, and perhaps also because the filling process of the pump chamber is largely dependent on suction, it is not surprising that successive combustion cycles in the same cylinder at the same steady load are not identical. A draw card taken slowly to show a succession of cycles will show this visibly, and a maximum pressure indicator will show by feel, or otherwise, that maximum pressures vary from cycle to cycle perhaps by 5 bar or more. (Any governor corrections will have an effect too, but progressively over several cycles.) These variations are a fact of life, and not worth worrying about unless they are 20 bar or more. It does mean that if indicator cards are being taken as a check on cylinder performance, several cycles per cylinder should be taken to ensure that a properly representative diagram is obtained.

Differences will also occur between cylinders for reasons of component tolerance (unless compensated adequately by the setting method), and because of the dynamic behaviour of the camshaft, which deflects in bending due to cam loading and torsionally due to the interaction of the cam loadings of adjacent cylinders. These will characteristically fall into a pattern for a given engine range and type.

Whilst engine builders should (and usually do) allow for variations in cylinder behaviour to occur, it is obviously good practice to minimise these to achieve maximum economy with optimum stresses. However, it would be wise to bear in mind that it is inherently very difficult to make pressure and temperature measurements of even the average behaviour of combustion gases in the cylinders.

To achieve, without laboratory instrumentation techniques, an accuracy which comes within an order of magnitude of the standard of accuracy to which fuel injection components are manufactured and set is impossible. This is particularly true of exhaust temperature as a measure of injection quantity, notwithstanding that this practice has been generally accepted by many years of use. Fuel injection settings on engines should not therefore be adjusted without good reason, and then normally only on a clear indication that a change (not indicative of a fault) has occurred.

Fuel quantity imbalance is far from being the only cause of a variation in the pattern of exhaust temperatures, nor timing of firing pressure. Consider, for instance, variations in compression ratio, charge air mass, scavenge ratio, not to mention calibration errors and the accuracy of the instruments used.

Fuel injection pumps (and injectors) have their own idiosyncrasies. The relationship between output and control rod position (the calibration) is not absolutely identical in different examples of pumps made to the same detail design. The output of a family of similar pumps can be made sensibly identical at only one output, namely at the chosen balance point. At this output a stop, or cap, on the control rod is adjusted to fit a setting gauge (or, in older designs, a pointer to a scale), so that it can be reproduced when setting the pumps to the governor linkage on the engine.

Obviously the calibration tolerances will widen at outputs away from the balance point, which is usually chosen to be near the full load fuelling rate. As the tolerances will be physically widest at idling, and will also constitute a relatively larger proportion of the small idling quantity (perhaps 35%) the engine may well not fire equally on all cylinders at idling, but this sort of difference is not sufficient to cause any cylinder to cut out unless another fault is present.

When a pump is overhauled it is advisable that its calibration should be checked, and the balance point reset. This ideally entails the use of a suitable heavy duty test rig, but in the largest sizes of pump it is sometimes more practical to rely on dimensional settings.

D.A.W.

6 Sulzer engines

Slow-speed marine diesel engines manufactured by Sulzer Brothers Ltd., of Switzerland, and their many worldwide licensees, are traditionally of the single-acting two-stroke turbocharged type employing loop scavenging. However, at the end of 1981 Sulzer announced a completely new family of crosshead engines designated as the RTA series and marking a complete changeover from loop to uniflow scavenging in acknowledgement of the superior thermodynamic performance of the latter coupled with a constant pressure turbocharging system. The original loop scavenged RD type was followed by the RND, RND-M, and RLA types, and the production programme in 1982 was centred on the RLB while development work with the RTA led to its introduction in 1983. Apart from the use of a centrally mounted exhaust valve in the cylinder cover, RLB and RTA components are in many respects similar; but a major characteristic of the newest RTA is the 'super long stroke', about 3:1 ratio of piston stroke to bore, in order to obtain both a better thermo-efficiency and a lower shaft speed to suit the gain in propulsive efficiency from a large diameter slower turning propeller.

RL TYPE ENGINES

A completely new engine, the RLA56, a small bore two-stroke engine, was introduced in 1977 to incorporate the basic design concept of the highly successful RND and RND-M series, but to extend the power range at the lower end.

This small engine, of comparatively long stroke design, was the first model of the RLA series. It retains many of the design features of the most recent economical loop scavenged RLB type, with both engines utilising many of the design features of the earlier RND-M series.

A number of RND-M engine features have been retained for the RLA and RLB type, namely:

1. Constant pressure turbocharging and loop scavenging.
2. A bore cooled cylinder liner and one piece bore cooled cylinder cover and water cooled piston crowns.

SULZER ENGINES

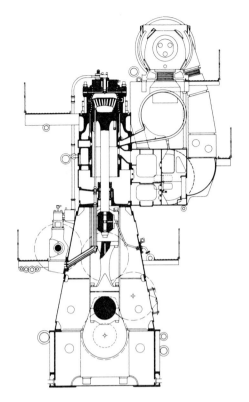

Figure 6.1(a) Cross section of Sulzer RLB90 engine

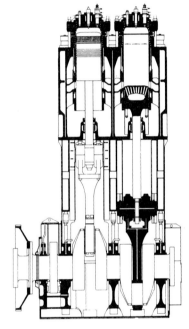

Figure 6.1(b) Longitudinal section of RLB90 engine

3. Double guided crossheads.
4. New cylinder liner lubrication system with accumulators for the upper liner part and thin-walled aluminium–tin crosshead shell bearings.

Major new design features peculiar to the RL types are:

1. A new bedplate design with an integrated thrust block; a new box type column design.
2. A semi-built or monoblock type crankshaft without a separate thrust shaft.
3. Location of the camshaft gear drive at one end for engines of four to eight cylinders.

4. Multiple cylinder jackets.
5. A bore cooled piston crown.
6. Modified crosshead.
7. A new design of air receiver.

Bedplate with thrust block

For the RL type engine bedplate a new concept of great simplicity has been applied. Both the cross girders and the longitudinal structure are of single-wall fabricated design giving very good accessibility for the welded joints. The central bearing saddles are made of cast steel and only one row of mounting bolts on each side is used to secure the bedplate to the ship's structure.

The completely new design feature is the method of integrating the thrust block into the bedplate (Figure 6.2) allowing an extremely compact design and saving engine length. The first crankshaft bearing on the driving end is a combined radial-axial bearing.

The bedplate is a one-piece structure for all engines of from four to eight cylinders, but if required it can be bolted together from two halves.

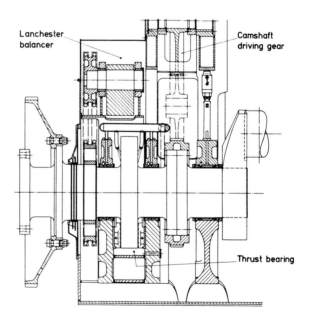

Figure 6.2 Integrated thrust bearing and camshaft drive

Thick-walled white-metal lined bearing shells are used to support the crankshaft and guarantee an optimum safety for the running of the crankshaft.

Crankshaft

In addition to the traditional semi-built type crankshaft, a monoblock continuous grain flow forged type can also be used. From four to eight cylinders, the crankshaft is a one-piece component with an integrated thrust collar section. Only one journal and pin diameter is used for all engines up to eight cylinders. For RLA engines larger than the RLA56, the crankshaft is semi-built, and the thrust shaft is separately bolted to the crankshaft.

Engine frame

For the small RLA56 engine a new method of frame construction is used. Cast iron central pieces, on which the crosshead guides are bolted, are sandwiched between two one-piece fabricated side-frame girders. This replaces the traditional construction of 'A' frame bolted to the bedplate with side plates attached with access doors forming the enclosure. Larger RL type engines use 'A' shaped columns of fabricated double wall design, assembled with longitudinal stiffening plates to constitute a rigid structure between the bedplate and cylinder blocks.

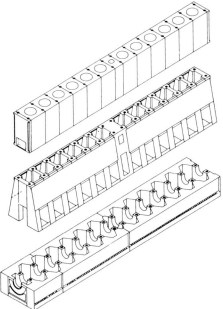

Figure 6.3 Structural arrangement of bedplate, columns and cylinder jackets for 12-cylinder Sulzer engine

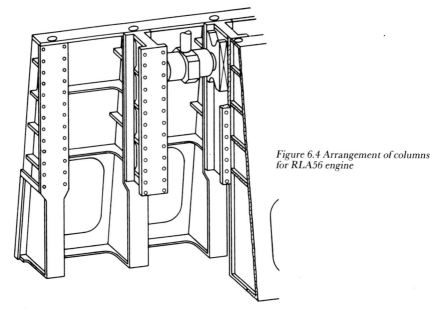

Figure 6.4 Arrangement of columns for RLA56 engine

Cylinder jackets

The cylinder jackets are made of fine lamellar cast iron produced as single block units bolted together in the longitudinal plane to form a single rigid unit, and held on top of the frame section by long tie bolts secure in the bedplate. For the small RLA56 engine, multi-cylinder blocks consisting of two or three cylinders are standard and provide great rigidity of the structure. The arrangement of the cooling water passages in the jackets ensures forced water circulation and optimal water distribution around the exhaust canal.

Because of the steadily rising charge air pressures, the design of the air receiver has been modified to allow the use of automatic welding techniques. Instead of a rib-stiffened plain side plate, a semi-circular pressure containment has been fitted with an integral air inlet casing. The auxiliary blower is mounted on the front end of the receiver, thus eliminating inclined ducts.

Combustion chamber components

The combustion chamber of RL type engines is principally of the same shape as that of the RND-M. The cylinder cover is basically a one-piece steel block with cooling bores, while an identical arrangement of bore cooling is applied to the upper collar of the cylinder liner which is made from lamellar cast iron, a material of good heat conductivity and wear

SULZER ENGINES

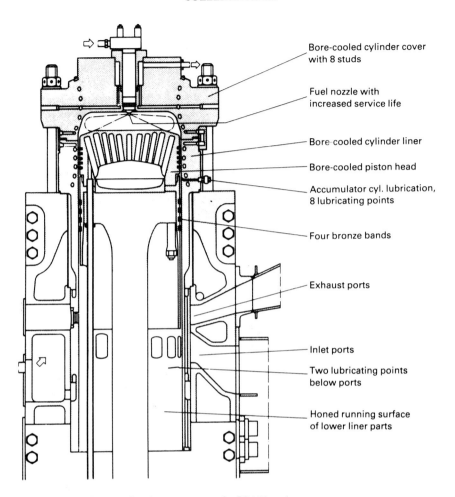

Figure 6.5 Combustion chamber components for RLA90 engine

resistance. Figure 6.5 shows the arrangement with cover, cylinder liner and piston.

A new type of piston crown was introduced for the RL series engines. This combustion chamber component similarly uses a bore cooling arrangement (Figure 6.6) as previously only applied to the liner and cover. Water cooling is retained but the piston bore cooling uses a somewhat different mechanism to the force flow system of the liner and cover.

The cooling space of the piston crown is approximately half filled with cooling water and as a result of piston acceleration and deceleration, a 'cocktail shaker' effect is produced to provide excellent heat

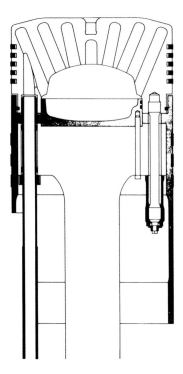

Figure 6.6 Detail view of piston with bore cooling

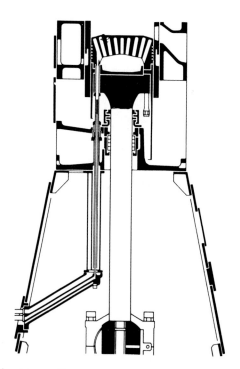

Figure 6.7 Arrangement of telescopic pipes for piston cooling

removal. This effect is capable of ensuring efficient heat transfer under all prevailing load conditions in order to keep the vital temperatures on the piston crown and around the piston rings within suitable limits.

As a result of this new design the crown temperatures are lowered compared with the previous construction. Forged steel blocks have been used for piston crown manufacture but cast steel versions are also used.

Cylinder lubrication

Lubrication of the cylinder is through six quills mounted in the upper area of the liner just above the cylinder jacket, as shown in Figure 6.8. The oil distribution grooves have a very small angle of inclination to avoid blow-by over the piston rings and small but regular quantities of cylinder oil supplied by an accumulator system. Two further lubrication points are provided below the scavenge ports on the exhaust side with the necessary oil pumps positioned on the front side of the engine above the camshaft drive.

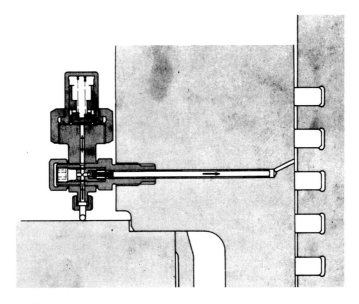

Figure 6.8 Cylinder lubricator

Crosshead design

The crosshead is similar to that used for RND engines and has double guided slippers. The pin size has been increased for safety and thin-walled half shells of the aluminium–tin type are used. The pin itself is of forged homogenous steel of symmetrical design, and can therefore be turned around in case of damage. The piston rod is connected to the pin by a single hydraulically tightened nut, while the slippers, made of cast steel and lined with white metal, are bolted to the ends of the pin. The cast iron double guide faces are fitted to the engine columns.

The connecting rods of traditional marine type have a forged normalised steel bottom end bearing, lined with white metal, held in place by four hydraulically tensioned bolts. Compression shims are provided between the bottom end bearing and the palm of the connecting rod.

Piston assembly

The piston (see Figure 6.6) consists of water cooled cast steel crown, a cast iron skirt with copper bandages and a forged steel piston rod, with the piston rings fitted in chromium plated grooves.

The water cooled piston has proved very reliable when running on heavy fuels and the use of water cooling has resulted in practically negligible system oil consumption. With oil cooling, oil is consumed usually as a result of thermal ageing on hot piston walls. Oil leaks from oil cooled pistons may also occur on other engine types. The two-part gland seals for the piston rod and telescopic piston cooling pipes can be inspected while the engine is running and can be dismantled without removing the piston. The double-gland diaphragm around the piston rod completely separates the crankcase from the piston undersides, preventing contamination of the crankcase oil by combustion residues or possible cooling water leakages. In addition, the fresh water piston cooling water system is served by an automatic water drain-off when the circulating pumps are stopped. This avoids leakages when the ship is in port.

Turbocharging arrangement

Sulzer RL engines all employ the constant pressure turbocharging system and with the latest high efficiency Brown Boveri series 4 turbochargers include provision for automatic cleaning. The layout of the turbocharging arrangement is shown in Figure 6.9.

The use of the piston undersides to provide a scavenge air impulse eliminates the need for large electrically driven auxiliary blowers. At higher loads a simple flap valve opens to cut out the piston underside pumping effect with a consequent improvement in fuel consumption,

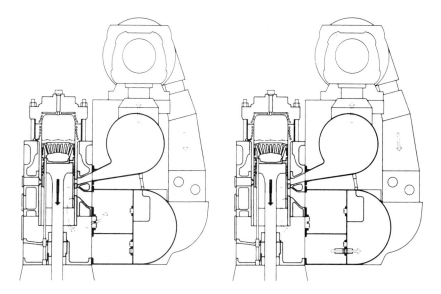

Figure 6.9 Constant pressure turbocharging and operation of under-piston supercharging

whilst a small auxiliary fan incorporated in the scavenging system improves the smoke values at the lowest loads. The piston underside scavenge pump facility allows the engine to start and reverse even with total failure of all turbochargers. The auxiliary fan and engine will even operate at up to 60% load, thus giving a 'take home' facility.

The turbochargers are mounted on top of the large exhaust gas receiver with the scavenge air receiver which forms part of the engine structure beneath. The charge air is passed down through seawater cooled intercoolers which are mounted accessibly alongside the scavenge air receiver. Flap type valves which operate according to the scavenge air pressure direct the air inside the three compartment air receiver to piston underside (at low scavenge air pressure) or direct to the scavenge ports when the higher air pressure from the turbochargers keeps the underside delivery flap valves shut.

For RL type engines equipped with only one turbocharger, a separate turbocharger/air intercooler module is available as a standard option. This unit can be located adjacent to the engine at either the forward or aft end thus considerably reducing the mounting height and width of the main engine. The turbocharger module consists of a base frame onto which the turbocharger, charge air cooler, air ducts and cooling water pipes are solidly mounted with flexible connections to the engine.

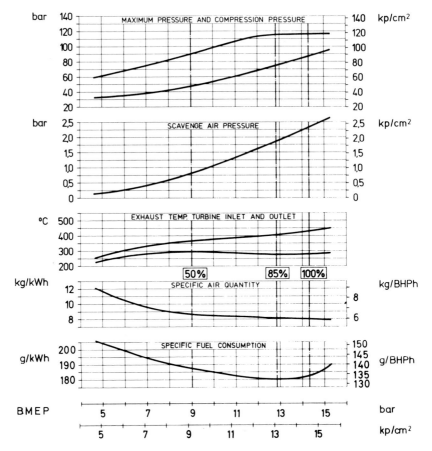

Figure 6.10 Typical performance curves for 6RLB90 engine

Controls

RL type engines use a pneumatic control system based on a Woodward PGA58 pneumatic governor. This is placed on the middle platform and driven by the same gear system as the fuel injection pumps. The engine manoeuvring stand is a small pneumatic unit designed for mounting in a control room desk and connected to the engine by a few 8 mm diameter copper pipes. This unit incorporates the reversing and speed setting levers, as well as starting, emergency stop and run buttons. The telegraph receiver can also be built into the stand with the reply lever directly connected to the engine reversing lever. Emergency manual controls are situated on the middle platform near the main governor.

The engines are provided with an electro-pneumatic overspeed and shut down device which acts directly on the fuel injection pumps.

Standard bridge control system

Remote control of the main engine from the bridge is nowadays considered essential for practically all ships, and Sulzer provides their own design SBC-7 system for their engines. The system (Figure 6.11) has been developed by the engine builder to form an integral part of the engine and to match its requirements.

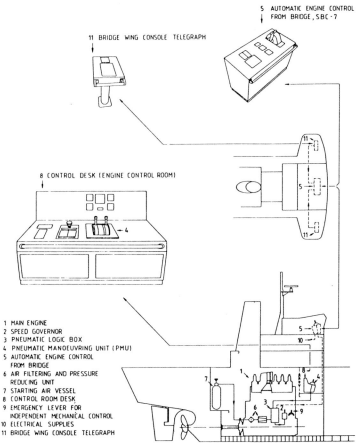

Figure 6.11 Engine and bridge control system

The SBC-7 bridge control system is basically electro-pneumatic in operation with pneumatics extended up to the wheelhouse so that any electronics faults will not prevent the engine from being manoeuvred from the bridge.

The system provides fully automatic engine direction selection, and starting and speed control from the telegraph level with full interlocks.

There are pre-set engine speeds for each telegraph order position, and additional push buttons are provided to allow fine speed adjustment up or down from the pre-set manoeuvring speeds. An automatic speed increase programme is provided to allow slow acceleration of the engine from 'Full ahead' manoeuvring speed up to any pre-selected service speed, and vice versa for when approaching port. A Sulzer telegraph specially designed for pneumatic remote control of the engine can also be used as a conventional telegraph to transmit orders to the engine room.

RTA TYPE ENGINES

At the end of 1981, Sulzer announced the introduction of a completely new series of two-stroke crosshead engines for the changing market of the 1980s. The new engine type, designated the RTA 'super long stroke', employs the large stroke-to-bore ratio of around 3:1 and — most significantly — a radical changeover from the loop to the uniflow scavenging system. It is likely that for the foreseeable future, this engine will dominate Sulzer's production programme because of its superior thermodynamic performance despite being offered as a companion to the RLB types.

Figure 6.12 First Swiss built Sulzer 7RTA58 on testbed

SULZER ENGINES

Table 6.1

Engine type	Bore/stroke (mm/mm)	Standard ratings	Rating point	Speed n (rev/min)	Mean piston speed (m/s)	Mean effective pressure (bar)	Mean effective pressure (kp/cm²)	Engine power P (kW/Cyl)	Engine power P (BHP/Cyl)	Specific fuel consumption* +3% 100% P g/(kWh)	Specific fuel consumption* +3% 100% P g/(BHPh)	90% P g/(kWh)	90% P g/(BHPh)	85% P g/(kWh)	85% P g/(BHPh)	Number of cylinders
RTA 84	840/2400	Engine maximum continuous rating	R1	87	6.96	15.35	15.67	2960	4030	173	127	170	125	170	125	4—10, 12
		Economy ratings	R2	82	6.56	13.64	13.90	2480	3370	169	124	167	123	167	123	
			R3	74	5.92	15.30	15.59	2510	3410	171	126	169	124	169	124	
			R4	70	5.60	13.53	13.78	2100	2850	167	123	166	122	166	122	
RTA 76	760/2200	Engine maximum continuous rating	R1	95	6.97	15.31	15.62	2420	3290	173	127	170	125	170	125	4—10, 12
		Economy ratings	R2	90	6.60	13.56	13.83	2030	2760	169	124	167	123	167	123	
			R3	81	5.94	15.22	15.53	2050	2790	171	126	169	124	169	124	
			R4	76	5.57	13.61	13.88	1720	2340	167	123	166	122	166	122	
RTA 68	680/2000	Engine maximum continuous rating	R1	105	7.00	15.34	15.64	1950	2650	174	128	171	126	171	126	4—8
		Economy ratings	R2	99	6.60	13.60	13.83	1630	2210	170	125	169	124	169	124	
			R3	89	5.93	15.31	15.59	1650	2240	173	127	170	125	170	125	
			R4	84	5.60	13.67	13.94	1390	1890	169	124	167	123	167	123	
RTA 58	580/1700	Engine maximum continuous rating	R1	123	6.97	15.31	15.64	1410	1920	175	129	173	127	173	127	4—9
		Economy ratings	R2	116	6.57	13.59	13.82	1180	1600	171	126	170	125	170	125	
			R3	105	5.95	15.27	15.55	1200	1630	174	128	171	126	171	126	
			R4	98	5.55	13.63	13.90	1000	1360	170	125	169	124	169	124	
RTA 48	480/1400	Engine maximum continuous rating	R1	150	7.00	15.32	15.63	970	1320	178	131	175	129	175	129	4—9
		Economy ratings	R2	141	6.58	13.61	13.86	810	1100	174	128	173	127	173	127	
			R3	127	5.93	15.29	15.52	820	1110	177	130	174	128	174	128	
			R4	120	5.60	13.62	13.91	690	940	173	127	171	126	171	126	
RTA 38	380/1100	Engine maximum continuous rating	R1	190	6.97	15.44	15.76	610	830	181	133	178	131	178	131	4—9
		Economy ratings	R2	179	6.56	13.70	13.90	510	690	177	130	175	129	175	129	
			R3	162	5.94	15.44	15.81	520	710	179	132	177	130	177	130	
			R4	152	5.57	13.61	13.76	430	580	175	129	174	128	174	128	

* for net calorific value 42 707 kJ/kg (10 200 kcal/kg) and ISO Standard Reference Conditions:

Total barometric pressure	1.0 bar	1.02 kp/cm²
Suction air temperature	27°C	27°C
Charge air cooling water temp.	27°C	27°C
Relative humidity	60%	60%

Bunker fuel now constitutes the major proportion of total ship operating costs and this has caused much effort to be made in reducing the fuel consumption of propulsion machinery. To assist in this aim the RTA series has been developed to provide much lower shaft speeds for the best propulsive efficiency with a simple direct drive installation and the lowest possible specific fuel and lubricating oil consumption. The super long stroke engine is said to represent a 10% cut in fuel cost of a propulsion plant compared with existing long stroke engines such as the RL types.

A major reason for the change to uniflow scavenging is the inherent thermodynamic characteristics of the two scavenging systems in relation to stroke/bore ratios. For traditional Sulzer loop scavenged engines the minimum specific fuel consumption can be obtained at stroke/bore ratios of around 2.2:1. At higher ratios the diminishing scavenging efficiency would lead to increased fuel consumption. However, for larger stroke/bore ratios, such as 3:1, the uniflow scavenging system is superior and as propulson engines are very competitively marketed on the basis of the specific fuel consumption, the changeover to uniflow was necessary.

The RTA series consists of six different bore sizes covering an output range from 2320 to 35 520 bhp at shaft speeds ranging from 190 down to 70 rev/min, and specific fuel consumption values ranging between 123 and 133 g/bhph, according to bore size and rating of the engine. The series is introduced with four standard ratings: maximum continuous rating (R1) and three lower economy ratings R2, R3 and R4, according to Sulzer's now standard output layout methods.

Engine design features

Sectional sketches of the RTA series in outline are shown in Figures 6.13(a) and 6.13(b) and the major design details are as follows.

A sturdy engine structure is provided with low stresses, small deflections and with a lower crankcase section in many ways similar to the RL series. A single-wall bedplate structure with an integrated thrust block mounts large surface area shell type main bearings. A-shaped fabricated double-wall columns support the cylinder blocks and the three structural components are held together by Corey pre-tensional tie bolts. The single unit cast iron cylinder jackets are bolted together to form a rigid cylinder block, and cylinder liners are made from lamellar cast iron with a bore-cooled collar and a load dependent cylinder lubrication system as used on the RL engines.

The semi-built crankshaft is divided into two parts for larger bore engines with a high number of cylinders. The crosshead has a very large

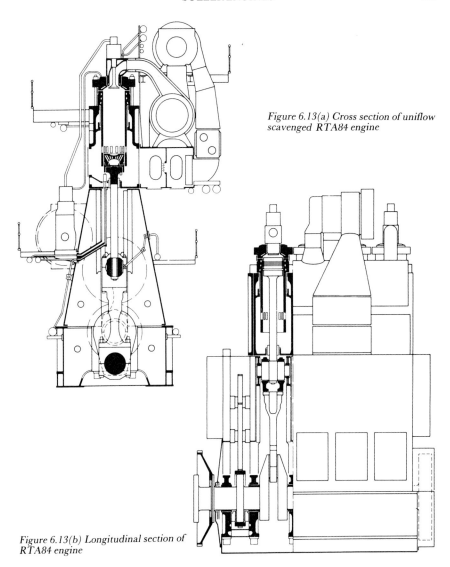

Figure 6.13(a) Cross section of uniflow scavenged RTA84 engine

Figure 6.13(b) Longitudinal section of RTA84 engine

surface area bearing with tin–aluminium type shells and double guided slippers. The piston is of bore water cooled type with a short skirt.

The camshaft drive gears are housed in a special double column space at the driving end of the engine, or at the centre for large bore engines with a larger number of cylinders. An option is an integrated power take off from the camshaft gear train for driving an alternator to generate auxiliary power when the engine is running. A further option is balancing gear for four cylinder engines.

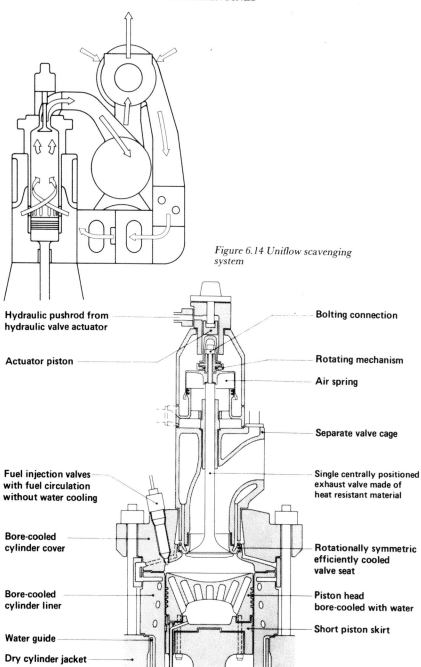

Figure 6.14 Uniflow scavenging system

Figure 6.15 Combustion chamber components

The single centrally placed exhaust valve is mounted in a cage fitted with an intensively water cooled seat and actuation of the valve is by an hydraulic system with the pressure plunger operated by a cam on the camshaft at half engine height position. The valve is made of high heat resistance material and is provided with a positive rotative drive. A compressed air spring system is built into the valve cage, eliminating return springs and avoiding vibrations and coil spring breakage. The camshaft drive is by three gear wheels from the crankshaft.

The fuel injection pump of Sulzer's standard double valve controlled type and the exhaust valve actuator are mounted in two cylinder casing set units on the camshaft.

Turbocharging is on the constant pressure principle with the scavenge box mounted alongside the engine and the large exhaust manifold sited on top of the scavenge air receiver. The turbocharger, of latest Brown Boveri VTR series 4 type, is mounted above the exhaust manifold. The auxiliary blower fitted for starting and low load operation automatically according to scavenge air pressure, is mounted in the scavenge air receiver.

FUEL PUMPS

All Sulzer engines are provided with a fuel pump of timed double valve type, driven by the camshaft (Figure 6.16).

Each pump consists of a plunger and guide bush and a driving piston. The roller is kept in contact with the cam by a powerful spring. A fuel pump can be taken out of service by a mechanical cut-out lever, which lifts the driving piston and roller clear of the cam. This lever can also be used for priming the injection system.

The fuel pump delivery is controlled by suction and spill valves. As long as the suction valve remains off its seat no fuel is delivered. When the fuel pump plunger is raised, the suction valve is lowered on to its seat. As soon as the suction valve closes, fuel is delivered. Normally the start of delivery is constant and the effective stroke is controlled by adjusting the spill valve. Retarding the spill valve increases the effective delivery stroke, while advancing reduces the effective stroke. These adjustments must be made when the engine is stopped and never when running. Ignition pressures are equalised only by turning the cams and not by altering the length of the pushrods.

The fuel charge for individual cylinders can be temporarily reduced, for instance when running in a spare part such as a liner or piston, etc. by fitting a spacer between the pushrod and the regulating rod. This raises the suction valve by 2.4 mm. When adjusting the fuel pump control valves no spacers should be fitted.

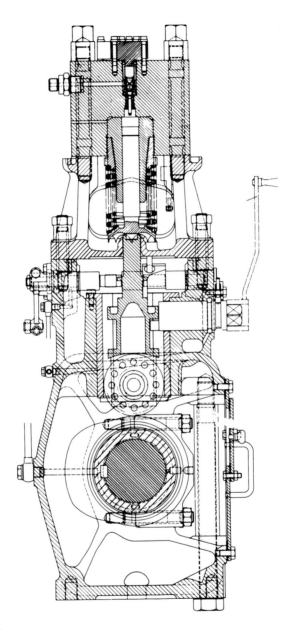

Figure 6.16 Sulzer fuel pump

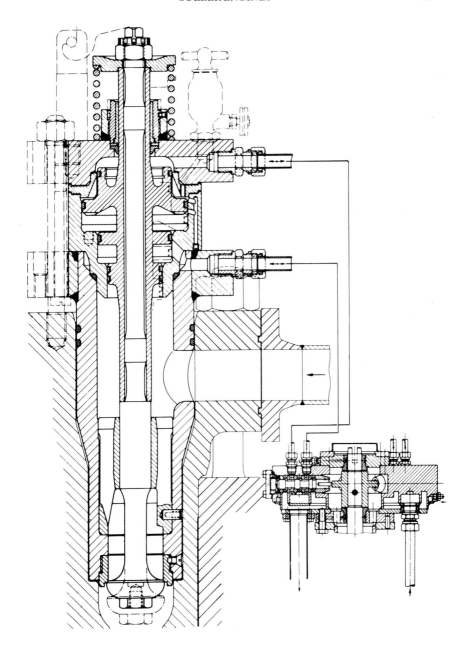

Figure 6.17 Starting valve

The effective fuel charge can be read off on the load indicator which has a scale graduated from 0 to 10. The load indicator serves as a guide for adjusting the fuel pumps, and for this reason its linkages should never be altered. The maximum fuel charge is made by a stop piece which is set during engine testing.

The pump bodies are safeguarded against excessive pressure by relief valves which are fitted on the backs of the fuel pump blocks and connected to the delivery passage along the plunger.

Variable injection timing

In the search for greater fuel economy Sulzer have introduced a variable injection timing (VIT) system to its RL and RTA engines. The specific fuel consumption is lowered by controlling the maximum permissible combustion pressure with differing loads through adjustment of the fuel injection timing. The reduction in fuel consumption is about 2 g/bhph at 85% engine load when carrying the maximum combustion pressure at its normal maximum value. The VIT mechanism automatically changes the injection timing according to load to maximum combustion pressure at engine loads between 85% and 100%.

A separate lever of the VIT mechanism allows the engines to make a manual adjustment of the fuel injection timing and is known as a 'fuel quality setting' (Figure 6.18). Burning heavier fuels can result in an

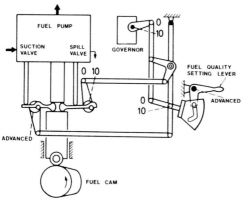

Figure 6.18(a) Variable injection timing mechanism

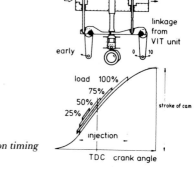

Figure 6.18(b) Variable injection timing fuel pump control

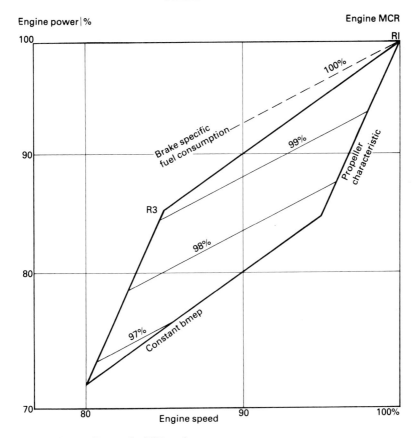

Figure 6.19 Layout diagram for RTA engine

ignition lag and consequent drop in the maximum combustion pressure which in turn increases the fuel consumption. By a simple readjustment of the fuel quality lever towards earlier injection, the pressure is raised and fuel consumption lowered to its best possible value. Peak pressure cards taken by the engineer after such adjustment will confirm that the engine is operating at the correct maximum pressure at any given load, and settings can be made for burning fuels of different quality and viscosity.

The simple VIT mechanism is linked to the governor load-setting shaft and a built-in cam system — which is positioned by the fuel quality setting lever — controls through linkages the simultaneous timing of the suction valve closure (beginning of delivery) and spill valve opening (end of delivery). Hence, the fuel injection timing (and combustion pressure) and the fuel delivery to the injectors (timing difference of both valves) are controlled load dependently.

C.T.W.

7 Burmeister and Wain engines

The long-established Danish builder of two-stroke engines, Burmeister and Wain, has for many years concentrated solely on the design and manufacture of crosshead type engines of high pressure turbocharged single-acting uniflow scavenged type with one centrally placed exhaust valve. The uniflow scavenging system gives effective complete scavenging of the exhaust gases in the cylinder in an even progressive upwards movement with low flow resistance out to the exhaust manifold through

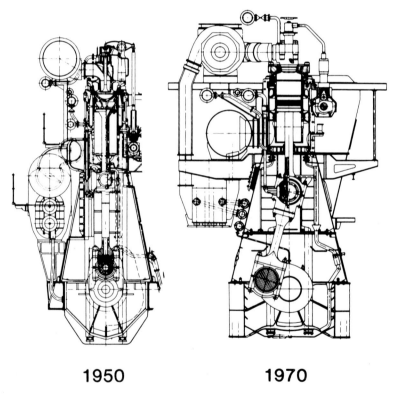

1950   1970

Figure 7.1 B & W uniflow scavenged engines of 1950 and 1970 with impulse turbocharging

the mechanically actuated large diameter sinngle exhaust valve of poppet type.

Until 1978 all B & W engines operated on the impulse turbocharging system (Figure 7.1) but the search for fuel economy during the last few years has meant a change to the constant pressure arrangement, giving a saving of about 5% in specific fuel consumption. Also during the last few years the company has brought out a whole series of engines, each very rapidly superseded by the next; namely, K-GF, followed by L-GF, then L-GFCA, followed by the L-GB, and the recent L-MC series 'ultra long stroke' engine.

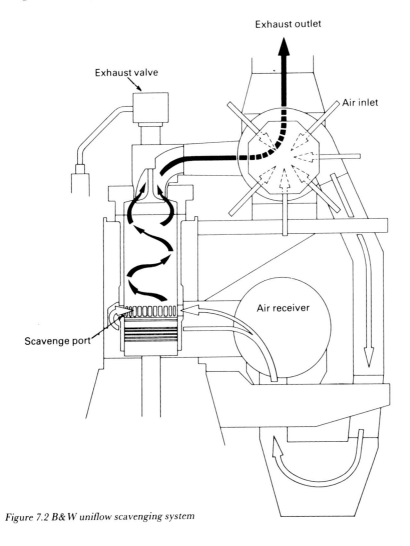

Figure 7.2 B&W uniflow scavenging system

All these recent designs are improved variants of the K-GF series which itself introduced many notable design innovations, such as box frame construction, intensively cooled cylinder components and hydraulically actuated exhaust valves.

A major change to the structure of the company followed its purchase by the German manufacturer MAN, which has resulted in Copenhagen becoming the centre for two-stroke engine development and all engines of either make now bearing the MAN–B & W trade mark.

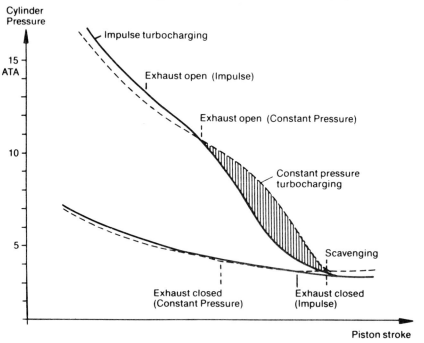

Figure 7.3 Working diagram

K-GF TYPE ENGINES

The current range of B & W two-stroke engines are all based on the K-GF series introduced about 10 years ago and featuring many new design features. Notable is the box-type construction of the crankcase and the use of hydraulically actuated exhaust valve instead of the conventional mechanical rocker arm system used in the earlier K-EF types. A sectional sketch of the K-GF is shown in Figure 7.4 and as this model has been rapidly superseded by the L-GF and L-GB models, only brief details of the K-GF construction are given here.

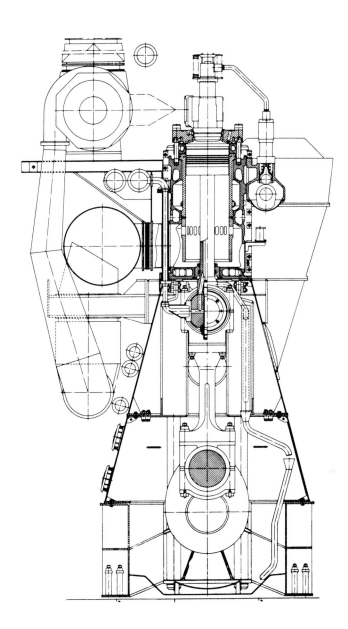

Figure 7.4 Cross section of K90GF engine

Bedplate and frame

The standard K-GF engine bedplate is of the high and fabricated design with cast steel main bearing housings welded to the longitudinal side frames. The standard thrust bearing is of the built-in short type and separated from the crankcase by a partition wall (Figure 7.5).

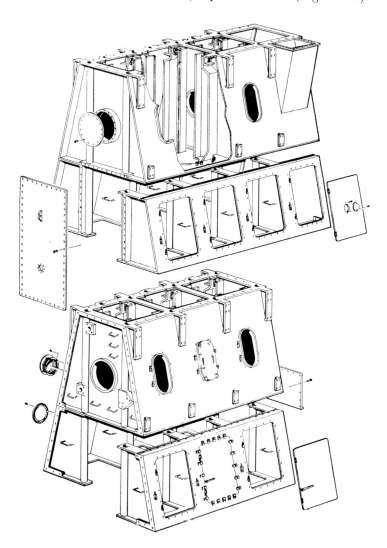

Figure 7.5 K-GF structural arrangement

The frame or entablature section consists of three units: a frame box of a height corresponding to the length of the crosshead guides, bolted together in the longitudinal direction in the chain drive section only, and two longitudinal girders, also bolted together in this section. This method of construction gives rigid units designed for easy handling and mounting, both on the test bed and when erecting the engine in the ship. The rigid design with few joints ensures oil tightness, and large hinged doors provide easy access to the crankcase.

The crosshead guides, consisting of heavy I sections, are attached at top and bottom in the frame boxes, with the lower attachment flexible in the longitudinal direction.

Cylinder jackets

Cast iron one-piece cylinder sections connected in the vertical plane by fitted bolts are held to the frame section by long tie-bolts which are secured in the main bearing saddles.

The cylinder liner is provided with a water cooled flange low down, to provide low operating temperatures in the most heavily loaded area. The liners are made of alloyed cast iron with ports for scavenge air and bores for the cylinder lubrication system.

Crankshaft

The crankshaft is semi-built for all cylinder numbers with cast steel throws for six to ten cylinder engines. Balancing of the engine is undertaken by varying the crankpin bore holes, thus entirely eliminating bolted-on counterweights.

Lubrication of the crank bearing is from the crosshead through the connecting rod, thus eliminating stress raising bores in the crankshaft.

Crosshead

The crosshead (Figure 7.6) is short and rigid and the bearings are so constructed that the bearing pressure between journal and bearing is distributed evenly over the entire length of the bearing. The bearing pressure is smaller than previously and the peripheral speed is higher, which improves the working conditions of the bearings.

Interchangeable bearing shells of steel with a 1 mm white-metal lining are fitted to the crosshead bearings. The shells are identical halves, precision bored to finish size and can be reversed in position during emergencies when a damaged or worn lower half can be temporarily used as an upper shell.

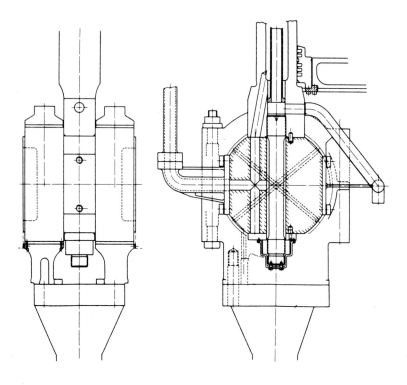

Figure 7.6 Crosshead arrangement

Piston

The piston is oil cooled and consists of the crown, cooling insert and the skirt. The cooling insert is fastened to the upper end of the piston rod and transfers the combustion forces from the crown to the rod. The crown and skirt are held together by screws while a heavy Belleville spring is used to press the crown and the cooling insert against the piston rod. The crown has chromium plated grooves for five piston rings while the rod has a longitudinal bore in which is mounted the cooling oil outlet pipe. The oil inlet is through a telescopic pipe fastened to the crosshead, and the oil passes through a bore in the foot of the piston rod to the cooling insert (Figure 7.7).

Cylinder cover

The cylinder cover is in two parts, a solid steel plate with radial cooling water bores, to which a forged steel ring with bevelled cooling water

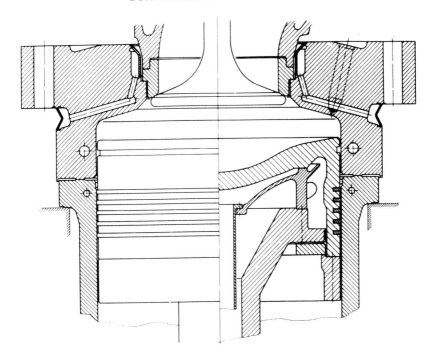

Figure 7.7 Combustion chamber and piston showing bore-cooling arrangement

bores is bolted. The insert has a central bore for the exhaust valve cage and mountings for fuel valves, a safety valve and an indicator valve.

The cover is held against the cylinder liner top collar by studs screwed into the cylinder block and the nuts are tightened by a special hydraulic tool to allow simultaneous tightening and correct tension of the studs.

Exhaust valve

The valve is similar to earlier types but with hydraulic actuation, a feature unique to the K-GF series (Figure 7.8). The valve consists of a cast iron cage and a spindle, with steel seats and the valve mushroom faced with Stellite for hard wearing. Studs secure the valve housing to the cylinder cover.

The hydraulic actuation system consists of a piston pump which is driven by a cam on the camshaft and the pressure oil is led to a working cylinder placed on top of the exhaust valve housing. The oil pressure is used to open the valve and closing is accomplished by a ring of helical springs.

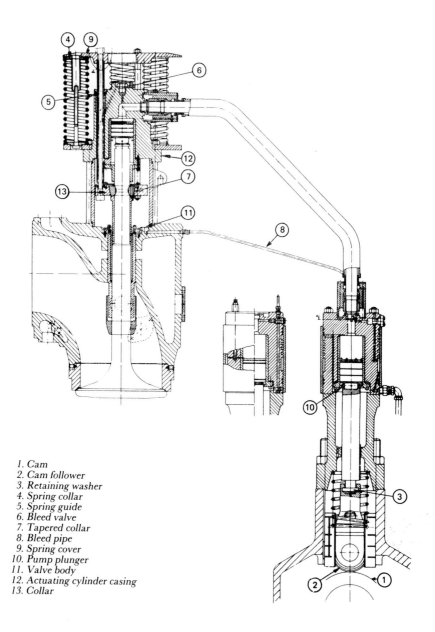

1. Cam
2. Cam follower
3. Retaining washer
4. Spring collar
5. Spring guide
6. Bleed valve
7. Tapered collar
8. Bleed pipe
9. Spring cover
10. Pump plunger
11. Valve body
12. Actuating cylinder casing
13. Collar

Figure 7.8 Components of hydraulically actuated exhaust valves

Camshaft

The camshaft is divided into sections, one for each cylinder, enclosed and suspended in roller guide housings with replaceable ready-bored bearing shells. The cams and couplings are fitted to the shaft by the SKF oil-injection pressure method and the complete camshaft unit is driven by chain from the crankshaft. For reversing of the engine, the chain wheel floats in relation to the camshaft which is driven through a self-locking crank gear; reversing crank pins are turned by built-on hydraulic motors.

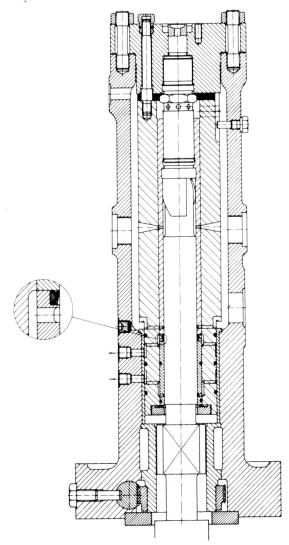

Figure 7.9 Fuel pump

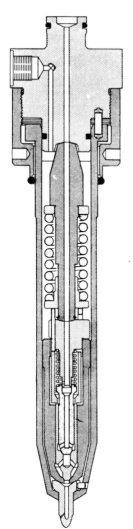

Figure 7.10 Fuel injector

Turbocharging

Until 1978 all engines were equipped with turbochargers operating on the impulse system, as was the case for the K-GF when introduced. The layout of chargers and intercoolers are seen in Figure 7.4. However, all B & W engines now have (including the K-GF) a constant pressure charging system in which the exhaust pulses from the individual cylinders are smoothed out in a large volume gas receiver before entering the turbine at a constant pressure. This system gives a fuel consumption which is some 5% lower. For part load operation and starting electrically driven auxiliary blowers are necessary. The arrangement of the constant pressure turbocharged K-GF engine is seen in Figure 7.4.

L-GF AND L-GB ENGINES

The oil crisis in 1973 and the resulting massive increase in fuel prices caused engine builders to develop newer engines of reduced specific fuel consumption. B & W's answer was the L-GF series combining constant pressure turbocharging with an increase in piston stroke: an increase in stroke of about 22% results in a lowering of the shaft speed by around 18%, leading to greater propulsive efficiency when using larger diameter propellers and around 7% increase in thermal efficiency.

The design of the L-GF series engine was heavily based on that of the K-GF and the necessary changes were mainly those relating to the increase in piston stroke and modifications to components to embody thermodynamic improvements (Figure 7.11). A major component design change is the cylinder liner, made longer for the increased stroke, and featuring cooling bore drillings forming generatrices on a hyperboloid to ensure efficient cooling of the high liner collar without cross borings with high stress concentration factors.

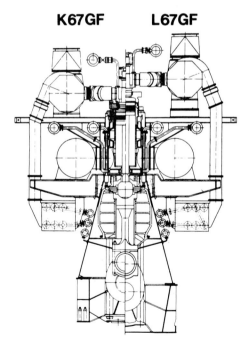

Figure 7.11 Short and long stroke 67GF engines

The cylinder covers are of the solid plate type used for the K-GF, while the pistons also are of the original oil cooled type but with somewhat improved cooling caused by the stronger 'cocktail shaker' effect resulting from the greater quantity of oil in the elongated piston rod. The design of the crankshaft, crosshead, bearings, and exhaust valve

Figure 7.12 12L90GFCA engine of 47 300 bhp at 97 rev/min

Constant pressure turbocharging

The increased stroke and thus reduced rev/min of L-GF engines compared with K-GF engines was not aimed at developing more power but an improvement in ship propulsive efficiency. The uniflow scavenging system has the advantage of a good separation between air and gas during the scavenging process, and the rotating flow of air along the cylinder contributes to the high scavenging efficiency and clean air charge.

As an engine's mean indicated pressure increases the amount of exhaust gas energy supplied during the scavenging period, relative to the impulse energy during the blow-down period, constant pressure turbocharging is advantageous; also for uniflow scavenged engines with unsymmetrical exhaust valve timing. Theoretical calculations for

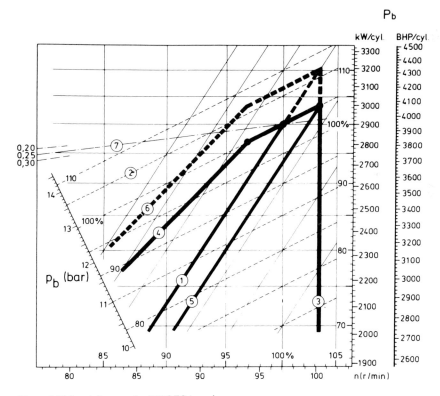

Figure 7.13 Load diagram for L90GFCA engine

improved fuel economy showed a possible gain of 5%–7% in specific fuel consumption by using constant pressure turbocharging.

The most obvious change with constant pressure turbocharging is that the exhaust pipes from each valve body led to a common large exhaust gas receiver instead of to the turbochargers as in the impulse system. When the cylinders exhaust into a large receiver, the outflow of gas is quicker because the large gas impulse at the commencement of the exhaust period is levelled out in the gas receiver and the outflow of gas will not be retarded as in the case of impulse turbocharging where a pressure peak is built up in the narrow exhaust pipe before the turbocharger. The opening of the exhaust valve can be delayed about 15°, thereby lengthening the expansion stroke and improving the efficiency and reducing the fuel consumption. The energy before the turbine is less than for impulse turbocharging, but as the pressure and temperature before the turbine is nearly constant the turbine can be adapted to run at peak efficiency and the blower can supply sufficient air above 50% of engine load. The scavenging air pressure is increased and the compression pressure somewhat decreased compared with impulse turbocharging.

A small auxiliary blower is necessary for satisfactory combustion conditions at loads up to about 50%. Two blowers, of half capacity, are used for safety, and even with one of these out of action the other with half the capacity is satisfactory for starting and load increase. The engine will still run at down to 25% load but with a smoky exhaust.

The change from K-GF to L-GF resulted in a 2% improvement in the specific fuel consumption and the lower speed accounts for 5% improvement in propeller efficiency. Constant pressure turbocharging adds a further 5% to the improvement, to result in a reduction by 12% in specific fuel consumption between the L-GF and K-GF types. The latest engine model of the constant pressure turbocharged engine is the L-GFCA, but an even newer L-GB type was introduced soon after.

L-GB TYPE ENGINES

A further improvement in specific fuel consumption was obtained by introducing the L-GB and L-GBE series engines. By using the optimum combination of longer stroke, higher output and higher maximum pressure and the newest high efficiency turbochargers, much lower fuel consumption rates are achieved. The L-GB series has a mep of 15 bars at the same speed as the L-GFCA to give an increase in power of 15% and an increase in the firing pressure from 89–105 bar. Accord-

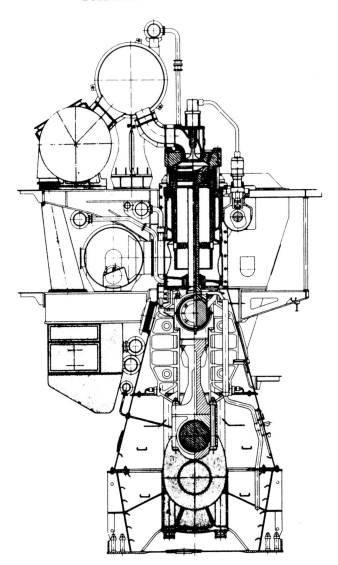

Figure 7.14 L-GB/GBE engine cross section

ingly the important Pmax/mep ratio is almost the same but the specific fuel consumption is some 3 g/bhph lower than the L-GFCA series. A further economy rating is obtained with engines of the L-GBE type, by holding the Pmax at reduced engine output, or so-called 'derating' as outlined in Chapter 2. The practice of offering propulsion engines in both normal and derated versions has now been adopted by many engine manufacturers, even for four-stroke medium-speed engines.

Bedplate and main bearing

The bedplate consists of high, welded longitudinal girders and welded cross girders with cast steel bearing supports. For the four and five cylinder engines the chain drive is placed between the aftermost cylinder and the built-in thrust bearing. For the six to twelve cylinder engines the chain drive is placed at the assembling between the fore and aft part. For production reasons, the bedplate can be made in convenient sections. The aft part contains the thrust bearing. The bedplate is made for long, elastic holding-down bolts tightened by hydraulic tool.

The oil pan is made of steel plate and welded to the bedplate parts. The oil pan collects the return oil from the forced lubricating and cooling oil system. For about every third cylinder it is provided with a drain with grid.

The main bearings consist of steel shells lined with white metal. The bottom shell can, by means of hydraulic tools for lifting the crankshaft and a hook-spanner be turned out and in. The shells are fixed with a keep and long, elastic studs tightened by hydraulic tool.

Thrust bearing

The thrust bearing is of the B & W-Michell type. Primarily, it consists of a steel forged thrust shaft, a bearing support, and segments of cast iron with white metal. The thrust shaft is connected to the crankshaft and the intermediate shaft with fitted bolts.

The thrust shaft has a collar for transfer of the 'thrust' through the segments to the bedplate. The thrust bearing is closed against the crankcase, and it is provided with a relief valve.

Lubrication of the thrust bearing takes place from the system oil of the engine. At the bottom of the bearing there is an oil sump with outlet to the oil pan.

Frame section

The frame section consists for the four and five cylinder engine of one part with the chain drive located aft. The chain drive is closed by the end-frame aft. For six to twelve cylinder engines the frame section consists of a fore and an aft part assembled at the chain drive. Each part consists of an upper and a lower frame box, mutually assembled with bolts.

The frame boxes are welded. The upper frame box is on the back of the engine provided with inspection cover for each cylinder. The lower frame box is on the front of the engine provided with a large hinged door for each cylinder.

The guides are bolted onto the upper frame box and have possibility for adjustment. The upper frame box is on the back side provided with a relief valve for each cylinder and on the front provided with a hinged door per cylinder. A slotted pipe for cooling oil outlet from piston is suspended in the upper frame box.

The frame section is attached to the bedplate with bolts. The stay bolts consist of two parts assembled with a nut. To prevent transversal oscillations the assembly nut is supported. The stay bolts are tightened hydraulically.

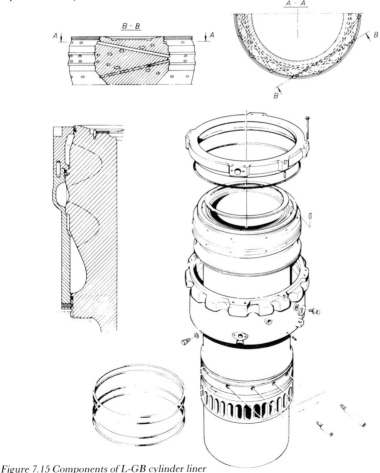

Figure 7.15 Components of L-GB cylinder liner

Cylinder frame, cylinder liner, and stuffing box

The cylinder frame unit is of cast iron. Together with the cylinder liner (Figure 7.15) it forms the scavenging air space and the cooling water space. At the chain drive there is an intermediate piece. The stay bolt pipes and the double bottom in the scavenging air space are water cooled. On the front the cylinder frame units are provided with cleaning cover and inspection cover for scavenging ports. The cylinder frame units are mutually assembled with bolts.

Housings for roller guides, lubricators, and gallery brackets are suspended on the cylinder frame unit. Further, the outside part of a telescopic pipe is fixed for supply of piston cooling oil and lubricating oil. At the bottom of the cylinder frame unit there is a piston rod stuffing box. The stuffing box is provided with sealing rings for scavenging air and oil scraper rings preventing oil from coming up into the scavenging air space.

The cylinder liner is made of alloyed cast iron and is suspended in the frame section with a low situated flange. The uppermost part of the liner has drillings for cooling water and is surrounded by a cast iron cooling jacket. The cylinder liner has scavenging ports and drillings for cylinder lubrication.

Cylinder cover

The cylinder cover is made in one piece of forged steel and has drillings for cooling water. It has a central bore for the exhaust valve and bores for fuel valves, safety valve, starting valve, and indicator valve (Figure 7.16).

The cylinder cover is attached to the cylinder frame with studs tightened by a hydraulic ring covering all studs.

Exhaust valve and valve gear

The exhaust valve consists of a valve housing and a valve spindle. The valve housing is of cast iron and arranged for water cooling. The housing is provided with a bottom piece of steel with Stellite welded onto the seat. The spindle is made of heat resistant steel with Stellite welded onto the seat. The housing is provided with a spindle guide. The exhaust valve housing is connected to the cylinder cover with studs and nuts tightened by hydraulic jacks. The exhaust valve is opened hydraulically and closed by a set of helical springs. The hydraulic system consists of a piston pump mounted on the roller guide housing, a high-pressure pipe, and a working cylinder on the exhaust valve. The piston pump is activated by a cam on the camshaft.

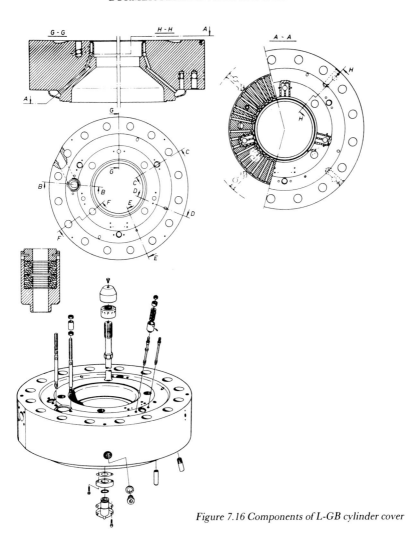

Figure 7.16 Components of L-GB cylinder cover

Cover mounted valves

In the cylinder cover there are three fuel valves, one starting valve, one safety valve, and one indicator valve.

The fuel valve opening is controlled by the fuel oil pressure and it is closed by a spring. An automatic vent slide allows circulation of fuel oil through the valve and high-pressure pipes and prevents the compression chamber from being filled up with fuel oil in case of possible sticking spindle at stopped engine. Oil from venting and other drains is led away in a closed system.

The starting valve is opened by control air from the starting air distributor and closed by a spring. The safety valve is spring loaded. The indicator valve is placed near the indicator gear.

Crankshaft

The crankshaft for four and five cylinder engines is made in one part, and for six to twelve cylinder engines it is made in two parts assembled at the chain drive with fitted bolts. The crankshaft is semi-built with forged steel throws.

The crankshaft has in the aft end a flange for assembling with the thrust shaft. The crankshafts are balanced exclusively by borings in the crank pins, though in some cases supplemented by balance weight in the turning wheel.

Connecting rod

The connecting rod is of forged steel. It has a Tee-shaped base on which the crank bearing is attached with hydraulic tightened bolts and nuts with Penn-securing. The L90GBE engine has shims placed between the base and the crank bearing, as this engine type needs a smaller compression chamber because of a higher compression ratio. The top is square shaped on which the crosshead bearings are attached with hydraulic tightened studs and nuts with Penn-securing. The bearing parts are mutually assembled with bolts and nuts tightened by hydraulic jacks.

The lubrication of crank bearing takes place through a central drilling in the connecting rod.

The crank bearing is steel cast in two parts lined with white metal. The bearing clearance is adjusted with shims. The crosshead bearings are of cast steel in two parts and provided with bearing shells.

Piston — piston rod — crosshead

The piston consists of piston crown, piston skirt, and cooling insert for oil cooling (Figure 7.17). The piston crown is made of heat-resisting steel and is provided with five ring grooves which are hard-chrome plated on both lands. The piston skirt is of cast iron. The piston rings are right- and left-angle cut and of the same height.

The piston rod is of forged steel. It is fixed to the crosshead with a hydraulic tightened stud. The piston rod has a central bore which, in connection with a cooling oil pipe and the cooling insert, forms inlet and outlet for cooling oil.

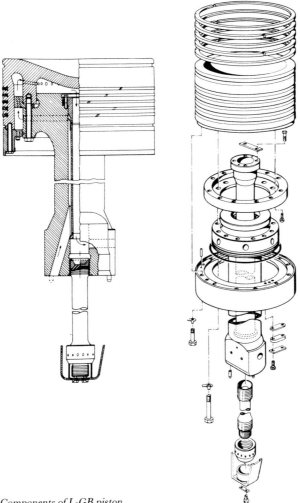

Figure 7.17 Components of L-GB piston

The crosshead is of forged steel and is provided with steel cast guide shoes with white metal on the running surfaces. A bracket for oil inlet from the telescopic pipe and a bracket for oil outlet to slit pipe are mounted on the crosshead.

Fuel pump and fuel oil high-pressure pipes

The fuel pump consists of a pump housing of nodular cast iron and a central placed pump cylinder of steel with sleeve and plunger of

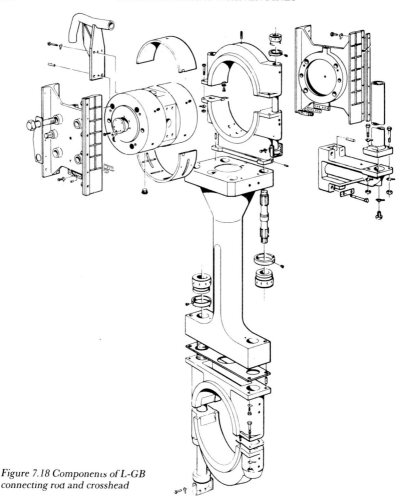

Figure 7.18 Components of L-GB connecting rod and crosshead

nitrated steel. The plunger has an oblique injection edge which will automatically give an optimum fuel injection timing. There is one pump for each cylinder. In order to prevent fuel oil from being mixed into the separate lubricating system on the camshaft, the pump is provided with a sealing device.

The pump gear is activated by the fuel cam, and the injected volume is controlled by turning the plunger by a toothed bar connected to the regulation mechanism. Adjustment of the pump lead is made with shims between top cover and pump housing.

The fuel pump is provided with a pneumatic lifting device: this can, during normal operation and during turning, lift the roller guide roller free of the cam.

The fuel oil high-pressure pipes have protecting hoses. The fuel oil system is provided with a device which, through the pneumatic lifting tool, disconnects the pump in case of leakage from the high-pressure pipes.

Camshaft and cams

The camshaft is divided into sections for each cylinder. The individual sections consist of a shaft piece with one exhaust cam, one fuel cam, one indicator cam, and coupling parts. The exhaust and fuel cams are of steel with a hardened roller race, and are shrunk on the shaft. They can be adjusted and dismounted hydraulically.

The indicator cams, which are of cast iron, are bolted onto the shaft. The coupling parts are shrunk on the shaft and can be adjusted and dismounted hydraulically.

The camshaft is located in the housing for the roller guide. The camshaft bearings consist of two mutually interchangeable bearing shells, which are mounted in hydraulic tightened casings.

Chain drive and reversing

The camshaft is driven from the crankshaft by two off $4\frac{1}{2}$ inch chains. The chain drive is provided with a chain tightener and guidebars support the long chain strands. The camshaft is provided with a hydraulic actuated reversing gear turning the camshaft to the position corresponding to the direction of rotation of the crankshaft.

Starting air distributor, governor, and cylinder lubricators are driven by separate chain from the intermediate wheel.

Moment compensator

Four, five and six cylinder engines are prepared for moment compensators, which can be fixed to fore and aft end of the frame section and are driven by the camshaft through flexible couplings. The moment compensator will reduce the second order external moments to a level between a quarter of the original figure and zero.

Governor

The engine rev/min is controlled by a hydraulic governor. For amplification of the governor's signal to the fuel pump there is a hydraulic amplifier. The hydraulic pressure for the amplifier is delivered by the camshaft lubricating oil system.

Cylinder lubricators

The cylinder lubricators are mounted on the cylinder frame, one per cylinder, and they are interconnected with shaft pieces. The lubricators have built-in adjustment of the oil quantity. They are of the 'Sight Feed Lubricator' type and each lubricating point has a glass. The oil is led to the lubricator through a pipe system from an elevated tank. A heating element rated at 75 watt is built into the lubricator.

Manoeuvring system (without bridge control)

The engine is provided with a pneumatic manoeuvring and fuel oil regulating system. This system transmits orders from the separate manoeuvring console to the engine.

By means of the regulating system it is possible to start, stop, reverse and control the engine. The speed control handle in the manoeuvring console activates a control valve, which gives a pneumatic speed-setting signal to the governor dependent on the desired number of revolutions. The start and stop functions are controlled pneumatically. At a shut-down function the fuel pumps are moved to zero position independent of the speed control handle.

Reversing of the engine is controlled pneumatically through the engine telegraph and is affected by means of the telegraph handle.

Reversing takes place by moving the telegraph handle from 'Ahead' to 'Astern' and by moving the speed control handle from stop to start position. Control air then moves the starting air distributor and, through the pressuriser, the reversing mechanism to the 'Astern' position.

Turning gear and turning wheel

The turning wheel has cylindrical teeth and is fitted to the thrust shaft. This wheel is driven by a pinion on the terminal shaft of the turning gear, which is mounted on the bedplate. The turning gear is driven by an electric motor with built-in gear and brake. Further, the gear is provided with a blocking device that prevents the main engine from starting when the turning gear is engaged. Engagement and disengagement of the turning gear is done by axial transfer of the pinion.

Gallery brackets

The engine is provided with gallery brackets placed in such a height that the best possible overhaul and inspection conditions are obtained.

Main pipes of the engine are suspended in the gallery brackets.

A crane beam is placed on the brackets below centre gallery manoeuvring side.

Scavenging air system

The air intake to the turbocharger takes place direct from the engine room through the intake silencer of the turbocharger. From the turbocharger the air is led via charging air pipe, air cooler and scavenging air pipe to the scavenging ports of the cylinder liner. The charging air pipe between turbocharger and air cooler is provided with a compensator and are insulated.

Exhaust turbocharger

The engine is as standard arranged with MAN or BBC turbochargers. The turbochargers are provided with a connection for Disatac electronic tachometers, and prepared for signal equipment, to indicate excessive vibration of the turbochargers. For water cleaning of the turbine blades and the nozzle ring during operation, the engine is provided with connecting branches on the exhaust receiver in front of the protection grid.

Exhaust gas system

From the exhaust valves the gas is led to the exhaust gas receiver where the fluctuating pressure will be equalised and the gas led further on to the turbochargers with a constant pressure. After the turbochargers, the gas is led through outlet pipe and out in the exhaust pipe system.

The exhaust gas receiver is made in one piece for every cylinder and connected to compensators. Between the receiver and the exhaust valves and between the receiver and the turbocharger there are also inserted compensators.

For quick assembling and dismantling of the joints between the exhaust gas receiver and the exhaust valves, a clamping band is fitted. The exhaust gas receiver and exhaust pipe are provided with insulation covered by a galvanised steel plate.

Between the exhaust gas receiver and each turbocharger there is a protection grid.

Auxiliary blower

The engine is provided with two electrically driven blowers which are mounted in each end of the scavenging air receiver as standard. The suction sides of the blowers are connected to the pipes from the air coolers, and the non-return valves on the top of the outlet pipes from the air coolers are closed as long as the auxiliary blowers can give a supplement to the scavenging air pressure.

The auxiliary blowers will start operating before the engine is started and will ensure complete scavenging of the cylinders in the starting phase, which gives the best conditions for a safe start.

During operation of the engine, the auxiliary blowers will start automatically every time the engine load is reduced to about 30%–40%, and they will continue operating until the load is again increased to over approximately 40%–50%.

In cases when one of the auxiliary blowers is out of service, the other auxiliary blower will automatically function correctly in the system, without any manual readjustment of the valves being necessary. This is obtained by automatically working non-return valves in the suction pipe of the blowers.

Starting air system

The starting air system contains a main starting valve (two ball valves with actuators), a non-return valve, a starting air distributor, and starting valves. The main starting valve is combined with the manoeuvring system, which controls start and 'Slow turning' of the engine. The 'Slow turning' function is actuated manually from the manoeuvring stand.

The starting air distributor regulates the control air to the starting valves so that these supply the engine with starting air in the firing order.

The starting air distributor has one set of starting cams for 'Ahead' and 'Astern' respectively, and one control valve for each cylinder.

L-MC TYPE ENGINES

The latest complete range of slow-speed engines designed by MAN–B and W is the 'ultra long stroke' L-MC series (Figure 7.19) which use a large stroke/bore ratio to exploit both the reduced fuel consumption rate achievable from the improved thermodynamic efficiency (around 50%) with uniflow scavenging and a stroke/bore ratio of 3:1 and the

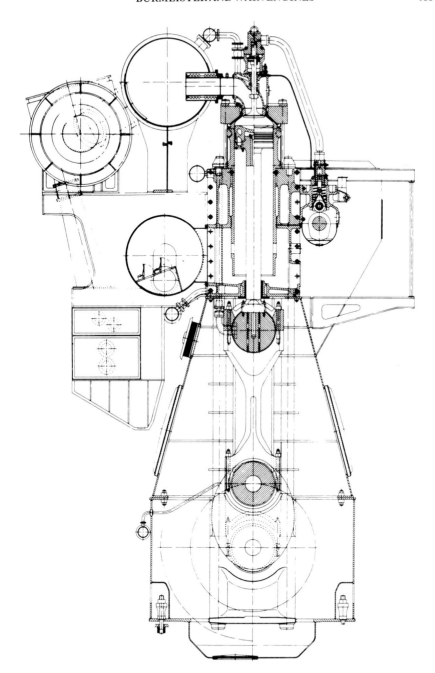

Figure 7.19 L80MC/MCE engine cross section

Cooled Bottompiece

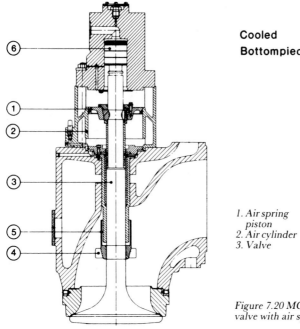

1. Air spring piston
2. Air cylinder
3. Valve
4. Valve rotation
5. Heat deflector
6. Actuating piston

Figure 7.20 MC/MCE type exhaust valve with air spring

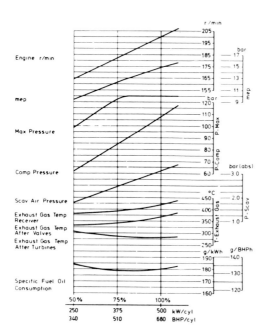

Figure 7.21 L35MC/MCE performance curves

Figure 7.22 First L35MC/MCE engine on testbed in Japan

lower shaft speed to improve propeller efficiency.

The first engine of this type to be built was a 350 mm bore unit manufactured in Japan. The L-35MC has a single exhaust valve in the cylinder head (Figure 7.20) as used on the L-GB type, and a single turbocharger is mounted at the aft end of the engine with the charge air cooler fitted above the integrated thrust bearing and coupling. This first engine of the new very long stroke MC/MCE series is shown in Figure 7.22 running initial trials in Japan in mid-1982.

C.T.W.

8 MAN Engines

Slow-speed marine diesel engines built by Maschinenfabriek Augsburg – Nurnberg AG (MAN) are single-acting two-stroke types of crosshead design. The KSZ type (Figure 8.1) was introduced in the mid-1960s, and in the intervening years KSZ-A, KSZ-B and KSZ-C types have all been introduced, each with improvements to promote greater reliability and fuel economy. The KSZ-C and KSZ-CL represent the most recent in MAN loop scavenged engine design technology. However, the fairly recent acquisition by MAN of the Danish company Burmeister and Wain, and the decision for B & W to concentrate on two-stroke engines and MAN four-strokes, places a question mark over future development of MAN two-stroke types. At present both MAN and B & W two-stroke, and four-stroke engines of both companies, are manufactured and marketed under the MAN–B & W banner.

MAN two-stroke engines now employ constant pressure turbocharging (Figure 8.2) and are scavenged according to the loop scavenging system. The cylinder exhaust ports are located above the scavenging ports on the same side of the liner, occupying approximately one-half of its circumference. The scavenging air is admitted through the scavenge ports, where it passes across the piston crown and ascends along the opposite wall to the cylinder cover, where its flow is reversed. The air then descends along the wall in which the ports are located, expelling the exhaust gases into the exhaust manifold. The piston closes the scavenging ports and then, on its further upward travel, also closes the exhaust ports compressing the charge of pure air in the cylinder.

When the KSZ series was introduced it followed the traditional two-stroke engine construction technique of cylinder blocks mounted on 'A' frames which sat on a cast bedplate and the three structural items held by a series of long tie-rods. However, this method of erection was abandoned when MAN introduced the KSZ-B type with the so-called box-type construction which has been retained for the most recent KSZ-C and KEZ-B types. This latter type is identical to the KSZ types but features electronic fuel injection instead of the traditional mechanical fuel pumps arrangement. In general design details the KSZ-B and KSZ-C engines are basically similar but for the longer piston strokes of

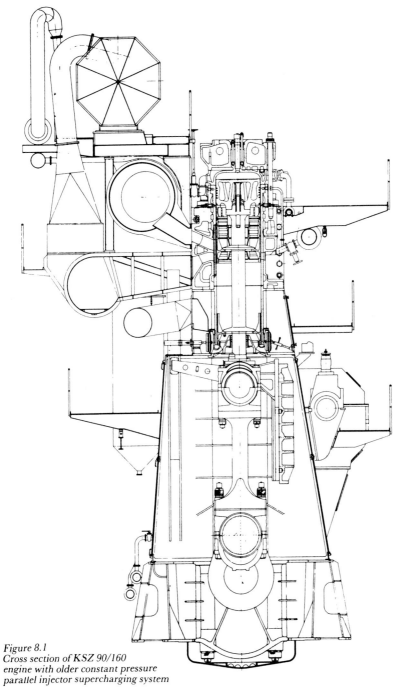

Figure 8.1
Cross section of *KSZ 90/160*
engine with older constant pressure
parallel injector supercharging system

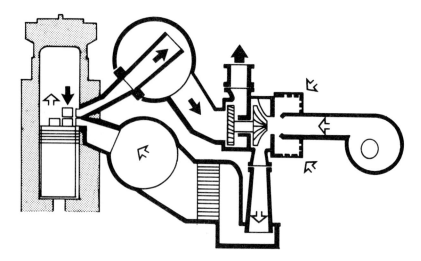

Figure 8.2 Constant pressure turbocharging of KSZ-B and KSZ C/CL engines

Figure 8.3 K8SZ 90/160 B/BL engine on the testbed

the latter series and other refinement to bring about a considerable reduction in the specific fuel consumption.

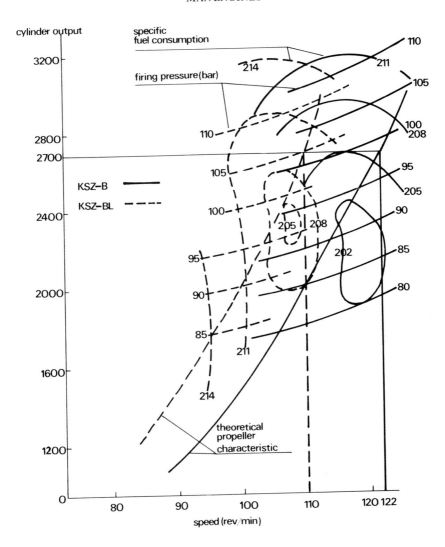

Figure 8.4 Performance curves for the K8SZ 90/160 engine

DESIGN PARTICULARS

MAN's KSZ-B/BL and KSZ-C/CL ranges are both of identical design, but with the later model giving a 10% reduction in the speed at the same horsepower rating, this being achieved by raising the cylinder mean effective pressure. Minor design changes are only where the increase in piston stroke has made this necessary.

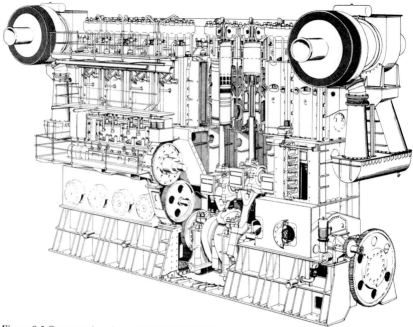

Figure 8.5 Cutaway drawing of K8SZ90/160B/BL engine

Engine structure

The significant constructional characteristic of the KSZ-B and -C type engines is the box-shaped longitudinal girders comprising a deep-section single-walled bedplate, a frame which is either in one piece or split (depending on the number of cylinders) on cast, high cylinder jackets. In the largest 900 mm bore engine the frame box is split horizontally (Figure 8.6) while smaller engines of 700 and 520 mm bore have the frame as a single-piece steel fabrication. For engines of seven to twelve cylinders the frames of engines of the two largest bore size are vertically split, whereas smaller engines with the camshaft gear drive located at the coupling end have no vertical split in the frame structure. The cylinder jackets are cast in either two or three units. The box frame design allows a high degree of stiffness and contributes to a gradual reduction in the deformations induced by the ship's double bottom and transmitted to the bedplate. These stiffness characteristics reduce the risk of uncontrolled transverse loading on the crankshaft main bearings, and cylinder liners at their clamping points are only slightly deformed by the outside structure. This contributes to keeping liner wear rates as low as possible.

MAN ENGINES

143

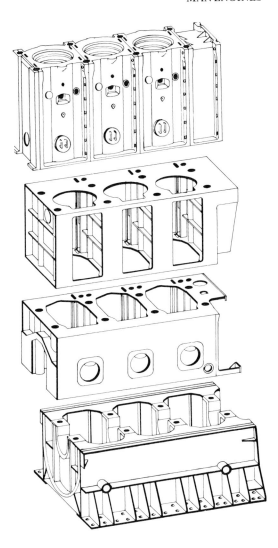

Figure 8.6 Construction method

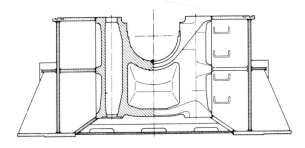

Figure 8.7 Bedplate design

Running gear

The crankshafts of the larger engines are of built-up type with forged steel crank throws pressed on to forged steel journals. Crankshafts for the smallest 520 mm bore engine have one-piece forged shafts up to the highest number of cylinders. The main bearings are of the so-called block bearing type comprising three-metal shells which can be replaced without subsequent fitting. The crankpin bearing is also of three-metal type with a comparatively thick white-metal layer.

The crosshead (Figure 8.8) features hydrostatic lubrication, a principle which ensures an adequate oil film between the crosshead pin and the single white-metal lined bearing shell. The high oil pressure required for this purpose is generated by double plunger high-pressure lubricators which are directly driven by linkages from the connecting rod to force oil under pressure into the lower crosshead bearing shell.

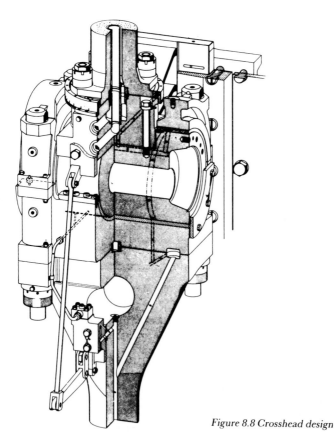

Figure 8.8 Crosshead design

Camshaft drive

The box shape construction of the frame forms a deformation resistant casing to accommodate the gear train of the camshaft drive. All engines have a gear train with only one intermediate wheel. The wear pattern and backlash on the intermediate wheel can be adjusted by draw-in and thrust bolts which act upon the frame at the bearing points of the intermediate wheel axle. The crankshaft gear wheel is clamped directly to the shaft as a two-piece component.

Combustion chamber components

In order to control thermal stresses, thin-wall intensively cooled components are used for the combustion chamber parts (Figure 8.9). The piston crown consists of a high-grade heat resistant steel and incorporates the MAN jet-cooled honeycomb system. An internal insert in the piston crown serves only to direct the cooling water flow and, in particular, to generate water jets which enter at the edge of each honeycomb element, thereby intensifying the conventional 'cocktail shaker' effect. The structural arrangement of the KSZ-C piston has the steel crown, insert, and piston rod and the cast iron skirt all held together by a single row of waisted studs. The fresh water for piston cooling (Figure 8.12) is fed into the piston rod with return through the co-axial bores to telescopic pipes.

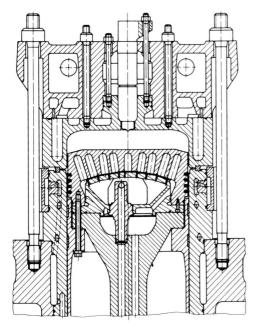

Figure 8.9 Cross section of combustion chamber

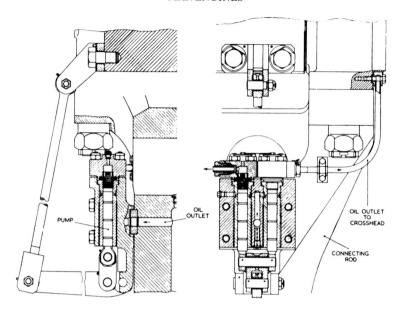

Figure 8.10 Crosshead oil pump

The cylinder cover comprises a stiff grey cast iron cap fitted with a thin-walled forged heat-resistant steel bottom. The pre-tension between the cap and bottom, which is necessary to prevent micro-motion, is achieved by an inner row of studs. An outer row of larger studs secures the cover to the cylinder block.

The cylinder liners are short single-piece units with intensive cooling of the upper part achieved by bore cooling of the liner flange (Figure 8.14). The vertical lands between the exhaust ports are also water cooled; the combined techniques ensure that sufficient heat is removed from the piston rings, the upper portion of the cylinder liner and the port lands to obtain a stable lubricating oil film, to reduce the wear rate and not to interfere with ring performance in way of the porting.

Turbocharging

An important feature of MAN engines is the use of the space beneath some of the pistons to supply charge air to assist the turbocharger at low loads and particularly on starting. This feature, known as 'injector drive turbocharging', was introduced on the KSZ-type engines but was replaced by a system of electrically driven blowers for the KSZ-A and subsequent engine models. These blowers cut in and out as a function of scavenge air pressure and usually remain in operation until the engine

MAN ENGINES

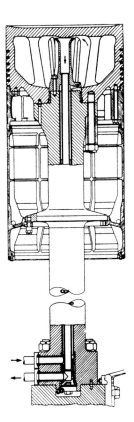

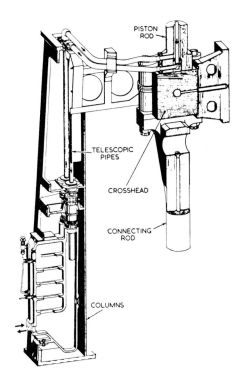

Figure 8.11 Piston and piston rod

Figure 8.12 Piston cooling arrangement

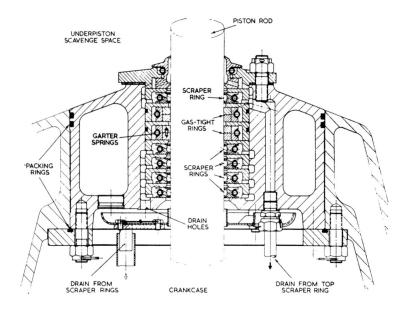

Figure 8.13 Piston rod scraper box

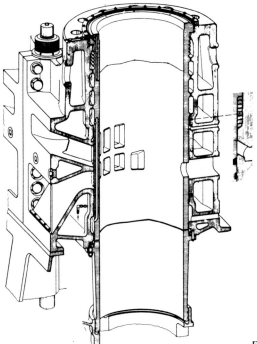

Figure 8.14 Cylinder liner and jacket

MAN ENGINES

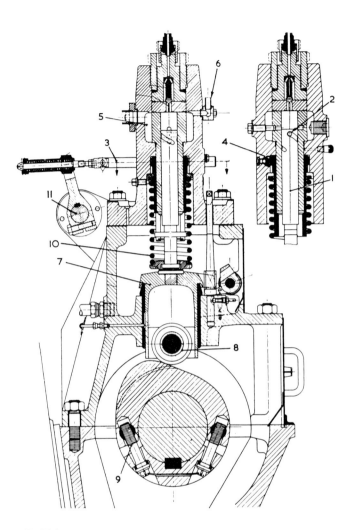

Figure 8.15 Fuel injector pump

reaches 50% load. The blowers are placed ahead of the turbocharger and need to be in operation before the engine will start.

The turbocharging arrangement is on the constant pressure system and embodies low flow resistance air and exhaust piping with suitable diffusers between the ports and manifolds. For engine outputs up to 8000 kW only one turbocharger is used and all engines above this power require two chargers, one at each end of the engine at the cylinder head level, though turbochargers have been mounted on special base frames together with the auxiliary blower adjacent to the engine.

ELECTRONIC INJECTION

A system introduced recently to MAN engines and no other slow-speed two-stroke engine types is electronically controlled fuel injection (Figure 8.16). The 'heart' of the electronic injection system is a microprocessor to which the actual engine speed and crankshaft position are transmitted as output signals. When the desired speed is compared with the actual speed, the processor automatically retrieves the values of injection timing and pressure for any load under normal operating conditions. In the event of deviations from normal operating conditions additional signals are transmitted to the microprocessor either manually or by appropriate temperature and pressure sensors. The result is an engine operating at optimum injection pressure and timing at all times. Adjustments to the system during operation are possible.

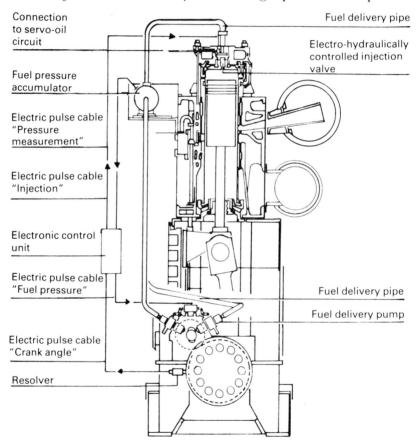

Figure 8.16 Components of electronic injection system

Corrections can be made for: adapting the injection system to varying conditions, such as ballast voyages, bad weather; matching the engine to different fuels, ignition lag, cetane numbers, etc; readjusting for operation in the tropics or in winter; controlling the injection characteristics of individual cylinders, such as when running in; reducing the minimum slow running speed, such as when in canals.

Injection system components

The most noticeable feature of a KEZ type engine with electronic injection is the absence of a camshaft and individual fuel pumps. The fuel is delivered to a high-pressure accumulator by normal fuel injection plunger pumps driven by gears and cams from the crankshaft. These pumps are located above the thrust bearing and, depending on the size and cylinder number of the engine, two or more of these pumps are used. To safeguard against failure of a pump or a pressure pipe, two

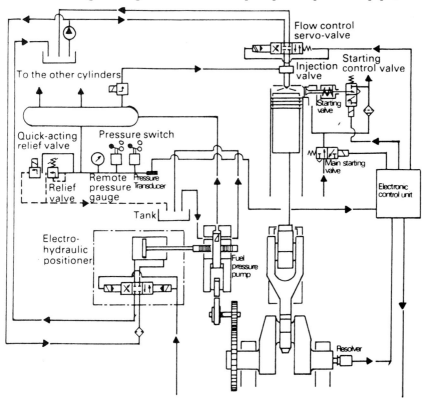

Figure 8.17 Schematic arrangement of electronic injection systems

independent fuel delivery pipes are placed between the high-pressure pumps and the accumulator.

The high-pressure accumulator generates a buffer volume to prevent the pressure in the injection valve from dropping below an acceptable limit when the fuel is withdrawn during injection into the cylinder. For the electronically controlled injection system, metering of the fuel volume for one cylinder is not affected by limiting the delivery volume per stroke of the injection pump plunger as in conventional systems, but by limiting the duration of injection. This is initiated by an electronic controller, i.e. by a low voltage signal which controls the opening and closing action of an electro-hydraulic servo valve. The electro-hydraulic valve is actuated by oil under pressure from two pumps and an air cushion-type accumulator.

Sequential control ensures that according to the engine firing sequence, the correct engine cylinder is selected in the proper crank angle range and that the electronically controlled injection valve is actuated at the exact point of injection.

For engine starting and reversing the correct injection timing is linked with the electronically controlled starting air valves. Since electronically controlled starting valves have no time delay between the switching pulse and opening of the air valve, and since there are no flow losses, due to the elimination of the control air pipes, the starting air requirement is reduced in comparison with the former conventional air start system.

The constant pressure maintained during fuel injection with the electronically controlled fuel injection system, and the fact that the injection pressure can easily be matched to each load condition, lead to lower fuel consumption rates over the entire load range. It also means satisfactory operation at very slow speeds, which contributes favourably to the manoeuvrability of a vessel; a K3EZ 50/105 C/CL engine on the test bed was operated at a dead slow speed of 30 rev/min, which has never been achieved before. The full load speed is 183 rev/min.

MAN engines of recent years have some very notable features to promote simplicity of design, ease of maintenance and low specific fuel consumption, not least the application for the first time to a marine diesel engine of electronic fuel injection. It is not known at this time whether electronic injection will be taken up by other engine builders and there is some doubt as to whether MAN two-stroke engines will still be produced after 1984. But it is now likely that electronic injection will live on in the future series of MAN–B & W engines designed in Copenhagen.

C.T.W.

9 Mitsubishi engines

Although Japan is the world leader in terms of annual production of slow-speed two-stroke engines, there is only one engine of Japanese design built in that country. That engine is the UE type built by Mitsubishi Heavy Industries Ltd. and licensees: it is a long stroke constant pressure turbocharged uniflow scavenged type for direct drive of the propeller. This make of engine has been in production since 1955 and has been successively improved and uprated. The latest model is the UEC-H type (Figure 9.1) which has a number of design changes compared with its immediate predecessor the UEC-E, one notable feature being the dropping of two-stage turbocharging (the only two-stroke engine so equipped) and the replacement of three exhaust valves in the cylinder cover with one centrally located unit of large diameter (on small bore models 370 and 450 mm only).

Figure 9.1 View of a 6UEC52/125H engine

Table 9.1 Mitsubishi UE Diesel Engine H & HA Series Principal Particulars

Engine type	No. of cylinders	Maximum output			Service output		
		Output PS (kW)	Engine speed (rev/min)	Specific fuel conservation g/PS-h (g/kW-h)	Output PS (kW)	Engine speed (rev/min)	Specific fuel conservation g/PS-h (g/kW-h)
UEC60HA	4	8,000 (5,885)	140	131 (178)	6,800 (5,000)	140	127 (173)
	5	10,000 (7,355)			8,500 (6,250)		
	6	12,000 (8,825)			10,200 (7,500)		
	7	14,000 (10,295)			11,900 (8,755)		
	8	16,000 (11,770)			13,600 (10,005)		
	9	18,000 (13,240)			15,300 (11,255)		
UEC52HA	4	6,080 (4,470)	170	132 (179)	5,170 (3,805)	170	128 (174)
	5	7,600 (5,590)			6,460 (4,750)		
	6	9,120 (6,710)			7,750 (5,700)		
	7	10,640 (7,825)			9,045 (6,655)		
	8	12,160 (8,945)			10,335 (7,600)		
	9	13,680 (10,060)			11,630 (8,555)		
UEC45HA	4	4,560 (3,355)	185	134 (182)	3,875 (2,850)	185	130 (177)
	5	5,700 (4,190)			4,845 (3,565)		
	6	6,840 (5,030)			5,815 (4,275)		
	7	7,980 (5,870)			6,785 (4,990)		
	8	9,120 (6,710)			7,750 (5,700)		
	9	10,260 (7,545)			8,720 (6,415)		
UEC37HA	5	3,250 (2,390)	210	135 (184)	2,765 (2,035)	210	131 (178)
	6	3,900 (2,870)			3,315 (2,440)		
	7	4,550 (3,345)			3,870 (2,845)		
	8	5,200 (3,825)			4,420 (3,250)		
	9	5,850 (4,305)			4,975 (3,660)		

Development of the Mitsubishi UEC-H has been in steps leading up to the recent introduction of the latest version, the UEC-HA model. A prelude to the development was the application of constant pressure turbocharging after many years of producing pulse turbocharged engines, the fitting of two-stage constant pressure turbocharging in 1975, and by 1980 the production of the UEC-H with single-stage constant pressure turbocharging, made possible by the fitting of the new Super MET turbochargers of higher efficiency.

Mitsubishi's aim was to design a new long stroke, slow-speed, constant pressure turbocharged engine of superior fuel economy to the previous UEC-E engine. The above design features, combined with uniflow scavenging, have enabled a reduction in specific fuel consumption by about 10% and the engine was designed specifically to burn heavy fuels, up to 5000 seconds Redwood being used in some cases.

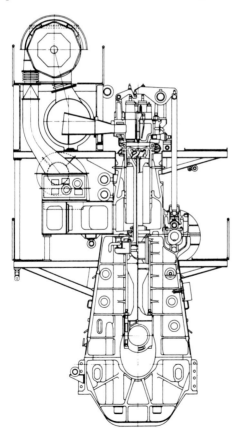

Figure 9.2 Cross section of the UEC52/125H engine

Table 9.2 Principal particulars of UEC-E type diesel engine

Item	Unit	UEC 85/180 E	UEC 65/135 E	UEC 60/125 E	UEC 52/105 E
Cylinder bore	mm	850	650	600	520
Piston stroke	mm	1,800	1,350	1,250	1,050
Maximum continuous output per cylinder	PS/cyl	3,800	2,300	1,900	1,330
Maximum continuous output (6 cyl)	PS	22,800	13,800	11,400	8,000
Maximum continuous engine speed	rpm	120	150	158	175
Brake mean effective pressure	kg/cm^2	13.95	15.40	15.31	15.38
Mean piston speed	m/s	7.20	6.75	6.58	6.13
Specific fuel consumption	g/PS h	152	153	155	155
Overall length (6 cyl)	mm	13,210	10,740	9,700	8,080

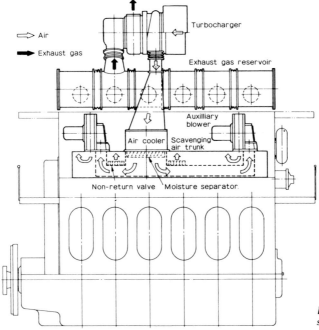

Figure 9.3 Exhaust and scavenge flow diagram

Engine structure

The UEC-H type engine is of cast iron construction with the three separate blocks of bedplate, entablature and cylinder block secured to one another by long tie-bolts. Unlike other designers of two-stroke engines,

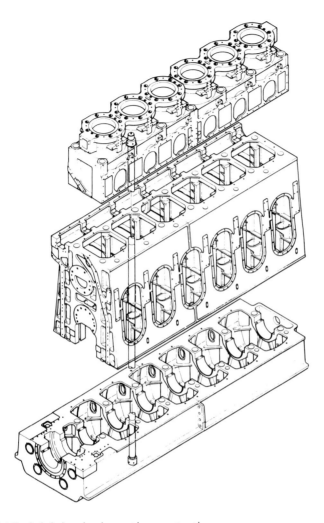

Figure 9.4 Exploded view showing cast iron construction

Mitsubishi has retained the solid cast iron construction (Figure 9.4) to achieve greater rigidity with less deformation in service and lower vibration and noise levels. The main bearings of shell type are supported in housings machined in the bedplate. The running gear is of straightforward conventional design using shell type or white-metal lined large end bearings and a crosshead of traditional table type (Figure 9.5) with four guide slippers and twin white-metal lined top end bearings and the piston rod bolted direct to the crosshead pin. The connecting rod itself is machined from forged steel.

Compared with the two-stage turbocharged E type engine the constant pressure turbocharged UEC-H engine has a delayed exhaust valve opening and consequently more effective work done each power stroke. There is also a resulting drop in the specific fuel consumption — at around 10% as mentioned earlier — while the long stroke design has brought additional gains said to amount to 4–7 g/bhph plus a 15% reduction in shaft revolutions.

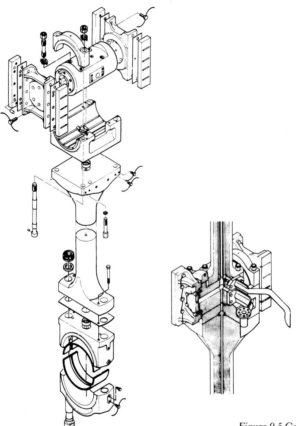

Figure 9.5 Connecting rod and crosshead

Exhaust and scavenging

Referring to the cross-sectional drawing Figure 9.2, the exhaust gases leave the engine by way of a diffuser type pipe to enter the exhaust gas manifold. Located above this manifold on a pedestal is the Super MET non-water cooled turbocharger, or two chargers in the case of a large number of cylinders. The air from this turbocharger is delivered to the

MITSUBISHI ENGINES

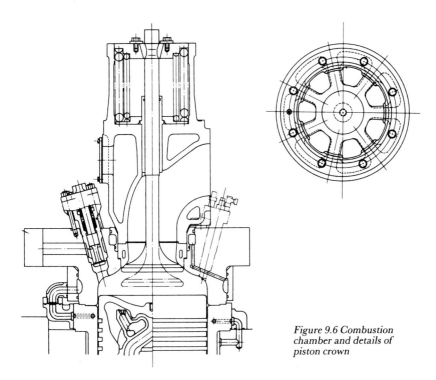

Figure 9.6 Combustion chamber and details of piston crown

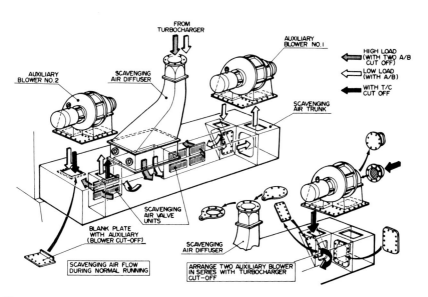

Figure 9.7 Scavenge air flow

intercooler and then the scavenge air trunk by way of non-return valves and then into the cylinder via ports uncovered by the piston. During low load operation the turbocharger air delivery is passed through two auxiliary blowers before entering the cylinders.

The diffuser type exhaust pipe is of a three branch construction where it connects to the engine cylinder. The gas flows from the three cylinder exhaust valves are merged into one to enter the exhaust manifold through the diffuser section. Figure 9.3 shows the arrange-

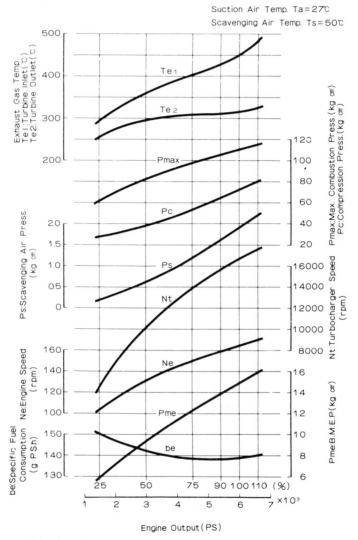

Figure 9.8 6UEC45/115H performance curve

ment of pipe connections. The exhaust manifold is actually made up of many short sections, each corresponding to one engine cylinder, joined together by flexible joints to accommodate longitudinal thermal deformation.

The scavenge air trunk, however, is divided into two compartments. Charge air from the intercooled, shown in Figure 9.3, enters the outer chamber and through the multi-cell non-return valves to the scavenge air entablature that surrounds the cylinder jackets.

The two motor driven auxiliary blowers consume approximately 0.6% of the rated engine output and as a matter of standard practice are automatically brought into operation when the engine operates at 50% load or less. Start up of the blowers is by a sensed drop in scavenge air pressure. An emergency feature for engines equipped with only one turbocharger is the facility to operate both auxiliary blowers in series to achieve the required scavenge air pressure for continuous engine operation. This emergency hook-up is a simple process of adjusting some partition plates in the scavenge box.

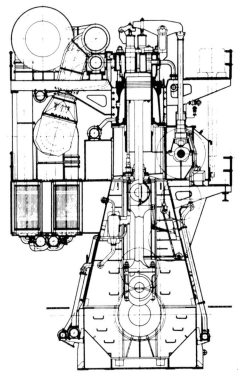

Figure 9.9 Cross section of UEC85/180E

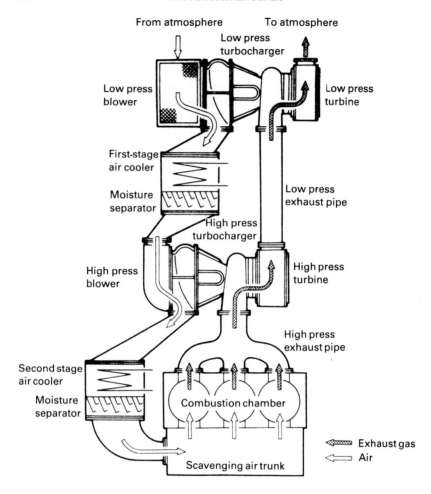

Figure 9.10 Two-stage turbocharging system

Two-stage turbocharged model

The distinctive feature of the Mitsubishi UEC-E type engine is the two-stage turbocharging system, introduced in 1977 to achieve a high output and bmep of 15 kg/cm^2. To match this increase in maximum cylinder pressures, the engine frame, bearings and combustion chamber components were made more rigid than those of the predecessor UE-D type engine, but still retaining the monobloc cast iron construction (three components held by tie-bolts). A new piston crown of specially reinforced molybdenum cast steel is employed with a combination of radial and circular ribs behind the frame plate, to withstand a rise in maximum pressure from 90–110 kg/cm^2. A sectional arrangement of the UEC-E type engine is shown in Figure 9.9.

However, this model was rapidly replaced by the current UE-H engine because of the much improved performance of the Super MET turbocharger (Figure 9.10) meaning that two of the older MET types in series were no longer necessary to achieve the high bmep. The two-stage turbocharged (impulse) UEC-E engine has an average fuel rate of around 155 g/bhph; the long stroke UE-H has achieved fuel consumption rates as low as 138 g/bhph and even lower for the most recent HA types.

C.T.W.

10 GMT engines

Grandi Motori Trieste (formerly Fiat) two-stroke crosshead engines are built in four basic cylinder bore sizes of 600 mm, 780 mm, 900 mm and 1060 mm, though few of the latter 'super large bore' type have been built in the last 10 years. These four engine sizes of the 'B' series are all basically similar, while 'C' series engines, and most particularly the unusual short stroke CC600 engine, are fairly recent developments with few examples actually at sea. The final development of the GMT two-stroke engine which will be worked through in the immediate future is the dropping of the lever-driven scavenging air pumps of each cylinder to result in a pure constant pressure turbocharged engine similar to other makes with electrically driven booster fans for low power operation and engine starting.

The GMT 'B' series of engines cover a power range from 6000 to 55 000 bhp and the characteristic feature is the two-stage air scavenging arrangement with turbochargers and mechanical scavenge pumps operating in series, the constant pressure turbocharging system and the

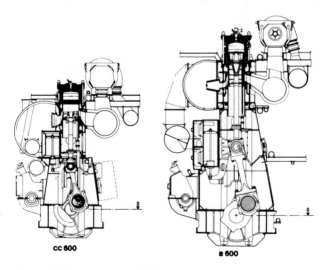

Figure 10.1 Cross section of GMT slow speed engines

GMT ENGINES

Table 10.1 Technical data for B600 engines

Cylinder bore (mm)	600	DIMENSIONS	
Piston stroke (mm)	1 250	Length at crankshaft (mm)	
Max continuous rating (bhp/cyl metric)	1 350–1 500	4 cylinder	6 670
		5 cylinder	7 770
Corresponding engine speed (rev/min)	145–160	6 cylinder	8 870
		7 cylinder	9 535
Bmep at mcr (kg/cm^2)	12	8 cylinder	10 635
		9 cylinder	12 485
Pressure charging/scavenging		10 cylinder	13 585
Type of scavenge	cross scavenge	Width at mounting (mm)	3 000
Scavenge air pressure (kg/cm^2)	1.55	Height above shafting (mm)	6 750
No. of turbochargers	1–2	Overhauling height above shafting (mm)	9 965 (9 045)*
Turbocharging system	constant pressure	Depth below shafting (mm)	1 100

cross scavenging arrangement in the cylinders brought about by air inlet and exhaust ports arranged on opposite sides of the cylinder liner. The reciprocating scavenge pumps are arranged aside each cylinder and driven through a rigid arm by the crosshead shoe. Unlike other engine designs, this design avoids using the lower part of the cylinder as a scavenging pump and results, instead, in being completely open to the atmosphere, easily to be inspected and cleaned of the sludge which

Figure 10.2 A ten cylinder B600 engine on test in Trieste

166 GMT ENGINES

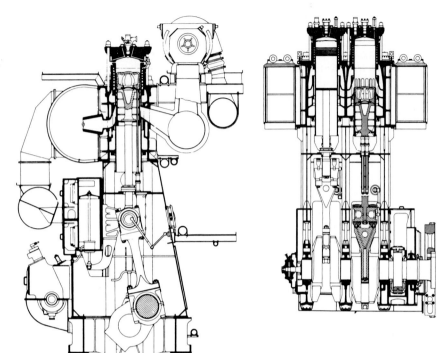

Figure 10.3 Cross and longitudinal sections of B600 engine

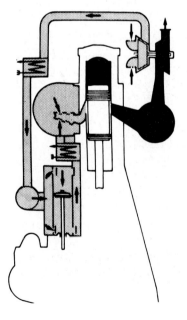

Figure 10.4 Scavenging and turbocharging arrangements

drains from the cylinder liner and collects on the upper crankcase wall. The solution with independent air pumps ensures the availability of a sufficient amount of scavenge air even when running at low speed, the avoidance of an auxiliary electrically operated blower and utilising the piston underside as a scavenge pump. The double-stage supercharging system has also the advantage of allowing operation of the engine as a naturally aspirated type, in case of failure of the first stage turbo-chargers, by using the second stage reciprocating pumps. Under such emergency conditions the available output is sufficient to ensure that the ship maintains a speed of about 80% of the normal value. But, as mentioned earlier, the most recent search for operating economy in the face of high fuel costs prompted GMT to redesign its engines in 1982 without mechanical scavenge pumps. This description is confined to the 'B' series engines, of which many examples are at sea, followed by the 'C' series which are installed particularly in Italian flag and built vessels.

DESIGN OF THE 'B' SERIES ENGINE

Bedplate and frame

For the engine types B600, C600, B780, C780, B900, C900 and B1060 the bedplate and the frame form, together with the cylinder blocks, a fixed structure, connected together by vertical steel tie-rods which put the whole structure under compression. The bedplate is of fabricated construction formed by cast steel cross members laterally welded to longitudinal girders of steel plate. One or more of these sections can be bolted together to form the bedplate. The frame is of box-type structure made from steel plates with welded-in pipe channels through which the tie-bolts are passed. At one side the crosshead guides are attached to the frame; this cast iron section also forms part of the scavenge pump cylinders.

Cylinders

The cylinder blocks are made in one-unit cast blocks joined together longitudinally by bolts and held down on the frame by the tie-bolts. The cylinder liners are seated at the top of each block and supported at both scavenge and exhaust ports and at the lower part. Each liner is made in two sections: the upper part is of high resistant cast iron, strong backed by an external steel ring with cooling channels cut within for surface cooling of the liner upper part.

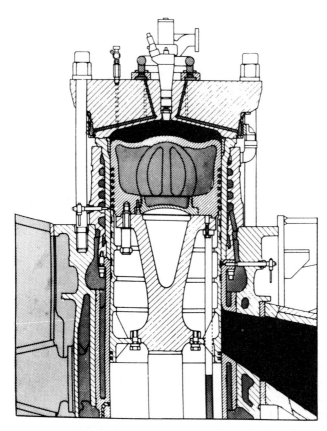

Figure 10.5 Combustion chamber components

Near to the combustion chamber the liner is protected by a thermal shield consisting of a stainless steel ring to reduce temperatures and stresses in the collar. The lower part of the liner is also made from high resistant cast iron and has the scavenge and exhaust ports separated by columns with internal water passages to avoid sludge and deposit formation around the scavenge ports.

Suitable ducts cast into the cylinder blocks convey the scavenge air from the pumps into the cylinders while air valves before the port openings prevent exhaust gases from entering the scavenge manifold, the valves opening only when the exhaust gas pressure drops below the scavenge air pressure, i.e. at the end of expansion and scavenging can take place.

The cylinders are lubricated through quills arranged on two different circumferences, one above and one below for lubrication of the piston skirt.

Cylinder covers

The cylinder covers consist of a simple forged steel body held in place by 12 studs around the circumference. A series of machined holes drilled radially close to the combustion chamber are provided for intensive cooling and strength to withstand thermal and mechanical stresses. A fuel injector is mounted in the centre of the cover together with relief and air start valves and an indicator cock.

Running gear

The crankshaft is of semi-built type with cast steel throws and forged steel main journals. The shaft is built in one section for engines on which the fuel pumps are driven at the flywheel side, and in two sections when the fuel pump drive is at the mid-length. The main bearings comprise steel shells with a white-metal lining.

The connecting rods are forged in steel in the traditional arrangement with a palm at the foot for the large end bearing. The crosshead pins are built to a specially patented design eccentrically machined so that intensive lubrication can be maintained at all conditions of load and speed while metal lined bearing shells allows easier maintenance. A single slipper is fitted to each crosshead (Figure 10.6).

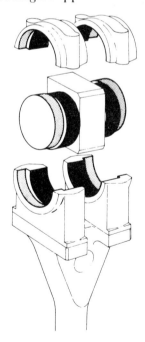

Figure 10.6 Eccentric crosshead bearing

The pistons consist of a cast steel crown bolted to a cast steel skirt, both components being connected to the piston rod via upper and lower flanges on the rod. The interior of the crown is provided with radial ribs for stiffness while the running surface has a special alloy coating to withstand corrosion and burning. Piston rings fitted are of the stepped type with reduced rubbing surface, and cooling of the crown is by water operating on the 'cocktail shaker' principle. The water is fed in and discharged through telescopic pipes running in boxes mounted between the engine frame and cylinders, with the outlet from the water boxes through a tall rate box mounted outside the engine. Separate fresh water systems with pumps, coolers, etc., are provided for piston cooling and cylinder jacket cooling. A water circuit is also provided for fuel valve cooling.

Fuel injection system

Fuel injection pumps are assembled in one or two bodies mounted externally on the engine and driven from the crankshaft by a train of three gear wheels located at the middle of the engine on eight to twelve cylinder models (Figure 10.8). The cam operated fuel valve has a timed spill valve which opens to cut off delivery to the cylinders, thus giving the fuel delivery quantity control. The fuel spill valve is actuated by a

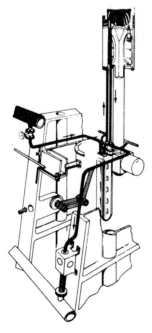

Figure 10.7 Piston cooling arrangement

GMT ENGINES

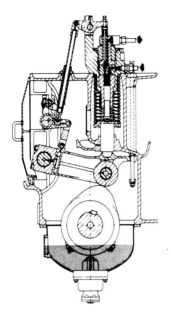

Figure 10.8 Fuel pump

rocker arm and push rod from the main plunger arm by way of an intermediate lever mounted on an eccentric shaft to vary the stroke and point of spill operation.

The fuel valves are of the screw-on nozzle cap type, cooled by fresh water, through drillings to a cooling ring inside the nozzle.

The CC600 engine

The CC600 was conceived in 1975 as a completely new engine of two-stroke crosshead type. Although it shares many features of the 'B' and 'C' series engines, this 600 mm bore by 800 mm stroke engine has a stroke/bore ratio of 1.33:1 instead of the 2:1 ratio used in the other

Table 10.2 Technical data for CC 600 engines*

Engine type	No. of cylinders	Output (bhp)	Output (kW)	Weight (tons)	Length (mm)
CC600.4	4	6600	4855	110	6310
CC600.5	5	8250	6065	132	7360
CC600.6	6	9900	7280	162	8410
CC600.7	7	11550	8495	184	8940
CC600.8	8	13200	9708	206	9990
CC600.9	9	14850	10921	228	11040
CC600.10	10	16500	12135	249	12090

* output	1650 bhp/cyl 1215 kW/cyl	bore stroke	600 mm 800 mm	speed mep	250 rpm 12,89 bar

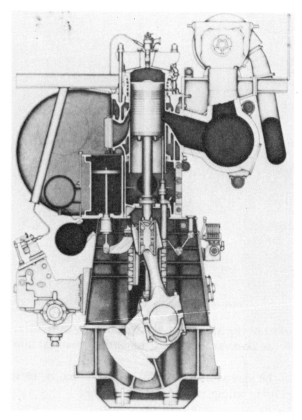

Figure 10.9 Cross section of CC 600 engine

models. This results in a higher speed engine of lighter construction making it suitable for geared applications in larger vessels or direct coupled in small ships. The output of the CC600 engine is normally 1650 bhp/cylinder at 250 rev/min.

The main structure consists of a bedplate and a frame both in fabricated steel with cast steel cross members welded to longitudinal girders. Cast steel cylinder blocks either as single units or multi-cylinder units are mounted on the frame, as are the cast iron scavenge pump bodies contained within the air manifold. The crankshaft is an alloy steel single-piece forging. The main bearings consist of a lower thick shell and a cap held by one strut, thus allowing the bearings to be easily inspected. A very light design of connecting rod has been used, similar to that of a four-stroke medium-speed engine of the same output. The crosshead shoes (Figure 10.10) are mounted at the ends of the crosshead pin.

The cylinder head (Figure 10.11) is similar in design to the longer stroke GMT engines, with bores close to the combustion chamber wall,

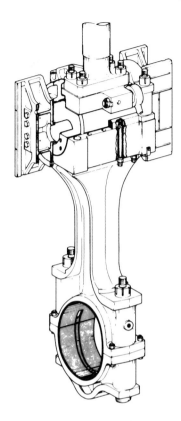

Figure 10.10 Crosshead and connecting rod

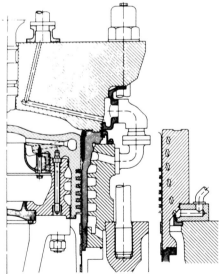

Figure 10.11 Combustion chamber components

while a similar design of cylinder liner has been retained. There is, however, a different type of piston fitted to the CC600 engine, the new one being of composite form featuring the forged steel crown sitting on an inner support which transmits the mechanical force to the piston rod and also conveys the cooling water through helical channels in the area near the piston rings. Radial holes near the piston top (with water inlet and return through the same holes) allow forced cooling of the hot surface of the thick crown top, whereas the thinner central section of the crown is cooled on the 'cocktail shaker' principle.

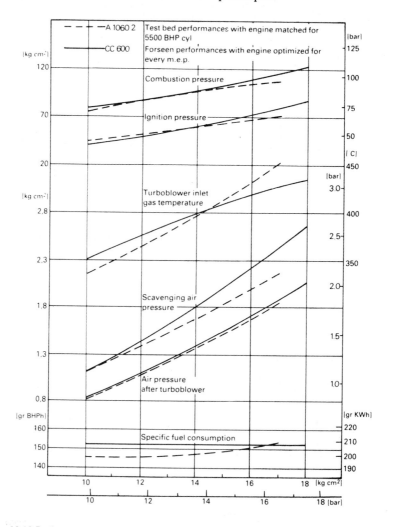

Figure 10.12 Performance curves for CC 600

GMT ENGINES

Figure 10.13 A CC 600H engine on the testbed

The new design of components in the CC600 have resulted in a most compact engine of high power with a weight-to-power ratio of only 13 kg/bhp. The specific advantages of this engine for marine propulsion plants are as follows:

1. Freedom to select the optimum propeller speed and reduce the installed power and fuel consumption.
2. Ability to burn the poorest heavy fuels.
3. Reduced maintenance times and cost.
4. Suitable for vessels with limited head room, such as ro-ro ships.
5. Reduced engine room length by using two engines geared together.
6. Ideal for special vessels with higher propeller speeds suitable for direct coupling.
7. Universal application of the engine allows standardisation of propulson machinery in different types of vessels.

C.T.W.

11 Doxford engines

The Doxford engine was the only remaining British design of two-stroke slow-speed engine, but a decline in demand for this long-established and once popular direct drive propulsion engine has now resulted in the cessation of production with the engine works permanently closed. However, quite a number of Doxford J-type engines — and indeed some of the earlier P-type — are still in service as well as some examples of the three-cylinder 58JS3 model, the last new engine type to come from the old Pallion engine works. Major design features of the most recent J-type (Figure 11.1) and the small three-cylinder engine are discussed here.

Figure 11.1 A 76J4 engine on the testbed

DOXFORD J-TYPE

Doxford J-type engines were built in long and short stroke versions in bore sizes of 580, 670, and 760 mm from three to nine cylinders up to a maximum of 27 000 bhp from a single engine. The engine is a single-acting two-stroke opposed piston type with each cylinder having two pistons which move in opposite directions from a central combustion chamber.

The pistons in each cylinder are connected to a three-throw section of the crankshaft, the lower piston being coupled to the centre throw by a single connecting rod, crosshead and piston rod. Each pair of side cranks is connected to the upper piston by two connecting rods, cross-heads and side rods. As the pistons move towards each other, air is compressed in the cylinder and shortly before reaching the point of minimum volume between the pistons (inner dead centre) fuel at high pressure is injected through the injector nozzles into the combustion

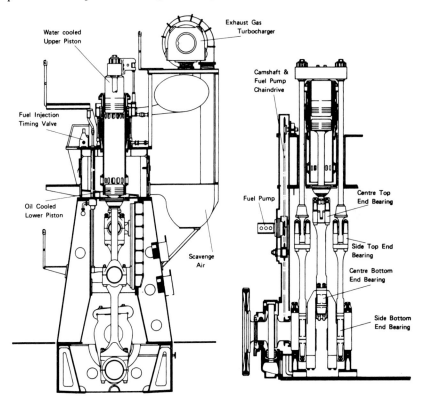

Figure 11.2 Doxford J-type arrangement

Table 11.1

	58J4	67J4	67J5	67J6	76J3	76J4	76J5	76J6	76J7	76J8	76J9
Engine type											
Number of cylinders	4	4	5	6	3	4	5	6	7	8	9
Cylinder bore (mm)	580	670	670	670	760	760	760	760	760	760	760
Stroke upper piston (mm)	480	500	500	500	520	520	520	520	520	520	520
Stroke lower piston (mm)	1 370	1 640	1 640	1 640	1 660	1 660	1 660	1 660	1 660	1 660	1 660
Combined piston stroke (mm)	1 850	2 140	2 140	2 140	2 180	2 180	2 180	2 180	2 180	2 180	2 180
Number of turbochargers	1	1	2	2	1	1	2	2	3	3	4
Weight, (excluding oil and water) tonnes	175	270	320	370	270	335	400	470	540	600	680
Ratings											
Max. continuous rating (MCR) bhp metric	7 500	9 000	11 000	13 500	9 000	12 000	15 000	18 000	21 000	24 000	27 000
Engine speed, rev/min	160	127	127	127	123	123	123	123	123	123	123
Brake mean effective pressure (bmep) bars	10.57	10.34	10.34	10.34	10.81	10.81	10.81	10.81	10.81	10.81	10.81
Recommended max service power (90% MCR)	6 750	8 100	9 900	12 200	8 100	10 800	13 500	16 200	18 900	21 600	24 300
Engine speed r.p.m.	155	123	123	123	119	119	119	119	119	119	119
Brake mean effective pressure (bmep) bars	9.82	9.61	9.61	9.61	10.06	10.06	10.06	10.06	10.06	10.06	10.06
Consumption											
Fuel consumption	150 to 153 g/bhp—h with H.V. fuel of 10,000 Kcal/Kg.										
Cylinder lubricating oil	0.45 g/bhp—h										

space. Due to compression the air in the combustion space is at high temperature, causing the fuel to ignite. During the first stages of combustion the pressure in the cylinder continues to rise until a maximum value is reached soon after the pistons begin to move apart. After combustion is completed, the hot gases continue to expand, thereby forcing the pistons apart until the exhaust ports in the upper liner are uncovered by the upper piston. As the exhaust ports open the hot gases in the cylinder, now at a reduced pressure, are discharged to the turbine of the turbochargers, so causing the pressure in the cylinder to drop to a level just below that of the scavenge air. At this point the air inlet ports in the lower liner are uncovered by the lower piston, so allowing air under pressure, which is delivered to the scavenge space by the turbocharger, to flow through the cylinder expelling the remaining burnt gases (Figure 11.3(a)).

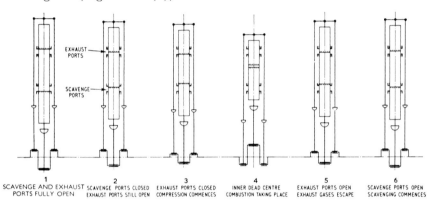

Figure 11.3(a) Exhaust and scavenge events

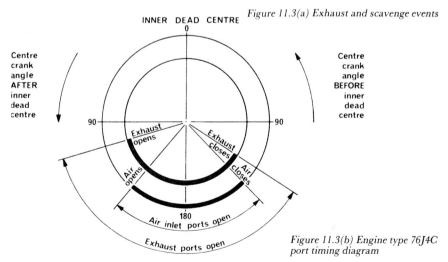

Figure 11.3(b) Engine type 76J4C port timing diagram

During the inward compression stroke of the pistons the air inlet ports are closed just before the exhaust ports. The air in the cylinder is then compressed and the cycle repeated. The opening and closing of the exhaust and air inlet ports are shown diagrammatically in the typical timing diagram (Figure 11.3(b)). This diagram is constructed with reference to the inner dead centre position, the unsymmetrical timing of the ports being due to the fact that the side cranks are at an angle of more than 180° from the centre crank. The angle of lead of the side cranks varies with the size of engine.

The upper piston of the Doxford engine requires its own running gear and crank throws. The advantages obtained with this principle are considerable. For equal mean indicated pressure, mean piston speed and cylinder bore, the Doxford opposed-piston engine will develop 30%–40% higher power per cylinder than a single-piston engine. The first order inertia force from the lower piston is balanced against the corresponding force from the upper pistons and a well-balanced engine is obtained. All the forces from the upper piston are transferred through the running gear. Long tension bolts, as used in single-piston engines to transfer the forces from the cylinder covers to the bedplate, are not required in the opposed-piston engine. The engine structure is therefore simple and relatively free from stresses. No valves are required in the scavenge-exhaust system and the scavenge efficiency of the cylinders is high. The flow areas through the exhaust ports are substantial, and fouling, which inevitably takes place in service, has little effect. The large port areas also make the engine well suited for turbocharging.

Engine construction

The bedplate is built up of two longitudinal box girders which extend over the full length of the engine. The transverse girders, which incorporate the main bearing housings, are welded to these longitudinal members. A semi-circular white-metal lined steel shell forms the lower half of each main bearing and is held in place by a keep secured by studs to the bedplate.

The entablature is also a welded steel box construction, arranged to carry the cylinders which are bolted to the upper face. It is bolted to the tops of the columns and to the crosshead guides. The entablature forms the air receiver from which the cylinders are supplied with air and its volume is supplemented by the air receiver on the back of the engine, the top face of which forms the back platform. The intercoolers are mounted on this receiver and the air deliveries from the turbochargers are connected to these coolers. Alternatively, in the case of engines with end mounted turbochargers, the air from the turbocharger is supplied

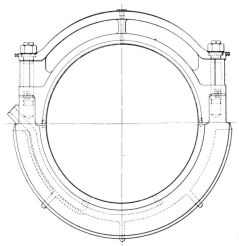

Figure 11.4 Main bearing assembly

directly to the entablature after passing through the end mounted intercoolers.

Crankshaft

Each cylinder section of the crankshaft is made up of three throws, the two side cranks being connected to the upper piston and the centre crank to the lower piston. The side crankwebs are circular and form the main bearing journals. Each centre crank is made of an integral steel casting or forging for semi-built shafts. For fully built shafts these units are made of two slabs shrunk onto the centre pin. The centre cranks are shrunk onto the side pins.

Lubricating oil is fed to the main bearings and through holes in the crankshaft to the side bottom end bearings. At the after end, the thrust shaft is bolted to the crankshaft. The thrust bearing of the tilting pad type is housed in the thrust block, which is bolted to the end of the bedplate. A turning wheel, which engages with the pinion of the electrically driven turning gear, is bolted to the after coupling of the thrust shaft.

Cylinders

The cylinder liner (Figure 11.5) is a one-piece casting and incorporates the scavenge and exhaust ports in the lower and upper sections respectively. A special wear resistant cast iron is used as cylinder liner material. An exhaust belt round the exhaust ports and a water cooling jacket are clamped to the top and bottom faces respectively of the com-

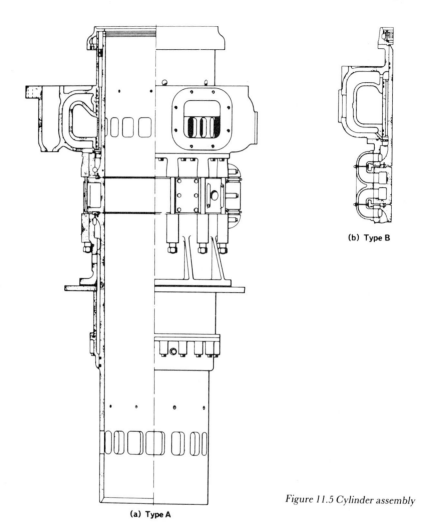

Figure 11.5 Cylinder assembly

(a) Type A
(b) Type B

bustion section of the liner by long studs. The cylinders are secured to the entablature by means of a flange on the jacket passing over studs and secured with nuts.

The cylinders are water cooled. The cooling water enters the jacket (Figure 11.6) and circulates round the top of the lower part of the liner. It then passes through holes drilled at an angle through the liner round the combustion chamber. Above the combustion chamber some of the water is taken directly into the water jacket of the exhaust belt and the remainder is passed through the exhaust bars first and then into the

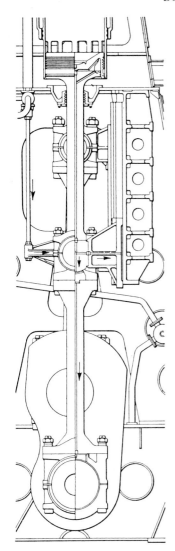

Figure 11.6 Lubricating and cooling oil circuits

exhaust belt water jacket. Finally the water is returned to the cooling water tank through sight flow hoppers.

Lubricator injectors are provided for the supply of lubricating oil to both the upper and lower cylinder liners (Figure 11.7). The lubricating points are equally distributed around the liners and supplied with oil from timed distributor-type lubricators.

The central combustion chamber section of the cylinder liner supports the fuel injectors, the air starting valve, the relief valve and the cylinder indicator connection.

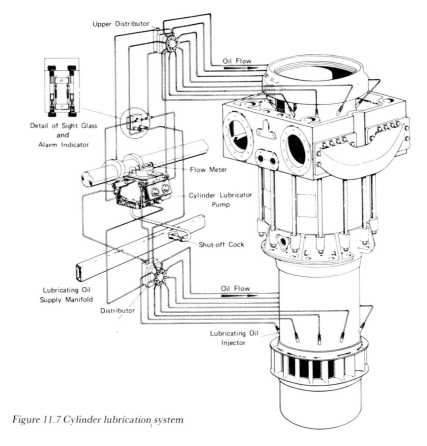

Figure 11.7 Cylinder lubrication system

The exhaust belts convey the exhaust gas from the cylinders to cast iron exhaust pipes which are connected to the turbochargers. Flexible expansion pieces are fitted between the exhaust belts and the exhaust pipes as well as between the exhaust pipes and the turbochargers. A grid is fitted at the inlet to prevent broken piston rings from damaging the turbocharger. A scraper ring carrier at the upper end of each liner prevents the passage of the exhaust gases past the upper piston skirt and scrapes the lubricating oil downwards.

Pistons

The upper and lower piston heads are identical. They have dish-shaped crowns to give a spherical form to the combustion space and are designed to be free to expand without causing undue stresses. The piston heads are attached to the rods by studs in the underside of the piston crowns so that the gas loads are transmitted directly to the piston rods. The under faces of the piston heads are machined to form cooling spaces between the piston heads and the upper faces of the piston rods, the cooling medium being supplied and returned through drilled holes in the rods.

A cast iron ring is fitted around each piston head to form a bearing surface. Four compression rings are fitted into grooves above this bearing ring and there is one ring below it to act both as a compression ring and as a lubrication oil spreader ring. The ring grooves are chromium plated to minimise wear of the surfaces in contact with the rings.

The lower piston rods have spare palms formed on their lower ends for bolting to the crossheads. The upper ends are cylindrical and form the faces to which the piston heads are bolted. Oil for cooling the lower piston is transmitted through the centre crosshead and up the piston rod to the piston head and is returned in a similar way (Figure 11.8).

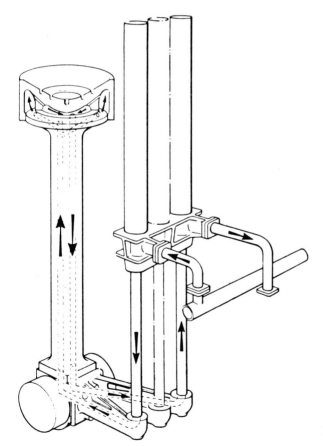

Figure 11.8 Lower piston cooling system

The centre crosshead bracket also carries the telescopic pipes for the piston cooling oil and lubricating oil to the centre connecting rod bearings.

Glands attached to the underside of the entablature, through which the lower piston rods pass, form a seal between the crankcase and the scavenge air space. These glands contain a number of segmental rings held to the body of the piston rod by garter springs. These rings are so arranged that the lower ones scrape oil from the rod back into the crankcase, whereas the upper ones provide an air seal and also prevent the passage of any products of combustion from the cylinders into the crankcase.

The upper piston rods are bolted to the upper piston heads and to the transverse beams which carry the loads from the pistons to the side rods. A cast iron skirt is provided around each upper piston rod to shield the exhaust ports and prevent the exhaust gases passing back into the open end of the cylinder.

For the upper pistons, water is used as the cooling medium and this is conveyed to and from the piston heads through holes in the upper piston rods. Brackets attached to the transverse beams carry telescopic pipes for the cooling water.

Connecting rods

The centre connecting rod has a palm end at the lower end to which the bottom end bearing keeps are bolted, whereas the upper end of the rod has an integral continuous lower half keep to which the upper half bearing keeps are bolted. The side connecting rod is formed with palm ends at both ends of the rod to which the top and bottom end bearings are bolted. The centre connecting rod top end bearings consist of continuous white-metal lined shells for the lower (loaded) halves, whereas two bearing keeps over the ends of the crosshead pins form the upper halves. The side connecting rod bottom end bearings are white-metal lined and are supplied with lubricating oil from the main bearings through holes passing up the connecting rods to the side top end bearings. The centre connecting rods also have cast steel white-metal lined bottom end bearings which are supplied with oil through holes in the rods from the top end bearings. The side top end bearings and centre top end bearings are fitted with white-metal lined thin-shell bearings.

The centre crosshead pins are made of nitriding steel and hardened by this process. At the top the pins are bolted to the palm end of the piston rods with the crosshead brackets sandwiched in between. Also, two long studs pass through the crosshead pins horizontally and secure the pins to the brackets at the back. The telescopic pipes for lubricating and cooling oil are supported by the crosshead brackets at the front of

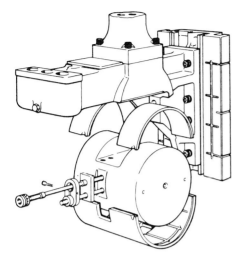

Figure 11.9 Centre crosshead assembly

the piston rods and the guide shoes are bolted to these brackets at the back. Each side crosshead (Figure 11.10) is made up of a steel casting into which is shrunk the crosshead pin, the side rod being screwed into the top of the casting.

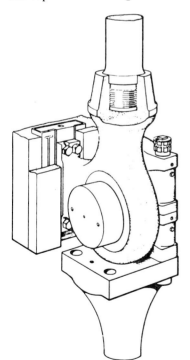

Figure 11.10 Side crosshead assembly

Camshaft

The camshaft is mounted on the top of the entablature and is driven through a roller chain from the crankshaft. It operates timing valves for controlling the fuel injection to each cylinder, cylinder lubricators and starting air distributors for controlling the starting air supply to the cylinders; it also drives the governor through a step-up gear. A drive to the fuel pump is also taken from the camshaft driving chain.

Fuel injection

The injection system operates on the common rail principle in which timing valves, operated by cams on the camshaft, control the injection of fuel from a high-pressure manifold through spring-loaded injectors to the cylinders. Fuel is delivered to the high-pressure main by the multi-plunger pump fitted at the after end of the engine, the pressure being maintained at the desired value by means of a pneumatically operated spill valve.

Two fuel injectors are fitted to each cylinder and these open when the timing valves are operated by the cams on the camshaft. The duration of opening of the timing valves and hence the period of injection of fuel into the cylinders is controlled by the governor. Either a centrifugal

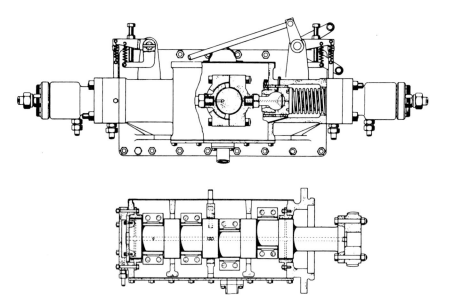

Figure 11.11 High pressure fuel pump

DOXFORD ENGINES

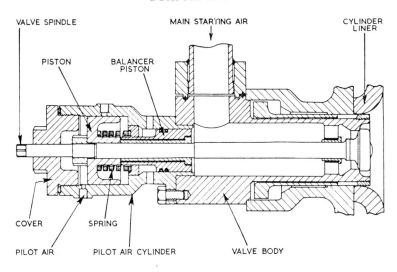

Figure 11.12 Air starting valve

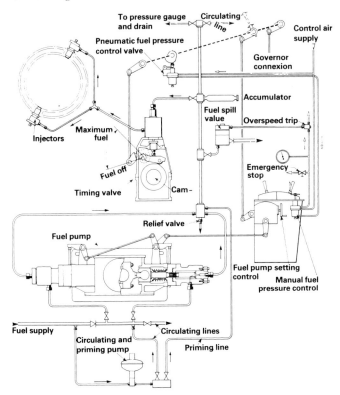

Figure 11.13 Air starting system

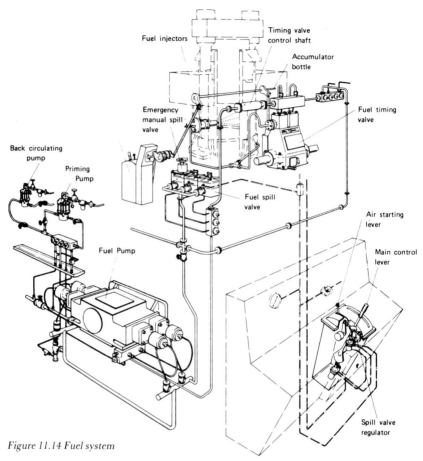

Figure 11.14 Fuel system

governor with a hydraulically operated output or an electronic governor with a pneumatic actuator can be fitted. In the former case the governor input is set pneumatically from the control station; in the latter case it is set electrically. In this way the speed of the engine can be adjusted as required. Means of adjusting the setting of the timing valves by direct mechanical linkage are also provided. The timing of injection is determined by the positions of the timing valve cams. Only one cam is required for each cylinder for both ahead and astern running.

Turbocharging

The latest J-type engines are turbocharged on the constant pressure system following a changeover from the original impulse charging system. Three and four cylinder engines have only one turbocharger, at

DOXFORD ENGINES

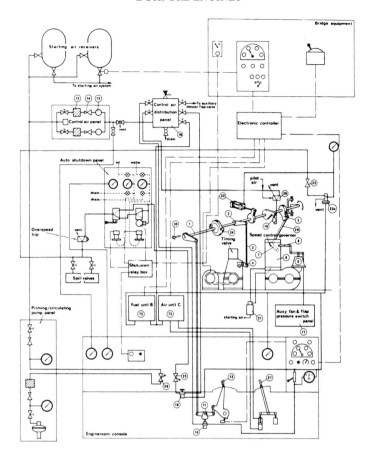

Figure 11.15 Diagrammatic arrangement of speed controls and other pneumatic controls, including remote control

the forward or aft end, while two or three chargers have been fitted to seven, eight and nine cylinder engines. Between each turbocharger and the engine entablature is a finned-tubed seawater cooled after cooler, and an electrically driven auxiliary blower is provided for slow running or emergency duties.

Starting of the engine is by compressed air, which is admitted to the cylinders through pneumatically operated valves, these being controlled by a rotary air distributor driven from the camshaft to govern the timing and duration of opening for starting the engine ahead or astern. Levers for starting and speed control are grouped in a control box near the engine and when bridge controls are fitted pneumatic valves to control engine movements are actuated by movements of the bridge telegraph.

THE 58JS3C ENGINE

The 58JS3C opposed-piston engine of 5500 bhp at 220 rev/min from only three cylinders was conceived as a direct challenge to medium-speed four-stroke engines used for the propulsion of smaller vessels (Figure 11.16). Several small container ships are powered by this three cylinder 580 mm bore two-stroke engine designed to burn cheaper low-grade residual fuels.

The design of the 58JS3C series is based on the Doxford J design, but because of the higher rotational speed and relatively short piston stroke, there are a few significant design features.

The engine bedplate, columns and entablature are of all-welded fabricated construction to produce a robust structure of optimum weight. The crankshaft is of single-piece construction with each cylinder section made up of three throws with the two side rods

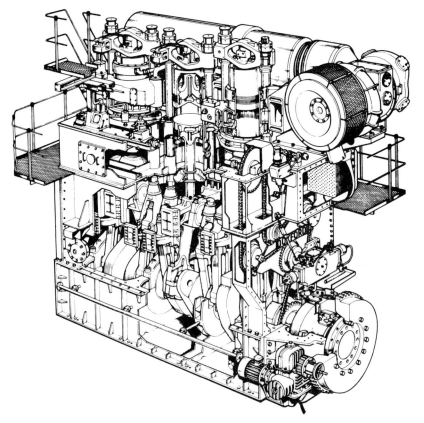

Figure 11.16 Cutaway drawing of a 58JS3C engine

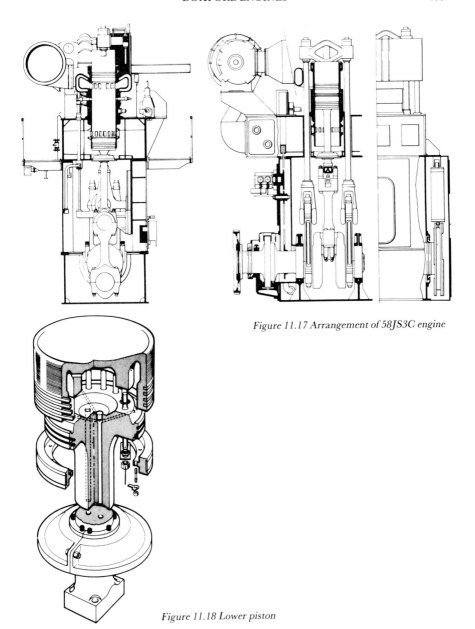

Figure 11.17 Arrangement of 58JS3C engine

Figure 11.18 Lower piston

connected to the upper piston and the centre crank to the lower piston. The side crank webs are circular and they also double as main shaft journals of large diameter. Each centre crank is an integral steel casting or forging and the centre crankwebs are shrunk on to the side pins.

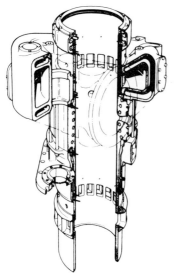

Figure 11.19 Cylinder liner

The cylinder liners (Figure 11.19) are one-piece special iron castings with the water cooling surface around the combustion area formed by bore holes drilled at an angle to the liner bore. The central part of the liner in the combustion area is fitted with fuel injectors, air starting valve and a combined relief and indicator valve. Scavenge ports are machined in the lower portion of the liner and these are encased in an air deflector to produce an air swirl. Exhaust gases pass from the cylinder through ports around the whole periphery at the top of the liner into exhaust belts which convey the gas from the cylinders to the

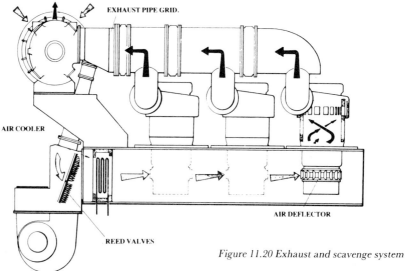

Figure 11.20 Exhaust and scavenge system

constant pressure exhaust gas manifold to which the turbocharger is connected (Figure 11.20). The cylinder liner is positioned in the entablature by a water cooled jacket which is bolted to the top face of the entablature and each cylinder unit assembly is cooled by distilled water. Equally spaced lubricating points are distributed around the liners and are fed from timed distributor-type lubricators.

Running gear

The upper and lower forged steel piston heads are interchangeable and have dish shaped crowns which at the inner dead centre position form a near spherical combustion chamber (Figure 11.18). The centre connecting rod has its lower end machined to take the upper half steel-backed and white-metal lined shell of the bottom end bearing whilst the top end is machined for shell bearings and the lower half shell and keep are bolted to the foot of the connecting rod (Figure 11.21). The centre crosshead bracket also carries a telescopic pipe for piston cooling oil and lubricating oil to the centre top end bearing is also supplied

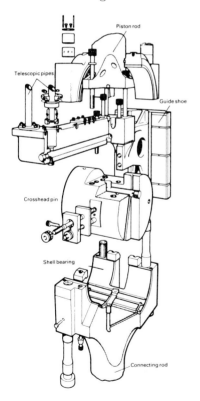

Figure 11.21 Centre crosshead assembly

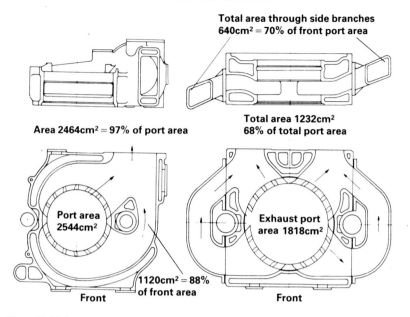

Figure 11.22 Arrangement of ports

through a telescopic pipe with a special valve to increase the oil pressure on the upward stroke of the connecting rod. The side rods are rigid castings designed to maintain the bearing shape, and long pre-stressed bolts take the load from the top end bearings to the side crankpin keeps.

The engine operates on the constant pressure turbocharging with one charger for normal running conditions, though for manoeuvring and slow running an electrically driven auxiliary blower is fitted. As in the case of the larger J type engines, the fuel system (Figure 11.23) operates on the common rail principle in which fuel is delivered to a high-pressure main by a multi-plunger pump and fuel injectors fitted to each cylinder are opened when timing valves are operated by cams on the camshaft, with opening and timing controlled by the governor. An unusual feature of the 58JS3C engine is the fitting of a starting air positioner to the crankshaft, which acts automatically to turn the shaft away from any dead spots so that starting can take place in the normal manner. However, according to reports from ships fitted with this pneumatically actuated device, it is rarely used.

C.T.W.

DOXFORD ENGINES

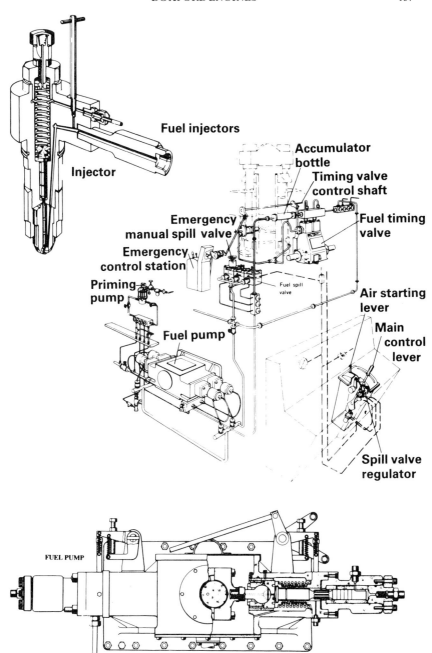

Figure 11.23 Fuel system

12 Götaverken engines

The two-stroke uniflow scavenged Götaverken engine has been out of production for a number of years, but since 1939 when the first vessel powered by a Götaverken engine ran her sea trials, more than 800 engines have been built, many of them still in service. Engines built between 1940 and 1950 were of naturally aspirated type, but a milestone for the Götaverken company in the 25th anniversary year of its own diesel engine in 1964 was the introduction of the highly turbocharged models built in three cylinder bore sizes of 850, 750 and 630 mm respectively. Early moderately turbocharged engines were introduced in the mid-1950s. Because few of the earlier turbocharged engines remain at sea, only the highly turbocharged types designated VGS-U are discussed here.

The application of high-pressure turbocharging to Götaverken slow-speed marine diesel engines commenced with the largest bore type of 850 mm originally designated 850/1700 VGA-U, but a revision of the engine design became necessary to accommodate the higher cylinder mean pressures and increase in output.

When redesigning the last type of engine Götaverken adhered to most of the main design features of the older engines. The uniflow scavenged, single-acting, crosshead types with one central exhaust valve in the cylinder cover proved to be reliable in service, economical in operation with low maintenance costs. The characteristic retained feature was the use of scavenge pumps actuated from the crosshead, operating in series with the turbochargers in a constant pressure system.

ENGINE STRUCTURE

The engine bedplate is of steel plate welded to form two sections assembled by fitted bolts to form two longitudinal girders with 'I' section cross members between these two side girders. Into these cross members are welded cast steel saddles for the main bearings, and with heavy bolt connections which transmit the forces from the entablatures directly to the bedplate. The thrust bearing is built into the after part of the bedplate and this consists of loose cast iron thrust pads faced with white metal. A sheet steel oil tray is welded to the underside of the bedplate.

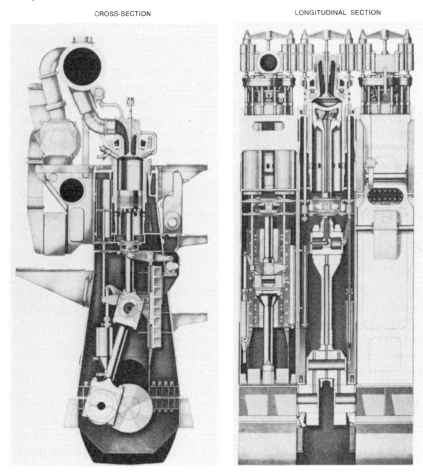

Figure 12.1 Arrangement of Gotaverken 850/1700 VGS-U engine

The entablatures are fabricated from steel plate to form closed units, one for each cylinder. A heavy steel casting forms the top of each entablature, into which it is butt welded. This top section is provided with heavy fitted bolts connecting each section to each other so that a strong longitudinal girder is formed. Furthermore, the upper part is provided with long studs to hold down the cylinder cover. The upper section of the entablature forms the scavenging air chamber, isolated from the crankcase by two horizontal plates between which a double stuffing box with scraper and sealing rings is fitted. All the scavenging air chambers are interconnected and form a common manifold for the engine. An open cofferdam formed between the two horizontal plates enables the piston rod to be examined while the engine is running. Furthermore, the cofferdam provides excellent protection in the event of fire in the scavenge trunking and eliminates overheating of the crankcase cover.

The lower part of the entablature is provided with large openings, covered by strong bolted doors, for easy access to the crankcase. These covers have automatic pressure relief valves: a small inspection cover is also fitted to each door. The guide face and keeps are of cast iron and secured to the entablature by fitted bolts, with adjustment by means of shims. The fixed telescopic pipe for the piston cooling which runs down one side of the entablature is of steel and connected with the engine oil cooling system. The upper part of this pipe is fitted with a gland with spherical bearings and a slotted steel discharge pipe is fitted on the opposite side of the crankcase and connected to an observation arrangement for control of the return oil from the working piston.

Crankshaft

In accordance with normal Götaverken practice the crankshaft is semi-built and made in two parts, flange coupled together by fitted bolts. The crankpins and webs are made of Siemens-Martin cast steel shrunk onto the journals. Generous fillets have been provided between the crankpin and web to reduce the concentration of stresses and give the largest possible bottom end bearing surface. The crankpin and main bearing journals are drilled to give passages for lubricating oil.

To obtain the smallest possible cylinder distance and engine lengths, especially large diameters are employed. The shaft diameter is 630 mm for engines with five to eight cylinders and 660 mm for nine and ten cylinder engines. The large turning gear wheel is mounted on the thrust shaft at the after end on the engine.

The connecting rod is of steel and drilled throughout its length to give passage for lubricating oil to the crosshead bearing and guide. The top end of the connecting rod is flanged to take the crosshead bearing, and the bottom end is forged in a T-form for connecting to the bottom end bearing. This latter has been designed with top and bottom halves of cast steel, white-metal lined, and provided with shims for adjusting the clearance. Two strong bolts of heat treated steel connect the two halves of the bearing with the connecting rod.

Crosshead

The crosshead is of cast steel incorporating a white-metal faced guide shoe. It has drilled supply passages for lubricating oil to the guide surface and is fitted with cast steel arms, bolted on, for driving the piston type scavenging air pumps. The crosshead pin is of special steel, carefully ground and polished and drilled throughout its length to reduce weight and provide passage for the cooling oil to the working piston. In addition, the crosshead pin is equipped with four studs to secure it to the crosshead and these also serve as connecting bolts to the connecting rod.

Attached to the crosshead pin is a cast steel arm to which is attached a telescopic pipe through which oil flows to the working piston. The cooling oil discharge is led through a steel outlet bend, which is also fitted to the crosshead pin, and connected with a steel pipe in the drilled hole in the centre of the piston rod.

The crosshead bearing has a bottom half and two top halves of cast steel, white-metal lined, and provided with shims for adjusting the bearing clearance. The bearing is shaped so that the entire length of the crosshead pin is carried in the bottom half where the maximum bearing surface is obtained. The bearing is fitted to the connecting rod by means of four heat-treated steel bolts, which also serve to secure the two upper half-bearings.

Piston and rod

The working piston (Figure 12.2) is of chrome–molybdenum steel and fitted with welded-in iron wear rings together with a special insert for

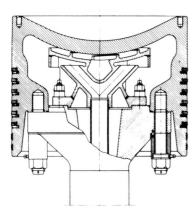

Figure 12.2 Cross section of 850 mm bore piston

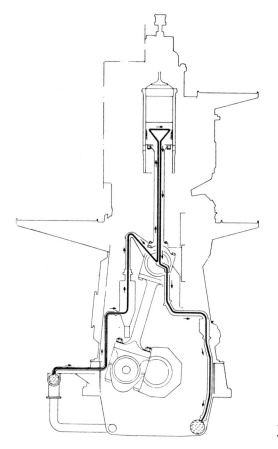

Figure 12.3 *Piston cooling oil charge and discharge*

cooling oil. The steel piston rod is drilled throughout its entire length to form a passage for the piston cooling oil, this passage housing a steel pipe through which the cooling oil return runs. The upper part of the piston rod has a flange with a ground face to which the piston is mounted: the bottom flange is square shaped and bolted to the crosshead.

Cylinder liner

The cylinder liners are of special cast iron with a high degree of wear resistance at high temperatures. The upper part of the liner is flanged, with a ground face between the cylinder head and the water jacket, the liner being sandwiched between these two. The cooling water jacket is fitted around the liner, the upper part forming a collar with ground faces sandwiched between the cylinder liner and the entablature top section. The jacket has cooling water connections and a stuffing box in the lower part allows for expansion of the liner. The jacket and liner are assembled as a complete unit, resulting in easy removal.

In the lower part of the cylinder liner there are a number of scavenge ports, equally spaced around the periphery, through which air from the scavenging air chamber flows into the working cylinder, when the piston uncovers the ports at the bottom of the stroke. A number of lubricating oil quills are fitted above the row of scavenge ports.

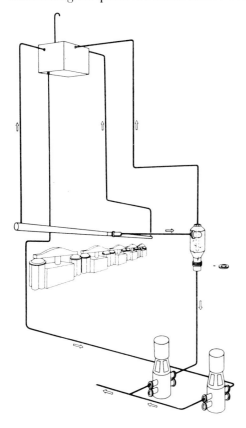

Figure 12.4 Air relief arrangement

Cylinder cover

The cylinder cover is in two parts, consisting of a cast iron water cooled lower half and cast steel non-water cooled upper half (with the exception of the 520 and 630 mm models which have covers in one piece and are cast iron). The lower half is sealed to the combustion chamber and contains all the valves; the exhaust valve is at the centre, the fuel, starting air and safety valves are located at the periphery of the cover. Four strong studs, which serve to hold together the two halves of the cover, also secure the exhaust valve cage. This valve casing is of cast iron with a replaceable seat of cast iron and the valve spindle is forged from heat resistant steel. The stem is clamped in a spherical bearing and mounted in the overhead cast iron actuating yoke.

The exhaust valves are actuated by cams, with the movement being transmitted to the yoke by twin rocker arms and side rods. Strong helical springs fitted between the yoke and the cylinder cover are used to close the exhaust valve.

The Götaverken engine operates on the uniflow scavenging principle; that is, air enters the cylinder through ports in the liner and the exhaust gas passes out through the single exhaust valve in the cylinder cover. The scavenge ports are shaped so that the air is given a swirling action during scavenging for complete gas removal and the air is supplied by turbochargers working in series with double-acting piston type scavenging pumps. These pumps are built into the entablature and driven by cast steel arms fitted on the crosshead. The suction and delivery valves for the air pumps are of simple design fitted for easy access in a casing on the entablature.

Camshaft drive

The engine has two camshafts: a lower one for the exhaust valve cams: the upper one for the fuel pumps drive (Figure 12.5). A chain system transmits movement from the crankshaft to the two camshafts with the chain drive in a special casing at mid-length of the engine. The sprocket on the crankshaft is in two pieces and keyed to the coupling flange: sprockets on each camshaft are secured by fitted bolts. The chain also runs over an intermediate sprocket wheel which drives the governor from its shaft, starting air distributor and tachometer, while the chain itself is tensioned by an adjustable jockey wheel carried in a steel support which is fitted on a movable arm inside the casing and provided with an adjusting screw and a pair of damping springs. The chain tension should be inspected under normal service conditions at 500 hour intervals, and the chain and sprockets carefully inspected at 1000 hour intervals for the first six months then at 3000–4000 hourly intervals thereafter. A smaller simplex chain drive connected to the lower intermediate shaft drives the governor, air distributor and tachogenerator; and the whole chain drive system is splash lubricated from a number of points coupled to the main engine lubricating oil system.

The two camshafts are fitted at the operating side of the engine extending the full length, and mount symmetrically shaped cams with the fuel cams adjustable within certain limits to adjust the fuel injection

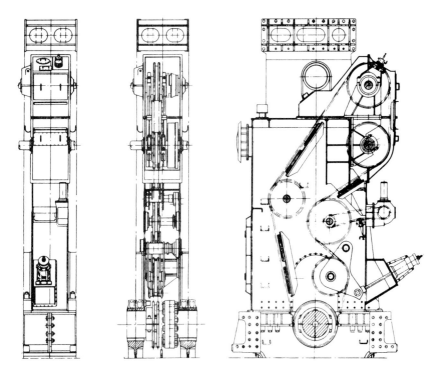

Figure 12.5 Arrangement of camshaft chain drive

timing to suit different grades of fuel oil. The chain drives the coupling sleeve on the upper camshaft which, through a dog clutch with a large tangential clearance between the dogs, transfers the torque to the camshaft. Clearance in the dog clutch enables the same cams to be used for ahead and astern running, while there is an air-operated locking sleeve on the camshaft which moves axially and locks the shaft to the driving sleeve when the engine is running.

Fuel pump and valve

The fuel pumps, one for each cylinder (Figure 12.6), are fitted above the upper camshaft and each is driven from its cam by a rocker, and through a spherically ended push rod transmits movement to the pump plunger. The pump barrel has a special hardened steel liner which is a press fit, into which the plunger is lapped. The pump housing contains a shock-absorbing cylinder with a lapped and spring-loaded plunger which has the effect of reducing the shockwave which occurs when high-pressure oil flows back to the suction side of the pump at the end of each pump stroke. The shock absorber cylinder is connected with the

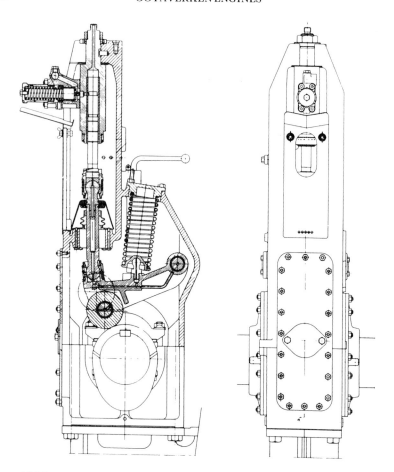

Figure 12.6 Fuel pump

pump casing through a drilled hole which serves as an inlet passage for the fuel oil. The pump plunger is of special steel and has two helical grooves which connect with the chamber above the piston through a vertically drilled hole. The quantity of oil delivered is regulated by the rotation of the plunger within the barrel, which makes the point of cut-off earlier or later in the stroke.

The fuel valves (Figure 12.7) are located in the lower cylinder cover and connected to the corresponding fuel pump by high-pressure piping. The valve is a spring-loaded needle valve type with a small lift, hydraulically operated and oil cooled. The valve needle is carefully ground into the atomiser, the seat being conical with an angle of 62° on the needle and 60° on the seat to obtain a narrow contact at the end of the seat, thus improving the seating and the functioning of the valve.

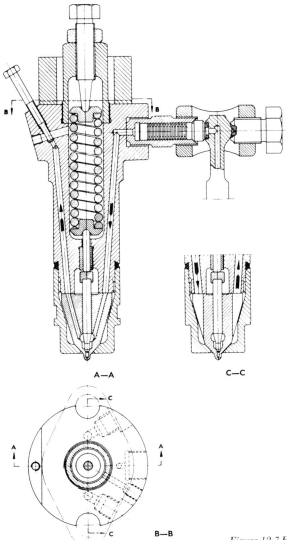

Figure 12.7 Fuel valve

Starting air system

Air for starting and reversing the engine is contained in two receivers, charged by motor driven air compressors. Air from the receivers is led to the starting slide valves and valves on the working cylinder. The pilot air from the air distributor is also taken from the air receivers.

For taking indicator cards the gas pressure is obtained from the working cylinder through a valve fitted in the cylinder cover and provided with a connection for the indicator. Movement for the indicator drum is provided by an eccentric cam mounted on the upper camshaft at each cylinder. The movement from this cam to the indicator is obtained through a spring-loaded pushrod, to which the indicator cord can be attached.

Lubricating oil system

The oil for the engine's bearings and piston cooling is stored in special tanks in the ship's double bottom, usually directly below the main engine sump and connected to it by drain pipes and shut-off valves. The lubricating oil is drawn up from the tanks by electric motor-driven pumps, which force the oil through filters and coolers to the various lubrication points on the engine. The oil afterwards runs into the sump and then back to the double bottom tanks. The pumps are of the screw type equipped with spring-loaded safety and control valves, and the oil coolers are of multi-tubular design provided with by-pass valves for both oil and cooling water.

The main inlet for the cooling oil and circulation lubricating oil is divided into two branches, one for the piston cooling, the other for lubrication of the bearings, etc. Separate valves are provided to allow adjustment of equal quantities of oil in each system. A tell-tale system on the manoeuvring side of the entablature shows the flow of oil through each piston as a return.

A pressure of 2.0–2.4 bar is required before the engine lubrication and cooling oil is considered normal, with a minimum pressure of 1.5 bar. The piston cooling oil outlet temperature is normally 45–50 °C, with a maximum of 55 °C.

Cylinder lubrication is by means of mechanical lubricators provided with sight feed glasses, each lubricator being operated by a pushrod from the indicator cam. There is a non-return valve for each lubrication point in the cylinder. The quantity of oil delivered to each lubrication point can be regulated by means of screws in the lubricator. A feed rate of 0.00055–0.00060 kg/kWh is recommended as reasonable, but this can be varied if piston ring conditions require a higher rate or a lower feed rate is thought possible.

Cooling water system

The main engine is cooled by fresh water circulated in a closed system by electric motor-driven centrifugal pumps (Figure 12.8). The pumps take their suction from the engine cooling water outlet and discharge the water through coolers to the cylinder jackets and then to the

GÖTAVERKEN ENGINES

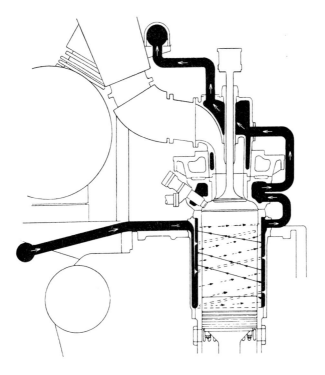

Figure 12.8 Cooling water system

cylinder covers and exhaust valves. An expansion tank located high above the main engine tops has a run down connection to the suction side of the water pumps, while the discharge from the engine cylinders passes through a cyclone separator. This separator acts as a deaerator in the line to the pumps to intensify the flow upwards of gas- or air-charged water from the deaerator to the expansion tank.

The fresh water coolers are either of tubular or plate type circulated with seawater from the main SW circulating pump. The same pump draws water from the ship's lower sea chest and delivers it to the lubricating oil cooler, turbocharger intercoolers and then the fresh water coolers before passing overboard. By-pass valves are used to regulate the quantity of fresh and sea water passing through the heat exchanger.

Fuel oil system

Fuel oil is kept in the ship's bunker tanks from where it is transferred to the settling tanks. From these the oil is passed to the heaters and centrifugal separators to be pumped into the daily service tanks as clean ready-to-use fuel. The daily service tanks are equipped with heating coils and from these tanks the fuel flows by gravity to the main engine fuel supply pump. This discharges through a filter to a steam heater which is fitted with a thermostatic control and a by-pass for the oil. From the heater the oil passes through a fine mesh filter to the engine fuel pumps. Surplus oil spilled back from the fuel pumps passes through a float valve back to the suction side of the fuel supply pump, with the float valve automatically venting the system. To safeguard the following of the main engine fuel pumps a spring-loaded check valve is fitted in the return pipe of the float valve.

The cooling system for the main engine fuel valves is completely separate from the engine fuel system. When burning lighter grades of heavy oil, diesel oil is normally sufficient as a coolant with the system comprising two circulating pumps which are connected to the suction side of the system. The heat exchanger is cooled by seawater but it can also, if necessary, be heated by steam. The coolant flow is in series through the fuel valve in each cylinder and then through the heat exchanger and back to the circulating pump.

Exhaust and turbocharging system

The exhaust gas manifold arranged at the top of the engine and running its full length is in direct connection with the turbocharger(s) inlet casing, thus operating on the constant pressure system with individual exhaust outlets from each cylinder cover fed into the one exhaust manifold.

Each turbocharger consists of a single-stage centrifugal compressor and a single-stage axial exhaust gas turbine assembled on the same shaft but within separate casings. The rotor shaft and turbine wheel are made of steel and the turbine blading of heat-resistant alloy. The compressor impeller is in the form of an open radial wheel and made of forged light metal alloy. Each end of the rotor is supported in ball bearings with a built-in lubrication system. The compressor casing is cast from light metal alloy and comprises a guide vane diffuser immediately after the impeller and a spiral diffuser which discharges into the air delivery volute. The turbine housing, of cast iron, is water cooled and divided into an inlet and outlet passage section. The inlet section contains the heat-resistant steel turbine guide vanes. Fresh water from the main engine cooling circuit is passed through the turbine casing for cooling.

Air from the turbochargers is blown through a welded trunk to an air cooler through which seawater — the cooling medium — is passed through tubes. These have copper fins brazed on to give a large cooling area and inspection doors are fitted so that the cooler can be inspected from the inlet and outlet sides. To simplify cleaning, the coolers have been equipped with two separate elements which can be removed without dismantling the connections to the air side.

After passing through the cooler, air in the outer scavenge manifold is delivered to the scavenge air pumps. This manifold is made up from a number of welded-plate casings, connected with the respective scavenge valve cover on each entablature. The covers are fitted with inspection doors to enable the scavenging air valves to be inspected, and are interconnected by means of flanges fitted with expansion bellows. The ends of the manifold are also provided with automatic non-return valves of the same type as those used for the scavenging air pumps. They remain in the closed position so long as the air pressure in the manifold is the same, or higher, than that in the engine room. If the turbocharger units are disengaged by locking of the rotors in case of failure, and the reciprocating scavenge air pumps draw air in, which is not pre-compressed, the scavenge air passes partly through these and partly through the stationary turbochargers.

Manoeuvring system

The main engine manoeuvring devices and instruments are delivered by the builders as standard equipment mounted in a console with pneumatic controls. By duplicating certain of the manoeuvring devices, control can be arranged from the bridge with the system incorporating a number of monitoring and alarm functions and a load control system to increase the engine revolutions in accordance with an adjustable programme. The pneumatic controls are backed up by a simple mechanical control system.

The manoeuvring system (Figure 12.9) is pneumatic with impulse air taken from the main starting air receivers through filters, driers, oil mist lubricating devices and reducing valves. The components of the system comprise the control box, direction of rotation monitor and control devices. The movement of the fuel regulating shaft and subsequent quantity of fuel oil injected is taken care of by a hydraulic speed governor which is driven off the starting air distribution shaft through a bevel gear drive. From a pressure regulator in the manoeuvring system the governor receives a pressure signal which is proportional to the control lever position. The governor controls the fuel supply so that a corresponding revolution pressure signal is received. When starting the engine, the manoeuvring lever is moved directly to the position which corresponds to the desired number of revolutions. A valve is actuated,

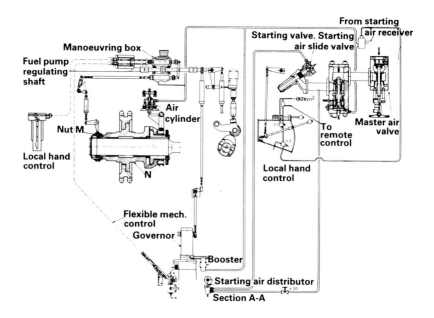

Figure 12.9 Diagram showing manoeuvring system

which releases air to the starting air distributor's reversing cylinder. If, however, the engine is to be reversed, the starting air distributor camshaft is moved in an axial direction and a locking cylinder is brought simultaneously into the blocking position. The fuel regulator shaft is now locked by two pistons, which are built into the control box and which also contains part of the fuel pump regulator shaft. When the starting air distributor camshaft is in the position corresponding to the engine's intended direction of rotation, impulse air is applied to an impulse valve, which then releases air to the starting air slide valve. This in turn opens and air is passed partly to the distributor and partly to the starting air valves in the working cylinders. The distributor opens the starting air valve in whichever cylinders have their pistons in a suitable position for starting in the required direction of rotation.

The engine should now rotate on air. The chain sprocket hub of the fuel pump camshaft turns 100° angular, but during this movement the camshaft itself remains at rest since the dog clutch is not locked. The fuel pump cams now come into the correct position relative to the crankshaft for the new direction of rotation. The relative movement between the camshaft and sprocket hub is transmitted through a threaded bush and linkage to a blocking piston, which is released when the 100° reversing angle is completed.

During this movement the direction of rotation indicator is also operating, and if the direction corresponds to the position of the manoeuvring lever, the blocking piston is uncoupled and the fuel pump control shaft is completely free for operating the engine on fuel.

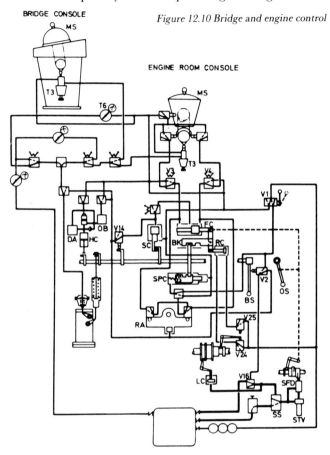

Figure 12.10 Bridge and engine control

When the engine is remote controlled the fuel pump control shaft is operated by the governor. The manoeuvring lever operates directly as a pressure regulator, which in turn gives signal air to the actuator for the governor. The latter contains a spring-loaded diaphragm, the movement of which is transmitted through rods and levers to the governor control shaft. The control panel is normally mounted in the engine room console and it contains a number of indicating lights which are illuminated by means of pressostats in the various air lines.

C.T.W.

13 Medium-speed engines

GENERAL

Some reference has already been made (in Chapter 2) to the factors governing the choice of medium- or slow-speed propulsion.

Apart from those cases where medium-speed engines are the natural choice, for example, vessels requiring an unobstructed deck near the water line, and smaller craft such as tugs, coasters, trawlers, etc., there are a substantial number of vessels whose owners have adopted medium-speed rather than slow-speed engines.

One major difference between slow- and medium-speed engines is that the medium-speed engine requires gearing (or other means) to reconcile the optimum speeds of engine and propeller, that is, indirect drive. Medium-speed engines operate at present at 400–600 rev/min in the larger sizes. Engines of up to 6000 horsepower and with speeds up to 1000 rev/min have been installed for main propulsion in particular cases. While most of this chapter is applicable to all such applications, the succeeding chapters, 14 to 22, will follow the presently accepted marine view of medium-speed engines as being those operating at speeds up to 600 rev/min.

Whatever the range of possibilities open to the engine designer when deciding how he will produce the necessary power, the propeller designer is constrained by efficiency considerations to keep the propeller speed low. A maximum speed of about 300 rev/min may be tolerated when absorbing 2000 hp, but this drops to 150 rev/min at 10 000 hp. In both cases maximum efficiency would be obtained at about two-thirds of these values. Indirect drives often allow lower speeds, nearer the optimum, to be used, and can achieve several percentage points improvement in efficiency compared with direct drive.

The choice of optimum propeller speed, however, is not always straightforward, (see *Marine Propellers*, by Emerson, Glover and Sinclair, in this series). Draught limitations dictate the maximum permissible propeller aperture, and if this does not permit the power for the desired vessel speed to be absorbed at the propeller speed for optimum efficiency, some compromise has to be made.

A higher propeller speed with a single screw may be easier to accept than twin or more screws. These make a more expensive installation and often forfeit some efficiency because of wake interference between them (the entry conditions for water reaching the propeller disc round the afterbody of the ship are important, and the builder has the option of using improvers such as nozzles).

Even if the propeller size imposes no limitations on the preferred power and speed, it may be that the implied size of final drive wheel in the gearbox, or of the carcass of the propulsion motor if electric transmission is used, cannot readily be accommodated in the machinery space. Moreover, in the case of gearing, if the ratio of engine and propeller speeds much exceeds 5:1, it may be difficult to obtain a standard range gearbox.

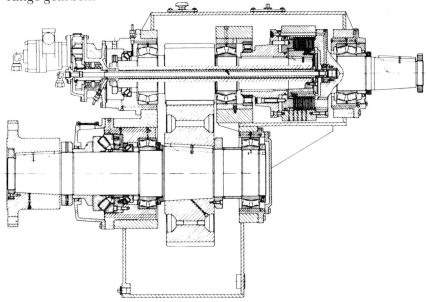

Figure 13.1 Typical reverse/reduction gearbox. The reverse train is an identical wheel and clutch assembly on a similar intermediate shaft also incorporating a pinion meshing with the final wheel (Reintjes) (Courtesy 'Marine Engineering Review')

Geared propulsion

The economic advantages and relative simplicity, allied with a wealth of successful experience, have made gearing the first choice for medium-speed installations. For most purposes it is now feasible to choose from a wide selection of makers a gearbox design suitable for almost any requirement. Gearboxes can be single reduction or twin, can accept from one to four inputs, can be adapted to drive power take off shafts for auxiliaries, and may or may not incorporate reversing mechanism. They may incorporate epicyclic and/or spur gearing, and they may have the output co-axial with the input, or stepped from it in any direction.

An extra reduction entails a slightly greater transmission loss, but is unavoidable with non-reversing engines driving fixed pitch propellers, or if a co-axial configuration is required. It may be necessary, too, because of gear design limitations, to enable a given reduction ratio to be achieved without an excessive diameter of bull wheel, which may create installation problems.

Typically on a reverse/reduction gearbox the input shaft(s) drive(s) the outer member of the clutch (Figure 13.1). Since it is normally essential that the gearbox has oil pressure as soon as the input starts to turn, the oil pump is driven from the input side of the clutch.

The input side of the clutch may incorporate a gear rim meshing with that of another clutch on an intermediate shaft, in order to provide a reverse rotation. Which is 'Ahead' and which 'Astern' will depend on the constraints of the engine and the installation, but usually the mode involving the fewer gears is preferred as the 'Ahead' configuration. Some designs have the ahead and astern clutches before the gear rims, so that with the clutches disengaged running the engine during maintenance does not turn any of the gearing. In either case engaging the ahead clutch turns the layshaft which carries the final pinion meshing with the bullwheel or final output wheel, while engaging the astern clutch does the same via intermediate shaft. With a stepped input and output configuration the input and output normally turn in opposite directions when running 'Ahead'. (If for any reason the engine or propeller cannot be adapted to do this, most (but not all) gearbox designs would permit operation via the normally 'Astern' train to be used for the 'Ahead' duty.)

If a co-axial configuration, or extra reduction, is required, this is achieved by incorporating an additional train before the clutch input shaft.

If reversing is possible by reversing the engine, or by the use of a controllable pitch propeller, the gearbox can be simplified to a single train (Figure 13.2), though it may retain a clutch if this is not fitted externally.

Internal clutches are invariably oil operated, usually of multi-plate design. Plates are of dissimilar materials, for instance, sintured bronze and steel. To permit prompt disengagement one set of plates (the steel)

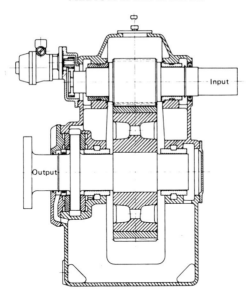

Figure 13.2 Typical plain reduction gearbox

will be produced in a concave form, so that releasing the engaging pressure enables them to push away from the flat bronze plates. The facility is usually provided to lock the clutches in the engaged position in the event of a hydraulic failure.

The gears themselves are usually cut to a shallow helix, which ensures a quieter, continuous drive. The teeth are cut to as large a modulus as can be accommodated and, increasingly in modern gearboxes, all are hardened. The helix angle is chosen to span the tooth pitch, so the angle only generates a modest end thrust, usually arranged to oppose the propeller thrust.

Pinions are usually integral with their shafts, but wheels are keyed or, increasingly, fitted by the keyless oil injection method, on an appropriately tapered seat.

Bearings in early gearboxes were usually plain, and the thrust bearing of classical Michell pattern. However, the attraction of roller bearings, namely lower friction, greater precision, higher load capacity and the ability to forgo pressure lubrication for a period, combined with their development and wealth of experience, have led to their growing acceptance in modern designs (Figure 13.3). The gearbox oil circuit passes from the pump to the filter and cooler before the relief valve which provides the clutch operating pressure. The spill is used for lubrication.

Gearboxes may be provided with additional trains to drive auxiliaries or power take-offs, or in some cases the latter can be coupled to the end

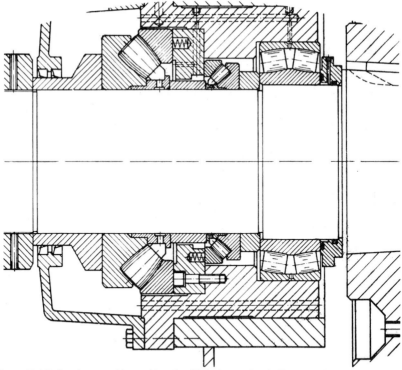

Figure 13.3 Roller thrust and journal bearing (Reintjes gearbox). (Courtesy 'Marine Engineering Review')

of one of the shafts, depending on the power required, and when they are to be used in relation to propulsion requirements.

Where a variable pitch propeller is used the gearbox oil circuit may be utilised to actuate the propeller pitch. If so, it is convenient to mount the oil distribution box for the pitch controls at the forward end of the final shaft in the gearbox. This lessens the sealing problem but it is not always possible. Otherwise, to get oil from the actuating pump to the hub, a distribution box has to be mounted against the tailshaft to lead pressure and return oil to and from the pitch operating cylinder in the hub.

With a non-reversing gearbox, it is good and most common practice to incorporate a disengaging clutch between the engine and the gear input, so that the engine can be run for maintenance purposes without turning the propeller (Figure 13.4). (The reversing gearbox already incorporates clutches which serve this purpose, as also do some non-reversing types.)

It is virtually universal to provide an alignment coupling at this point, and this is often the stage at which the two-node mode of torsional vibration can be most effectively controlled by choice of the

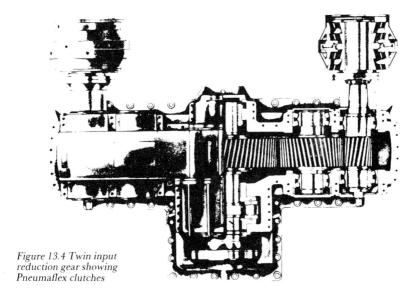

Figure 13.4 Twin input reduction gear showing Pneumaflex clutches

effective stiffness and damping characteristics of this coupling.

With vessels such as tugs, a separate, suitably rated slipping clutch may be used to limit damage if the propeller fouls a hawser or some other object.

In some specialised applications a fluid, or an electromagnetic, coupling is used, but for normal applications this is today a costly solution and may entail another 2–3% slip. However, designs of such couplings can now be contrived which can be positively locked at certain operating conditions.

There are geared vessels with direct reversing engines (two- and four-stroke) and fixed pitch propellers.

There is no unanimous view about the desirability of variable pitch propellers, despite their ability to be set always at the optimum efficiency. They are widely accepted for larger vessels (even with direct coupled engines) where fuel savings are very important, and the ability to stop more quickly is valued. There is also the attractive possibility of using the main engine to drive an alternator when on passage. (This applies to slow-speed engines too.) By contrast, they are at present less well accepted in the case of smaller vessels, partly because these tend to spend more of their time in estuarial waters where there is thought to be greater risk of wear or damage which would be more expensive to repair in the variable pitch case. Much will depend upon a particular operator's own experience.

It would be unfair to take sides, other than to say that ship operators will decide this question, like all others, by the best experience they can find. The answer is likely to be that which gives, over the life of their ship, the lowest costs.

The effects of the disparity in optimum engine and propeller speeds has already been mentioned. One factor which is influenced is torsional vibration (particularly in the one-node mode). Another is the transient behaviour. This is because gearing has the effect of increasing the effective inertias of the slower turning masses and reducing the effective stiffnesses of the slower turning shafts. To facilitate analytical treatment, correction is made from propeller speed to engine speed. To do this, the propeller inertia is multiplied by the square of the gear ratio. In other words, a propeller coupled through a 4:1 reduction gearbox behaves as if it had 16 times its real inertia. This produces a low one-node frequency, possibly below minimum engine speed, which is always satisfactory. On the other hand, its effective inertia may equal or exceed that of the engine (this is more of a risk with a lighter, higher speed engine). If that is the case, and a fixed pitch propeller is used, reversing, as in a crash stop, may cause the propeller to prevail over the engine. With large reductions, particularly if a highly rated engine is used, it is generally safer to use a variable pitch propeller.

DIESEL–ELECTRIC

In earlier times speed reduction was invariably achieved as part of the functions of a diesel–electric transmission, but nowadays this is very expensive to install compared with gearing. The transmission efficiency attainable is lower at maximum output (90–92% compared with 96–98% for a gearbox) though at partial outputs this could often be offset by using fewer engines so that they ran within their most efficient ranges of output.

Today, apart from semi-submersible multi-hulled offshore support vessels, where the engines usually have to be remote from the propellers, diesel–electric drives are used only for special purpose vessels where, for instance, very precise station keeping or speed control is essential. Examples include ice breakers, buoy tenders, oil exploration and diving/support vessels, and cable ships. Sometimes, a vessel requiring substantial power for non-propulsion purposes may find dual purpose electric transmission an advantage. For instance, in bulk carrying vessels requiring a high rate of self-discharge, it is advantageous to call upon the main engine power for the purpose.

In principle, the classification also extends to electrically driven transverse thrust units which are fitted to many vessels whose owners require better manoeuvrability. Diesel–electric drive affords a greater flexibility in laying out machinery, and there are advantages in keeping the propulsion motor close to the propeller. However, copper (for main bus bars) has become extremely expensive, and while this can be reduced by using higher voltages to reduce currents, these too create their own problems. Older diesel–electric installations were invariably designed for direct current operation, usually on the Ward/Leonard system or a derivative of it.

The Ward/Leonard system couples the armatures of the generator(s) and motor(s) in a loop. The motor field is held constant, and the generator field varied to control the voltage in the loop and therefore the speed. Usually for marine propulsion, the operation is simplified by controlling the voltage to give a constant current in the loop (it is then called the constant current loop) while the excitation of the motor is altered to give the required speed (Figure 13.5). Torque varies as field × current and therefore is directly proportional to the field, if the current is held constant. Operation, including reversing, is much simplified, and the number of motors and generators connected can be varied at any time.

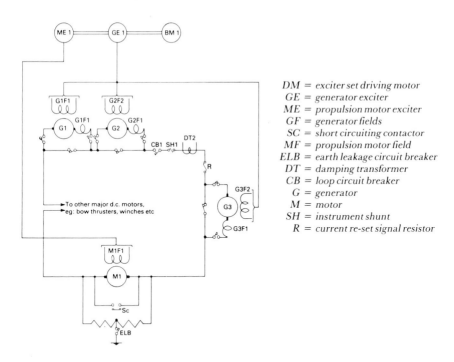

Figure 13.5 Basic circuit of constant-current control: the main propulsion loop

At present there is a discernible trend in diesel–electric installations towards alternating current generation, usually with thyristor control of the direct current propulsion motor. This permits cost savings because standard alternators can be adopted, and operating at up to 11 kV, which also reduces the size of conductor necessary. Further developments are possible using ac motors, which offer cost and weight saving, but they have not so far been taken up in any quantity.

It is important to note that torsionally, isolating the engine(s) from the propeller in this way does not eliminate torsional vibration arising in the generator sets, nor yet in the motor/propeller shaft, where propeller excited vibrations can still be significant.

Although for normal purposes there is little advantage in adding to an expensive electric transmission the cost of a variable pitch propeller, there is an advantage in doing so where it is important to have precise control down to very low speeds, at which the control of a propulsion motor is difficult.

Characteristics

Next to the type of drive, an important difference between slow- and medium-speed engines is in the build philosophy. The slow-speed engine is designed almost individually, to be selectively assembled and tuned, and to absorb the relatively rough handling imposed by the sheer size of its components. The medium-speed engine is usually one of a series, is much more likely to rely on standard settings, and will be built with finer tolerances and clearances. It will therefore require rather more sophistication and care in handling its albeit lighter components. It is likely to include in its tool kit jigs and fixtures to ease the labour and reduce the man-hours involved in replacing major components.

In operation, the initial applications of medium-speed engines, often drawing on design and service experience with relatively unattended stations on land (or in smaller sizes in locomotives), meant that they were better prepared for the advent of the unmanned engine room. However, the makers of direct coupled engines have not been slow to adapt their machines to operate without constant attendance.

Medium-speed engines, being built in greater numbers, tend to favour the philosophy of reliance on standard sizes and standard settings with controlled tolerance, whereas it is traditionally slow-speed engine practice to adjust to compensate for actual variations in each.

Medium-speed engines can afford to be four-stroke (and usually are) but slow-speed engines are invariably two-stroke (and have to be). Medium-speed engines with one or two (two-stroke) exceptions, are of trunk piston construction, whereas direct coupled slow-speed engines usually employ crossheads.

A less fundamental but very significant difference lies in the fuel injection arrangements. Size, and the absence in many cases of any substantial camshaft, have historically dictated to the slow-speed engine maker a preference for common rail systems, or more exotic solutions like the Archaiouloff injector activated by the combustion air pressure in the cylinder.

Builders of medium-speed engines approached the marine market with a long tradition of using jerk pumps. They have been fortunate in that the makers of this kind of fuel injection equipment have been able to keep pace not only with the considerable rise in injection pressures needed to obtain good fuel economy while ratings and sizes continued to increase, but also with the problems of handling inferior fuels.

Noise is often cited as a disadvantage of the four-stroke medium-speed engine, but it is probably fair to say that by the standards laid down by modern Health and Safety legislation no engine room is quiet enough to avoid potential damage to operators' hearing. The characteristics of the noise emitted by different types of engine differ, and subjectively, some kinds of noise are less objectionable than others. However, modern vessels all tend to adopt a soundproofed control room.

A generation ago this would have been a serious disadvantage in that a trained ear is one of the best diagnostic tools ever evolved for the care of machinery. Nowadays, however, the development of monitoring techniques on the one hand and the greater ability of engines to run without attention on the other, have conspired to reduce reliance on the engineer's senses.

Apart from their obvious differences, slow- and medium-speed (and high-speed engines for that matter) each require a characteristic maintenance philosophy. Examples can be quoted where an engine of one of these classes, acknowledged to be successful in its usual environment, has been a disappointment to its operators when used by staff more at home with a different class of engine. For instance, engineers used to the tightening torques and forces of very large engines have been known to find it difficult to adjust to the very much smaller fastenings on high-speed machines; and these can be overtightened in consequence.

A large engine has bearing clearances which will safely absorb relatively large foreign particles without damage (i.e. they need to be bigger to bridge the oil film). The smaller the engine the more necessary it is to be scrupulously clean during reassembly.

Psychologically a faster running machine may seem more strident and awesome than a larger slower one, and this has been known to discourage essential maintenance — or even to put it off.

It has usually been preferable, if not essential, to carry out the maximum possible amount of repair work *in situ* on a slow-speed installation, and designs still tend to facilitate this. By contrast, the medium-speed engine is designed to a large extent on the assumption of repair by replacement. The smaller engines lend themselves least well to improvised repair of other than detail.

These distinctions, however, are becoming increasingly blurred with widening experience, and are tending to depend as much on the balance between on the one hand the skill of the vessel's engineers or dockyard staff, and on the other hand, the thoroughness, cost and availability of the maker's spares and service organisation, as it does on the characteristics of the engine. Smaller engines tend to generate more widespread availability of spares worldwide, especially where the type has won acceptance in non-marine fields also. It should also be less costly in the last resort to airfreight spares.

However, in the marine context most repairs cost far less than the revenue lost while the ship is off-hire to execute them.

Acknowledgement

Figures 13.1 and 13.3 are taken from an article by Mr E. A. Jackson of European Marine and Machinery Agencies, and which appeared in *Marine Engineering Review*, Aug. 1979.

D.A.W.

14 SEMT Pielstick engines

Pielstick engines, as they are generally known, are licensed by Société des Études de Machines Thermiques (SEMT), a division of Alsthom-Atlantique, and built in over a dozen countries. They are numerically the best established range of medium-speed engines in marine service.

The Pielstick concept originated early in the 1950s as part of a family of monobloc multiple-crankshaft engines which were designated PC1. In the early 1960s further development of the PC1 engine produced the PC2, which has since sold extensively as a single-crankshaft engine of conventional form. PC engines are all medium-speed directly reversible four-stroke engines, turbocharged and intercooled.

SEMT have also designed and licensed two ranges of high-speed engines — the PA types — which are discussed in Chapter 23. This chapter, however, is concerned only with the PC type. These have been built in three basic sizes: the PC2 since the mid-1960s, the PC3 from 1971, and the PC4 since the late 1970s. Current ratings of these three sizes of engine are shown in Table 14.1 below. While PC2 engines have retained the same bore and stroke, the later stages in the evolution of this type have been accorded a suffix, viz: PC2–5 and now PC2–6. This is because the PC2–5 and its successor incorporate some major changes of scantling, which affect interchangeability.

Table 14.1 General characteristics

	PC2	*PC3*	*PC4*
Bore mm	400	480	570
Stroke mm	460	520	620
Rev/min	520	460	400
Max. cyl power kW	550	700	1215
No. of cyls in-line engines	6,8,9	6,7,8	6,7,8,9
No. of cyls Vee-engines	10,12,14,16,18	12,14,16,18	10,12,14,16,18

The maximum power attained by individual examples of an engine type depends on the state of development. In the case of the PC2, the cylinder power attained in the PC2–6 variant has nearly doubled that

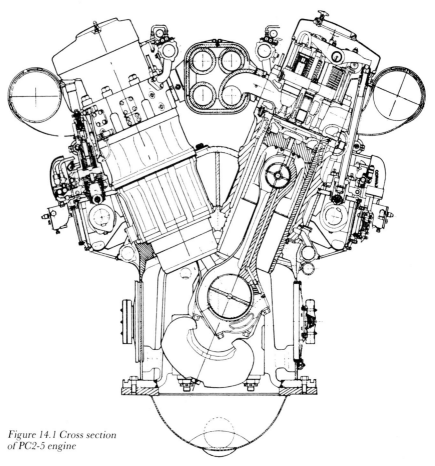

Figure 14.1 Cross section of PC2-5 engine

achieved by the first PC2 when it replaced the PC1. This means that there are many changes in component design, although the basic dimensions of each type have generally not changed.

The latest types are described and, since the PC family of engines share many common design principles and characteristics, in general the description is based on the PC2 size, but noting where necessary the essential differences which distinguish the two larger types. Figure 14.1 indicates a cross section of a PC2–5 Vee engine, which is characteristic of the philosophy of all PC engines. Compared with the earlier PC2 engines the scantlings have, generally, been increased and the older traditional parallel bolt design of large end, split at right angles to the cylinder axis, has been discarded in favour of the stiffer automotive type rod. Figure 14.2 shows a cross section of a PC4–2 engine. The Vee angle on all PC engines is 45°. The components are briefly described below.

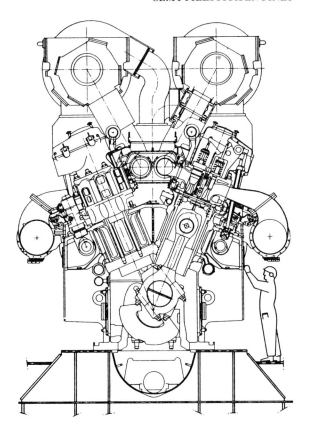

Figure 14.2 Cross section of PC4-2 engine

Crankshaft

Compared with earlier PC2 engines, the crankshaft scantlings of the PC2–5 and later engines have been increased. Balance weights, etc. have been adjusted to provide an increased oil film thickness despite the increase in cylinder pressures and output.

The crankshaft is manufactured from chromium-molybdenum steel by continuous grain flow or free forge methods, with an ultimate tensile strength of 80 kg/mm^2. The total output is transmitted by the flange on the coupling end, but as with PC2 engines, alternators can be driven from the flange at the free end. The timing gears and auxiliaries are driven by a special collar which eliminates the split centre disc linking, on the PC2 engines, the primary wheel with the crankshaft. The end of the crankshaft which is opposite to that having the timing gear is normally equipped with a Pielstick damper. The characteristics of this unit are identical to that used on the PC2, though its design has been slightly altered because of the increase in flange dimensions. The balance weights are fixed, as on the PC2 crankshaft, to the dovetail-shaped crankwebs and secured by jack-screws. This feature can be discerned on both Figures 14.1 and 14.2.

All crankshaft bearings are steel-backed thin-wall lead–bronze with a lead–tin overlay. The connecting rods on all Vee engines in the PC ranges run side by side on the crankpin so that the running parts are fully interchangeable with the in-line versions. The thrust bearing is adjacent to the aft coupling face.

Connecting rod

The connecting rod is die-stamped from steel having an ultimate tensile strength of 70–80 kg/mm^2, and the crankpin bearing is inclined at 40° to the rod to allow the passage of the rod through the liner. The joint face is serrated. The oil feed for the small end is taken from the cap half bearing, which is more lightly loaded, and can better tolerate the discontinuity in oil film associated with the transfer ports to the circumferential collecting groove behind the shell. The oil is taken from this groove to a drilling which communicates via a ferrule in the joint face with a single passageway drilled in the connecting rod to deliver cooling oil to the piston.

Pistons

The single-piece aluminium alloy pistons with cast-in cooling coil used in the early PC2 engines have given place in all current PC engines to composite pistons with 'cocktail shaker' cooling. These have a light alloy skirt and a steel forged crown containing 1% manganese, the two parts being secured by bolts; the tightness between crown and skirt is ensured by selected Viton joints. Cooling oil is led to the piston through a longitudinal drilled hole in the connecting rod and this oil lubricates the gudgeon pin, then penetrates the piston pin bush and eventually reaches the annular channel in the piston. From there the oil passes into a central chamber and finally to the oil sump by a drilled hole in the piston centre. To avoid subjecting the gudgeon pin bearings in the PC2–5 to excessive pressures, the pin retains its original diameter but is lengthened by 20 mm. The gudgeon pin covers have been eliminated — the piston rings being mounted above the pin — and replaced by two oval circlips (Figure 14.3).

The piston ring arrangement is composed of: one chromium-plated slotted top ring having a torsional chamfer of 8 mm thickness; three copper-plated compression rings each of 8 mm thickness; two spring-loaded scraper rings each of 10 mm thickness.

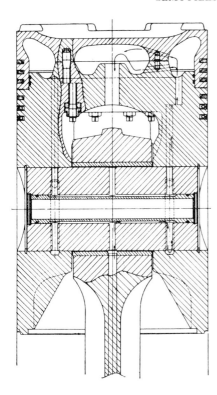

Figure 14.3 Section through a two-part piston – PC2-5 engine

The thickness of the rings has been reduced when compared to the PC2 engine rings to increase the piston skirt sliding area. Both top ring grooves undergo a high-frequency treatment, to allow a wear rate lower than 1/100 mm/1 000 h.

On earlier engines with coil cooled aluminium pistons and a conventionally split large end, the piston cooling oil was channelled in both directions up and down separate drillings in the connecting rod, and discharged to the crankcase just above the joint face of the big end.

Cylinder arrangement

While early PC2 engines had a plain flanged liner, the rating of current engines requires more effective cooling of the combustion area, particularly so where heavy fuel is to be burnt.

The bore-cooled liner adopted is made from FT25 centrifugal cast iron, and the internal bore is machined to obtain a finish of 80 to 120 µ in CLA. This liner, the upper parts of which are drilled with sloping channels at 40°, permits a bore temperature at top dead centre level of lower than 150 °C. The liner is centred in a cast iron water jacket, thus preserving the steel crankcase. Sealing between the liner and water jacket is ensured by elostomer rings.

The joint between the cylinder liner and cylinder cover is dry, the main transfer of water being made through an external connection on the camshaft side of the water jacket (i.e. the lower side on Vee engines), with a small by-pass port on the opposite side to obviate the formation of a steam pocket (Figure 14.4).

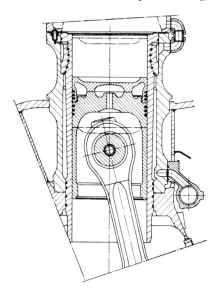

Figure 14.4 Sectional view showing cylinder liner cooling arrangements – PC2-5 engine

Cylinder cover

In all engines improvements have been made to the shaping of exhaust ports, and to the cooling of the seat in the valve cage both by water circuit and by tighter fit of the cage in the head.

Figure 14.5 illustrates the evolution of the PC2 valve. Note the valve cage seat cooling arrangement and the improved passage shape. Other PC engines follow a similar pattern. Some PC2 engines incorporated water cooled valve stems.

The cylinder cover is manufactured from FT25 cast iron, and for the PC2–5 engine retains the general design of the PC2, although increased mechanical and thermal stresses have necessitated greater reinforcement. With this in view, the cover height was not increased because this would require a modification of the valve gear. However, the upper section has been increased by 10 mm which stiffens the cover and permits better sealing for the valve cage, while the bottom of the cover has been made thicker and various ribs have been reinforced, especially

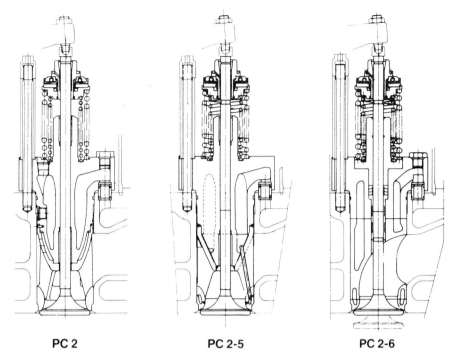

Figure 14.5 Exhaust valve basic design evolution

in the water passages between the inlet and exhaust valve housings. The aerodynamic design of the ducts has been modified to decrease the specific fuel consumption, and both inlet and exhaust ducts have been provided with a convergent orifice near the valve seats: the diameter of the exhaust valves has also been increased by 5 mm.

For diesel oil operation solid valves of chromium-silicon-molybdenum steel are used, the valve seat being made from hardened steel and supporting the valve body at its lower section to decrease the length exposed to dilation. The valve assembly, which can be withdrawn without dismantling the cylinder cover, is secured by long studs and distance pieces — which has eliminated the use of Belleville washers — and a locking ring linking the valve seat and body has been provided to facilitate seat extraction should they become difficult to remove because of carbon deposits. The shape of the valve mushroom has been modified, as well as the shape of the split collets, to decrease the level of mechanical stress. For operation on heavy fuel with a vanadium content below 150 ppm it is sufficient to add a Rotocap to the upper part of the springs.

Today as vanadium content is generally not known, SEMT has replaced the standard cage by a water cooled seat of a one-piece cage.

The spindles are made in two parts welded together, the lower part being in Nimonic 80A alloy. As with the PC2 engine, water cooled exhaust valves are employed for operation on heavy fuel of vanadium content systematically above 250 ppm.

Crankcase

Two types of crankcase have been designed for the PC2-5 engine; both consist of welded steel plates and elements of cast steel. The first design made from cast steel is similar to the PC3 engine (Figure 14.6, A Type). The eight cylinder tie-bolts are secured in the crank housing which supports the water jacket, and the top of the frame therefore no longer needs to be machined except for the entablatures supporting the camshafts. This arrangement eliminates the distance pieces between the crankshaft and the top of the frame; moreover, accessibility for welding is improved during fabrication and during weld inspection.

The second type of frame (Figure 14.6, B Type) has been decided upon because of economic factors. The monobloc crankhousing made from cast steel has been replaced by cast elements identical to the PC2 design, that is, one for each section of the Vee-form. In this case, the distance pieces between the top plate and crankhousing had to be retained because the water jacket can no longer rest upon the crankhousing, but on the top plate of the frame, as with the PC2 engine.

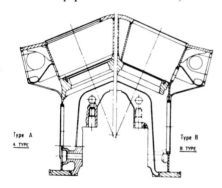

Figure 14.6 Two different crankcase sections. (Left half) cast steel monobloc; (Right half) fabricated

Both solutions can be chosen because, between two engines one having the first type of frame, and the other the second type of frame, the only difference is in the water jackets, which are not interchangeable. The second type of frame can also be used for PC2 engines provided that the main bearings are machined at a diameter of 290 mm.

The transverse wall is made from cast steel and the main bearings are similar to those of the PC2, the bearing body being clamped to the frame by two vertical tie-bolts which are hydraulically tightened, and by two lateral tie-bolts which are accessible from outside the frame. The bearing caps are tightened by a hydraulic jack which also serves to carry oil to the main bearings (Figure 14.7). Note that although the crankshaft is underslung, the design allows maintenance access to the main bearings from above.

The governor (usually Woodward) and the engine driven oil, water, and fuel transfer pumps are driven off the main phasing gear train via a spur gear, and are mounted at the flywheel end of the engine below the camshaft.

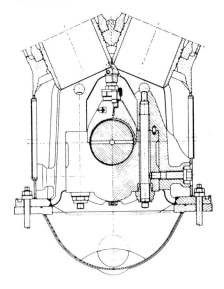

Figure 14.7 Crank housing with main bearing

Injection System

It is Pielstick practice that the fuel pump, which is of their own design, incorporates its own roller follower. In all recent engines the high-pressure fuel pipe from the injector to the pump has given way to a machined connection threaded through the cylinder head top deck to the injector stem. This shortens the connection and avoids the risk of fuel oil reaching the lubricating oil circuit if an external fuel pipe is damaged. It also facilitates the high injection pressures which SEMT consider now to be essential to the successful combustion of inferior fuel, as well as to improved economy.

The fuel valve is water cooled. See typically Figure 14.8 for the PC2-5 arrangement. Figure 14.9 shows a PC4 fuel pump. Note the barrel lubrication arrangement. The fuel pump control rods are ganged together along the length of a PC engine.

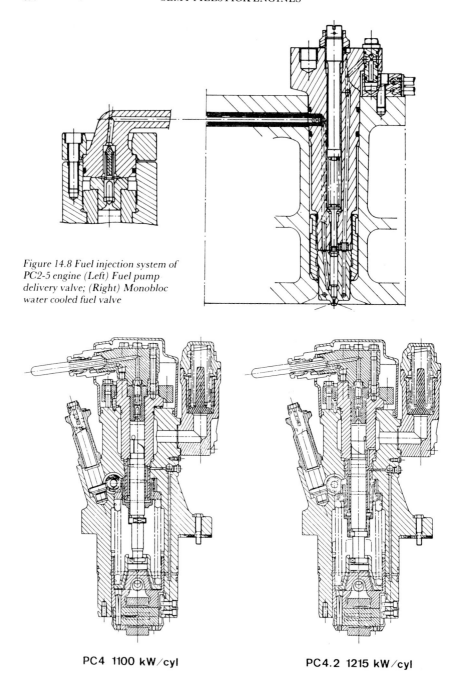

Figure 14.8 Fuel injection system of PC2-5 engine (Left) Fuel pump delivery valve; (Right) Monobloc water cooled fuel valve

PC4 1100 kW/cyl   PC4.2 1215 kW/cyl

Figure 14.9 Evolution of PC 4 engine fuel pump

The camshaft bearings are housed under the fuel pump flange, so that the stresses on the frame are minimised. The camshaft is easily withdrawn as a unit by releasing the bearing keeps.

Exhaust system

Earlier engines, in common with others of similar rating, used the conventional pulse system for connecting the cylinders to the turbocharger inlet. However, at the highest ratings involving bmep in excess of 20 bar, the constant pressure system gives better performance, though at the expense of performance at lower outputs.

In order to preserve flexibility of output to suit the requirements of marine propulsion, Pielstick have therefore evolved an exhaust branch layout with a venturi-shaped passage, or pulse converter, which preserves the pulse energy at the valve sufficiently to give good part load performance without prejudice to the high load advantages of the constant pressure system (Figure 14.11). It is termed the Modular Pulse Converter system and, as can be seen from Figure 14.10, it leads to a very much simpler layout.

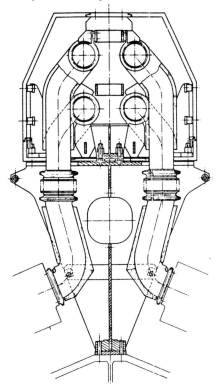

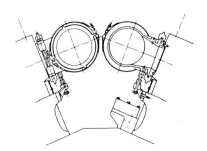

Figure 14.10 Exhaust system (Left) Pulse system (Right) Modular pulse convector system

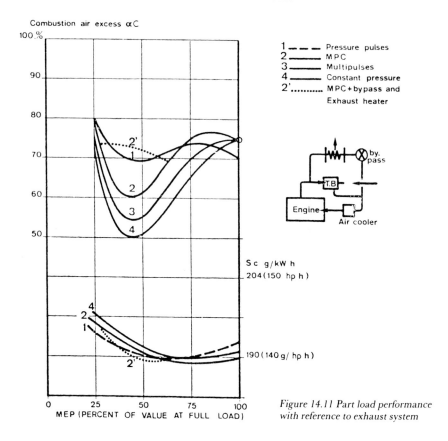

Figure 14.11 Part load performance with reference to exhaust system

In general, Pielstick engines are equipped with one turbocharger and one intercooler per bank of cylinders, irrespective of the number of cylinders. In common with the experience of other engine designers Pielstick have found that the optimum valve timings, particularly of the exhaust valve, need to be amended as cylinder output rises.

The PC2–5 and PC2–6 engines, for instance, have the exhaust valve opening advanced from 140° ATDC to 120°. The exhaust valve closing time is also advanced slightly. The inlet valve period of opening is extended by 10° by opening earlier and closing later.

Maintenance

The main advantages of a compact engine are lower costs and space requirements but, unfortunately, accessibility can be difficult and complicate maintenance operations. In designing the PC4 engine, a considerable effort has been made to eliminate this latter reputation.

Taking into account the maintenance schedule, which retains all the advantages of medium-speed engines — particularly piston overhauls every 10 000 to 17 000 running hours — the components which must be subject to special studies, with a view to reducing overhauling time and labour, are: the pistons; the main bearings and big end bearings; the plunger/barrel of the injection pumps.

For all these parts, the component arrangements and the overhauling tools have been designed in such a way that the man-hours for overhauling are, if anything, less than on the PC2 engines. This is due, for example, to the disposition of the inlet manifolds under the platforms, thus giving good accessibility to the cylinder head, valves and the valve

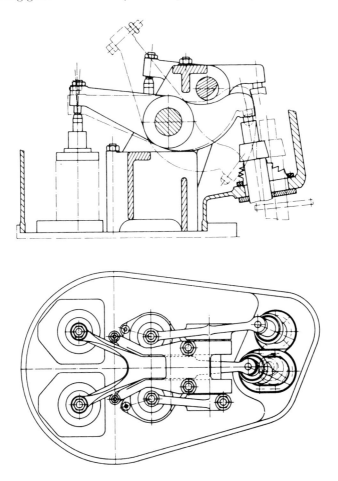

Figure 14.12 PC4 valve gear showing how the pushrod can be moved sideways after slackening the tappet screw, permitting the exhaust rocker to be tipped clear for withdrawing the valve cage

gear, as well as to the injection pump, while leaving sufficient height from the base so as not to impair accessibility to the main bearings and big end bearings.

Another example of the simplification of maintenance operations concerns the removal of the cylinder heads: a special device equipped with eight hydraulic jacks allows the tie-bolt nuts to be slackened with little manual effort.

The exhaust and inlet manifolds are connected to the cylinder head by easy-to-dismantle and refit connections using a collar tightened by only one screw easily accessible for the exhaust manifold, and a rubber joint on the admission side, ensuring automatic tightness.

To remove the main bearings and big end bearings, hydraulic jacks are used and oil/pneumatic tools allow their refitting without manual effort or loss of time.

It is possible to transport and erect in place complete cylinders (connecting rod, piston, cylinder liner, water jacket, and cylinder head) while the principal pipework is arranged in the form of a bundle together with the inlet manifolds, thus making a sub-assembly of the platform-support type at each side of the engine.

Acknowledgement

The editors would like to acknowledge the assistance of Alsthom-Atlantique in the preparation of this chapter.

D.A.W.

15 MAN four-stroke engines

Although MAN have made four-stroke propulsion engines for many years, they first entered the higher output medium-speed market in the mid-1960s with the VV40/54 engine, followed in the late 1960s by its derivative the VV52/55. (The numbers designating all MAN four-stroke engines are respectively the bore/stroke in centimetres.)

These engines were built as in-line or as 45° Vee engines with from six to eighteen cylinders. They had cast iron frames, with the crankshaft installed into the bedplate from above, and the Vee engines used an articulated connecting rod. Their general features can be appreciated in Figure 15.1, which is a cross section of a Vee form 40/54 engine. Composite oil cooled pistons were employed, and all valves were caged, with water cooling for the exhaust seats.

Some engines, particularly where low output operation was significant, were fitted with tandem fuel pumps to achieve finer metering control at these outputs.

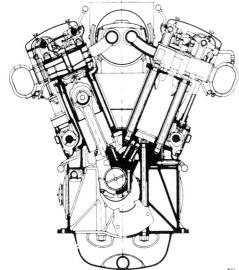

Figure 15.1 Cross section of VV40/54 engine

During the 1970s this programme was modified by adding the 32/36 cylinder size, and by reducing the stroke of the two earlier engines to produce, in the redesigned form, the 40/45 and 52/52 ranges. All of these designs share many basically similar features. MAN have licensed a number of builders around the world to produce these and other of their engines.

The current ratings of the principal engines are shown in Table 15.1. In-line engines are designated by the prefix L (e.g. 6L 32/36), and Vee engines by the prefix V (e.g. 12V 52/52).

Table 15.1 General characteristics

Engine type	32/36	40/45	52/52
Max. speed rev/min	750	600	514
Cyl/output kW	370	550	885
No. of cyls in-line engines	6,8,9	6,7,8,9	6,7,8,9
No. of cyls Vee-engine	12,14,16,18	10,12,14,16,18	10,12,14,16,18

These designs are described in order, the larger engines usually in terms of their differences from the smaller ones.

THE 32/36 ENGINE

A cross section of a Vee engine is shown in Figure 15.2.

Engine frame

The engine frame, rising from the mounting girders located well below the crankshaft centre line, right up to the top surface of the cylinder blocks and extending over the entire length, is a monobloc casting. This one-piece frame incorporates spaces and passages to house timing gear, camshafts with bearings, injection pumps and inlet and exhaust valve tappets.

The force generated by firing is transmitted directly from the level of the cylinder head bolts down to the main bearings. The transverse members of the frame in way of the flow of force are spaced in such a way as to minimise material stress and structural deformation. The bearing caps for the underslung crankshaft are cross bolted, producing a complete enclosure round the bearings: this helps to distribute the forces uniformly.

The fillet between the horizontal seating surface of the bearing cap and the vertical fitting surfaces, generally considered critical, is sub-

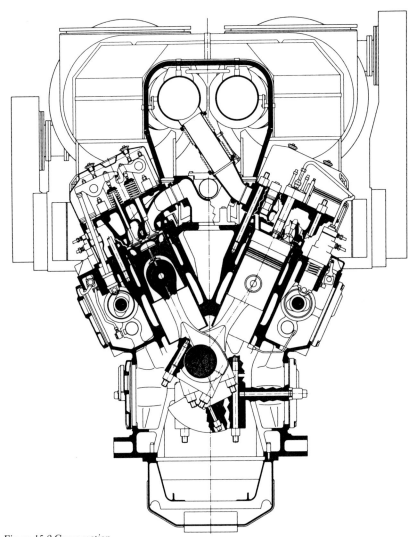

Figure 15.2 Cross section of V32/36 engine

jected to initial compressive stressing so that the dynamic stress amplitudes, which are relatively small anyway, can be controlled more easily by the rigid frame structure.

The frame features box-shaped cross sections running the entire length of the engine, which enclose the crankshaft and cooling water spaces around the cylinder liners, and are the seat of the engine on the foundation. This lends the engine extreme longitudinal rigidity.

There is an outboard bearing at the coupling end of the crankshaft, and this also provides thrust location (Figure 15.3).

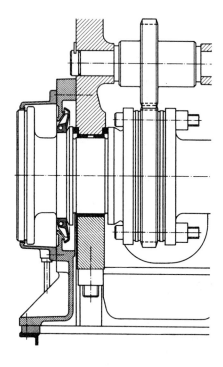

Figure 15.3 Outboard and thrust location bearing, 32/36 engine

Connecting rods

The connecting rods, which are angled split at the big end with a ground serration pattern joint face, are particularly resistant to deformation, as Figure 15.2 shows. The length of the small end bearing on the lower side, that is, in the direction that carries the gas forces, is larger than on the upper side which carries only the inertia forces. The converse applies to the gudgeon boss areas in the piston. The connecting rod bolts are tightened hydraulically and the tightening torque is checked by measuring the elongation of the bolts. To counteract any increase in stresses likely to occur in the threads of the bolts as a result of resistance to deformation in the relatively massive foot of the connecting rod, the bolt ends are hollowed to promote conformability.

Piston

The composite piston used for medium-speed engines of this size burning heavy fuel oil follows standard MAN practice. The aluminium skirt, which is maintained at a relatively low temperature, has a double oval shape and hence adapts itself well to the cylinder liner surface during the stroke.

Liner

The liner, whose wall thickness equals 8% of its diameter, is of sturdy construction. Deformation is kept to an absolute minimum, thanks to the short supporting length between the thick-walled top sections of the frame and the liner, and to intensive cooling of the collar only (Figure 15.4).

The liner is radially supported in the top section of the frame, keeping deformations by temperature small and equal over the whole circum-

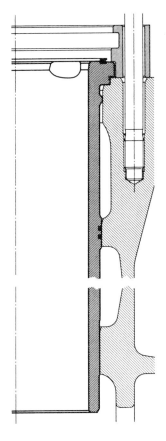

Figure 15.4 Liner and frame detail 32/36 engine

Cylinder head

The cylinder head is handy in terms of both weight and size. With the aid of hydraulic tools, it can quickly be dismantled with the valves still in place and without removing the tandem rocker arms which can be swung clear. Valve cages have, therefore, not been used. Special importance has been attached to good air and exhaust flow paths. The exhaust valve seats are intensively cooled via the water cooled seat rings. The seat rings for both the inlet and exhaust valves are of wear-resistant material.

Thanks to the sturdy deck plate of the cast iron cylinder head, to which the gas forces are directly transmitted vertically, the flame plate can be made particularly thin. This, plus the intensive flow of cooling water to the centre of the combustion area, due to the intermediate deck, yields such low stress values that material strain is very slight.

A collar is used to locate the bottom of the cylinder head, and to secure the flange of the cylinder liner. Six bolts on each cylinder head, with hydraulically tightened nuts, hold the cylinder head symmetrically to the collar and the frame, so that the liner remains perfectly round during operation.

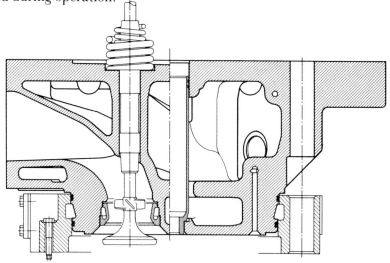

Figure 15.5 Cylinder head section of 32/36 engine

As in the larger four-stroke engines, the exhaust valves are rotated by vanes fitted to the stem, and driven by the gases expelled from the cylinder. This feature can be seen in Figure 15.5. The main advantage of this rotator is that the valve still has sufficient momentum to turn as the head comes to touch the seat, thus scraping off the thin deposits formed, particularly during heavy fuel operation.

Turbocharging system

This is arranged on the constant pressure system as with the 40/45 engine.

Camshaft

Figure 15.6 indicates a diagrammatic view of one camshaft section. Each camshaft is made up of several such sections, equal in length to the distance between two cylinders. The camshafts are mounted in passages in the frame running the entire length of the engine. Sections of the camshaft can easily be removed and changed by unfastening the 10 coupling bolts at each end and exploiting the axial travel (provided to achieve direct reversing) to disengage the spigot.

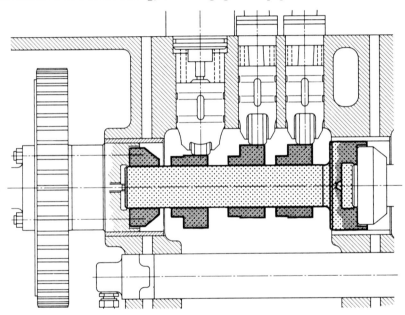

Figure 15.6 Arrangement of camshaft section 32/36 engine

THE 40/45 ENGINE

A cross section of in-line engine is shown in Figure 15.7.

Engine frame

The philosophy of the 40/45 engine frame is exactly the same as the smaller 32/36 and size has not prevented the adoption of a monobloc casting extending from the deep foundation to the cylinder jacket, cam-

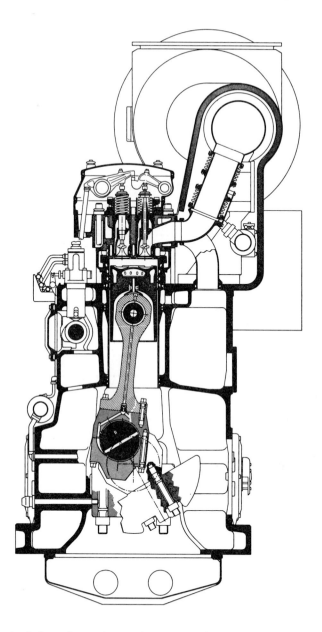

Figure 15.7 Cross section of L40/45 engine

shaft housing, air receiver, cam followers, etc., and incorporating a housing for the vibration damper at the free end.

Tie-rods are not used, the firing stresses being carried directly from cylinder head studs to main bearing studs by means of generous ribbing. The bearing caps of the underslung crankshaft are vertically and horizontally bolted to the engine frame, giving a very rigid assembly. The design ensures a low level of material stress and of structural deformation.

It has not been found necessary to adopt serrations or any similar treatment to eliminate fretting movement between the bearing keep and the main bearing housing, this being amply achieved by the location geometry of the assembly.

The horizontal tie-bolts ensure that the critical fillet between the horizontal seating surface and the vertical fitting surfaces is subjected to an initial compressive stress to control the dynamic stress amplitudes. The choice of cast iron, with its large masses and inherent damping capacity, has a favourable influence on noise and vibration.

Vibration damper

A MAN sleeve spring vibration damper is fitted at the free end. This design was chosen because it combines long service life with constant detuning. Maintenance requirements are minimal (Figure 15.8, half view).

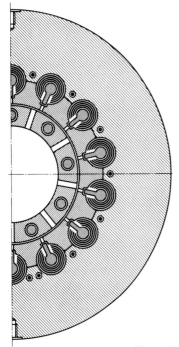

Figure 15.8 Vibration damper for 40/45 engine

Connecting rods

The connecting rods of the Vee 40/45 engine have been developed not so much with an extremely short engine or particularly small masses in mind, but sturdiness of construction combined with ease of maintenance. The big end is therefore horizontally divided twice. For normal maintenance purposes only the upper connection is loosened, leaving the bearing undisturbed on the crankshaft. Running-in procedures usually required after the bearing has been opened up can thus be discarded. Less space is needed overhead for piston withdrawal, thus permitting a lower engine room height. As can be seen from Figure 15.9 (right half) the cooling and lubricating oil for the small end and piston is collected from the cap half near the joint face to avoid discontinuities in the loaded area of the rod half bearing shell. The oil is then led through internal passages to enter the central drilling in the connecting rod shank via a non-return valve. The alignment of each joint is preserved by a pair of dowel pins. On Vee engines the connecting rods run side by side on the crankpin.

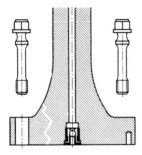

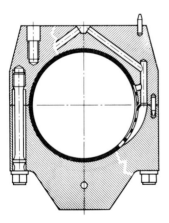

Figure 15.9 Arrangement of bottom end of connecting rod of 40/45 engine

Pistons

The composite pistons, as on all current MAN engines, consist of a forged aluminium skirt bolted to a forged steel crown which has hardened ring grooves. The first ring has a wear resistant running layer applied by a plasma spray technique. This is designed to enhance its life when burning inferior fuels.

Cooling oil is fed via the small end, as shown in Figure 15.10, and returns via the gudgeon bosses to the crankcase. The difference in length between the inertia loaded and the gas loaded halves of both the small end bearing and the gudgeon bosses can be discerned in Figure 15.10. In this regard the 40/45 engine is similar to the 32/36 engine; and, as on the 32/36 engine, the piston skirt has a double oval shape to conform to the cylinder liner surface throughout the stroke.

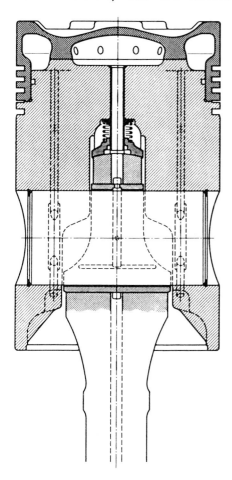

Figure 15.10 Composite section through piston in plane of gudgeon pin of 40/45 engine

Cylinder head

As with other MAN engines, the thin, uniformly cooled flame plate absorbs the thermal load, while a sturdy deckplate supports the head against the firing forces. This can be seen from Figure 15.7. The radial flow of cooling water between the two decks from the circumference to the centre where the thermal loads are most intense, as well as the progressive decrease in wall thickness towards the middle of the head, combine to yield uniform temperature gradients and good control of temperatures (Figure 15.11).

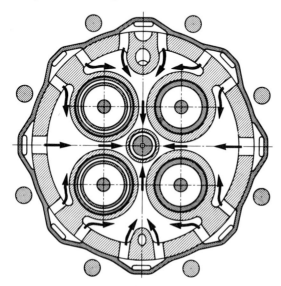

Figure 15.11 Cooling water flow in cylinder head of 40/45 engine

The firing forces are evenly transmitted from the top deck plate to the eight cylinder head bolts with nuts, which hold the cylinder head, liner and engine frame together with little deformation. For an engine of this size exhaust valve cages are considered most suitable. The cages are seated on the intensively cooled flameplate of the cylinder head and can move with it, as they are not attached to any other point. Valve seat and valve guide are thus properly co-ordinated at all load conditions. The provision of a conical gap between the underside of the upper flange of the cage and the lower plate of the cylinder head enclosure, through which the cage passes, makes it easier to remove the exhaust valve cages even after long periods of service. The cages are intensively cooled at the valve seats, yet the ducts are allowed to remain hot enough to suppress corrosion. When the cages are being removed for servicing the water inlet and outlet connections from the rest of the cooling system can be sealed by a slide valve.

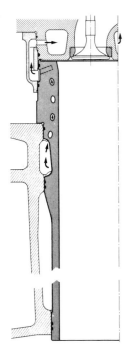

Figure 15.12 Cylinder liner section showing coolant flow, 40/45 engine

Cylinder liner

As Figure 15.12 shows, the liners have extra thick walls in the combustion zone which make for rigidity. Bore cooling, of the collar opposite the combustion zone only, and less intensive cooling down to the intermediate location where the water enters the jacket space, makes for uniform distribution of temperature over the entire length of the liner. The liner is only cooled by air below the collar. These features ensure that the liner remains almost perfectly round during operation. The liner is radially supported at three levels to curb deformation of its longitudinal axis. Due to the rigidity of the liner itself the transfer collar for conveying water from the liner to the floor of the cylinder head is of relatively modest scantling.

Valves

The seats of the exhaust valve cages are armoured and cooled, as is usual on MAN engines. The seat of the valve has a wear resistant contact surface. The exhaust valve head is rotated by vanes fitted to the stem and driven by the exhaust gases as they are expelled from the cylinder. This rotator is simple and effective, as on the 32/36. The intake valves are rotated by Rotocaps.

The material of the cages is particularly insensitive to corrosion. As Figure 15.13 shows, the continuous, single-piece valve guides are sealed at the top end, from which the oil is evenly fed. This does much to prevent stem corrosion.

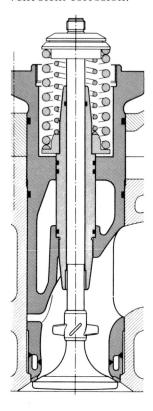

Figure 15.13 *Valve cage section of 40/45 engine*

Turbocharging system

All MAN engines are normally fitted with turbochargers of MAN design. The engine is constant pressure turbocharged. Thanks to the charge air pipe being integrated in the engine casing on both the Vee and the in-line engines, the pipe arrangement is simple and silencing is efficient. Air is admitted to the cylinder heads through pipebends secured to the cylinder heads by quick acting couplings. Above the charge air pipe one exhaust manifold per cylinder bank is arranged. Each manifold leads to a turbocharger at the end of the engine.

To improve the efficiency of the system some of the outlet velocity of the exhaust gas is converted into static pressure in the conical adapters between the cylinder head and the flexible sections of the branch pipes, which serve in this respect as a form of pulse converter. The exhaust gas pipes are also secured to the cylinder heads by quick acting couplings.

As in the case of the other constant pressure turbocharged medium-speed MAN engines, the exhaust manifolds are interconnected by a transverse pipe at the end opposite to the turbochargers in the case of Vee engines so that, in the event of one turbocharger failing, the engine can still develop an output higher than that of a naturally aspirated engine. The exhaust gas system is completely jacketed for heat and sound insulation. The individual parts of the exhaust gas pipe and its jacketing can be readily dismantled in sections.

Because of the relatively long flexible branch pipes, it is not necessary to make further allowance for expansion in the common manifold for each bank, which therefore bolts directly to the turbocharger.

Camshafts

On the 40/45 engine the camshafts are in a single unit for each bank of cylinders, and are removed sideways having removed the bearing keeps. The cams themselves are hydraulically press fitted. The cams are replaceable, but two part muff type cams are available to facilitate urgent repairs should the need arise.

Injection system

Features of the 40/45 system are basically the same as those of the 32/36 engine, whose pump is shown in cross section in Figure 15.14.

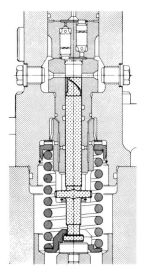

Figure 15.14 Section through fuel injection pump of 32/36 engine.

The engine is fitted with individual pumps designed for high pressures. Each plunger features not one but two offset control edges, so that fuel is released in stages after the end of delivery and the peak pressures in the delivery pipe are reduced, thus preventing flow cavitation and erosion of the plunger barrel and pump casing. The pump barrel has three angular grooves: the top groove reduces the pressure between plunger and barrel to that of the suction space; the bottom groove conducts lubricating oil; the middle groove permits the mixture of fuel and lubricating oil formed, to be discharged without pressure, thus preventing contamination of the lubricating oil by fuel. Ample plunger lubrication is assured.

The high-pressure fuel line connects with the injector stem through a passage in the cylinder head, so that dilution of the lubricating oil by fuel is extremely unlikely.

THE 52/52 ENGINE

Since the size of this engine does not lend itself to the monobloc philosophy of the two smaller ones, it has been decided to retain for the 52/52 the maximum commonality of components with the LV52/55 engine. The newer engine can therefore draw on the successful service experience of some 300 units of the older engine. This enhances reliability. A cross section of the Vee engine is shown in Figure 15.15.

Engine frame

The casing of the 52/52 engine is a two-part design, consisting of engine frame and separate cylinder blocks for the Vee engine, while the in-line engine has a bedplate and cylinder frame. Each cylinder bank of the Vee engine has a continuous cylinder block connected to the engine frame by tie-rods. The tie-rods retrieve the cylinder block of tensile stresses occurring during combustion. The cylinder block incorporates camshaft bearings, injection pump and the inlet and exhaust valve tappets. The cylinder frame of the in-line engine is a monobloc casting and incorporates the supports of the above components. This frame too is tightened to the bedplate by means of long tie-rods.

In both cases the engine casing is designed for the crankshaft to be introduced from the top. The firing forces pass through a steel transverse member — in the case of the Vee engine — which links the forces from the cylinder head studs to the tie-rods securing the cylinder blocks to the bedplate.

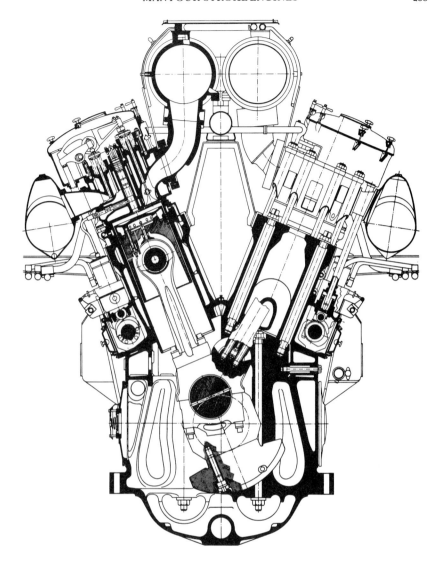

Figure 15.15 Cross section of V52/52 engine

The crankcase of the Vee engine and the bedplate of the in-line engine are made of nodular graphite cast iron, whereas the cylinder blocks and the cylinder frame are made of grey cast iron.

Figure 15.16 illustrates the features of the engine frame design of both the in-line and the Vee engines. Sturdy seating beams are formed in the bedplate to connect the engine structure to the foundation.

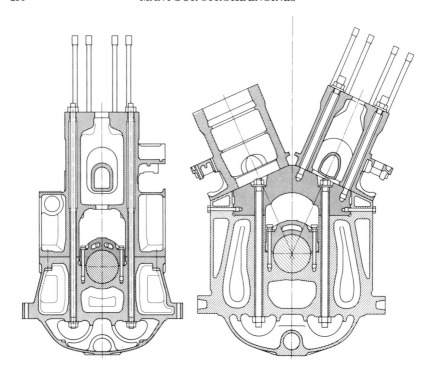

Figure 15.16 Frame construction of L and V 52/52 engine

The engine frame incorporates the control drive gear at the coupling end and at the free end space is provided for the torsional vibration damper. In this case the design does not lend itself to incorporating the air manifold in the block, and a separate air manifold is used made up of separate sections for each cylinder.

Connecting rod

In the Vee engine the die-forged connecting rods consist of master rod and slave rod, articulated to the connecting rod big end by a slave pin (Figure 15.15).

For maintenance work on the in-line or Vee engines, the connecting rods can be removed from the big end without disturbing the connecting rod bearing. In this regard the 52/52 engine is similar to the 40/45. The lubricating passages and also the small end design are also very similar.

All the bolts in the assembly are designed for hydraulic tightening. The bolt ends are recessed to ensure uniform contact over the entire thread length.

Other components

The piston is very similar in principle to that in the smaller engines. Figure 15.15 shows on the master rod cylinder a section across the gudgeon pin, with an insert (upper left) to show the resilient bolt securing the two parts of the piston to each other.

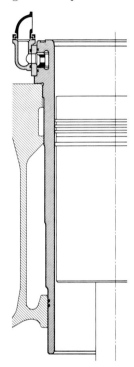

Figure 15.17 Cylinder liner section of 52/52 engine

The cylinder liner cooling is differently arranged, as shown in Figure 15.17. The 52/52 liner has a particularly rigid collar which is intensively cooled by an incorporated ring space. This not only ensures good cooling and uniform temperature distribution but also makes for small stress amplitudes under firing forces. Water is transferred from the jacket to this collar, and from the collar to the head by separate external connections.

The cylinder head, turbocharging and injection system are, in principle, the same as the 40/45 engine. Once again the exhaust branch adapters are designed to act as pulse converters.

The exhaust gas pipes are coupled to the cylinder heads by turn-lock fasteners.

The camshaft is again a single-piece design, removed sideways. It has shrunk-on valve cams, but a split fuel injection cam, clamped on to the shaft.

Maintenance

Reference has already been made in the descriptive text to the use of hydraulically tightened nuts on the major threaded fastenings, and to the use of quickly detachable clamps for the air and exhaust connections to the cylinder heads.

The constant pressure exhaust system is inherently simpler and easier to handle, and this simplification lends itself to easier maintenance.

Acknowledgement

The editors would like to acknowledge the assistance of MAN in the preparation of this chapter.

D.A.W.

16 Stork Werkspoor engines

Stork Werkspoor Diesel was formed in 1954 by the amalgamation of Stork and Werkspoor and also included other components of the Dutch diesel engine industry.

Not long after the company was formed it decided, in view of the growing market for high-powered medium-speed engines for the propulsion of ocean going ships, to develop their own design of medium-speed engine, which was introduced in the later 1960s as the TM410.

The engine was designed on conservative lines and quickly established a reputation for reliability. Since its inception it has been uprated some 50% with some design detail changes. In addition, in the mid-1970s, the larger TM620 engine has been introduced and, to a very large extent, this is a scaled-up version of its smaller stable companion.

The description which follows is based therefore mainly on the TM410, but with a note of the differences which apply in the case of the TM620.

SWD have adopted a cautious policy in licensing other builders to produce their designs. So far, relatively few licences have, in fact, been concluded, and most of the SWD engines are built in Amsterdam.

Table 16.1 The current outputs of the TM410 and TM620 engines

Engine	TM410	TM620
Max speed rev/min	600	425
Cyl output kW	564	1348
Bore mm	410	620
Stroke mm	470	660
Mep bar	18.2	19.34
No. of cyls in-line engines	6,8,9	6,8,9
No. of cyls Vee engines	12,16,18,20	12

Both engines are directly reversible four-stroke designs and the Vee angle is 40°. Cross sections are shown of the TM410 in Figure 16.1 and of the TM620 in Figure 16.2. These show very clearly the family likeness of the in-line versions of the two engines, and the cross section of the Vee form TM410 (Figure 16.3) is likewise typical of both sizes.

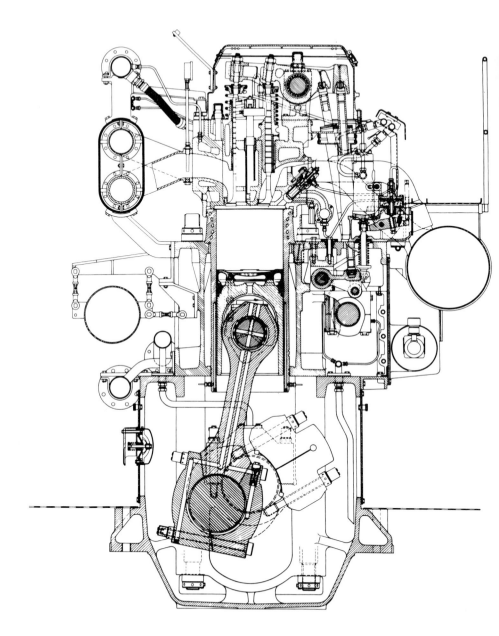

Figure 16.1 Cross section of in-line TM410 engine

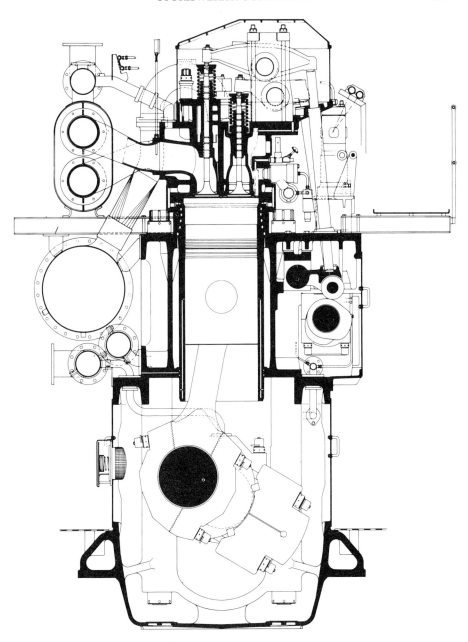

Figure 16.2 Cross section of in-line TM620 engine

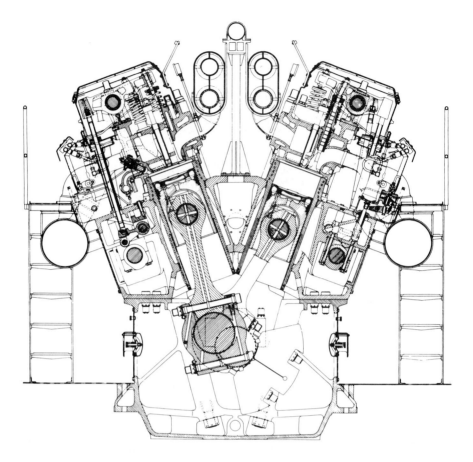

Figure 16.3 Cross section of Veeform TM410 engine

Bedplate and main bearings

Figure 16.4 shows the bedplate of the TM410 to be a U-shaped strong iron casting, in which the main bearing caps are fitted with a serrated joint to form, with the bearing saddles, a rigid housing for the thin-wall steel main bearing shells. For the TM620 what is, in effect, a single tooth serration is used in the main bearing cap joint. That is to say, the joint faces are inclined towards the abutment (Figure 16.5).

The bearing shells have copper–lead lining and lead–tin plating. Both upper and lower shells can easily be removed by lifting the bearing cap (Figure 16.6). Large inspection openings in the bedplate enable easy access to the crankcase. The inspection covers are fitted with relief valves.

STORK WERKSPOOR ENGINES

Figure 16.4 Bedplate and crankshaft — in-line TM410 engine

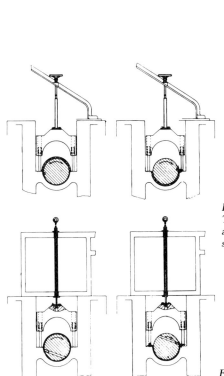

Figure 16.5 A major design change with the TM620 engine is the main bearing cap seating arrangement, to obtain the lowest possible stress

Figure 16.6 Main bearing removal

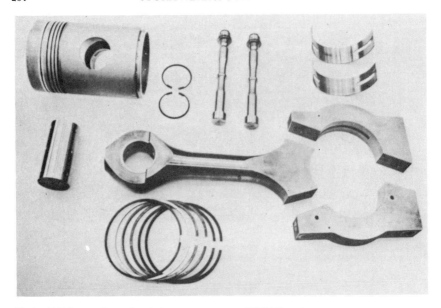

Figure 16.7 Connecting rod and piston components of TM410 engine

Integrally cast columns in the in-line engine bedplate accommodate alloy steel tie-rods, to connect bedplate and cylinder block tightly together in a rigid construction and to prevent high tensile stresses from cylinder pressures being transmitted to these cast iron components. In the Vee type engine also, two rows of tie-rods are applied. As these rods are in an oblique position, a number of short bolts are fitted in addition in order to prevent relative movement of cylinder block and bedplate. These can be seen in Figure 16.3. The thick upper bridge between pairs of cylinders on opposite banks also contributes to this rigidity.

Crankshaft

The crankshaft is a fully machined high tensile one-piece continuous grain flow forging. The large diameter journals and crankpins are provided with obliquely drilled holes for the transmission of lubricating and piston cooling oil. See Figure 16.4 for general construction of the crankshaft. Counterweights are fitted on all crankwebs, being secured by means of two hydraulically stressed studs and two keys.

Cylinder block with camshaft

The cylinder block is a rigid iron casting and includes the cooling water jackets and the camshaft space.

For the inspection of cams and rollers, the cylinder block is provided with covers on the camshaft side. Three roller levers are fitted per cylinder, one each for the inlet and exhaust cams and one for the fuel pump drive in the TM410. The TM620 has only two — for the valves.

The pushrod passages are fitted with seals to prevent oil leakage from entering the camshaft space. The camshaft is ground to a single diameter over its entire length. The cams are of hardened steel, hydraulically shrunk-on with the aid of tapered bushes.

When necessary, the complete camshaft can be removed sideways. Cams can be exchanged without removing the camshaft from the engine, as they can be removed from the shaft at the forward end of the section. The camshaft is driven from the crankshaft by nodular cast iron gearwheels and runs in thin-wall bearings. The direct reversible engine is provided with double cams with oblique transition faces and a pneumatically controlled hydraulic reversing gear, to move the camshaft in an axial direction.

Connecting rod and piston

As will be seen from Figure 16.7 the connecting rod has an extremely heavy big end, to form a rigid housing for the copper–lead lined lead–tin plated thin-wall steel bearing shells.

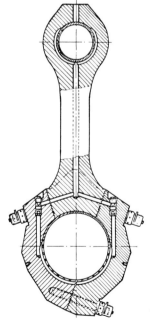

Figure 16.8 Connecting rod of TM620 engine

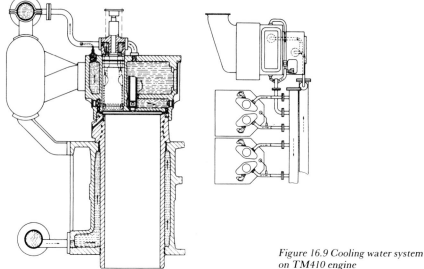

Figure 16.9 Cooling water system on TM410 engine

To allow the rod to be removed through the cylinder liner, the big end is split by serrated joints in two planes. The advantage of the design is the low headroom required, together with easy dismantling. The big end bolts are stressed hydraulically.

The TM620 connecting rod is similar in principle to that of the TM410, as Figure 16.8 shows, but uses three pairs of studs, each set normal to the face of the joint it closes rather than bolts at right angles to the axis of the rod.

The lower rated TM410 engines have a light alloy piston with a cast-in top ring carrier, and a cast-in cooling oil tube. (This can be seen in Figure 16.1.)

The TM620 and the higher rated TM410 engines now embody a two-piece piston of the usual layout. (This can be seen, for instance, in Figure 16.3.)

Both engines use one chrome-plated top ring, three compression rings with bronze insert, and an oil control ring above the gudgeon. The TM410 uses also a second oil control ring at the base of the skirt when the single-piece piston is fitted.

The gudgeon pin is fully floating, and the small-end bearings have different widths so as to provide a greater area to withstand better the combustion loads. (This can be seen in Figure 16.7.) The small end bearing material is to the same specification as the large end.

The small end and piston cooling oil supply passes through drillings in the connecting rod; the spent oil from the pistons and gudgeon pins is expelled from the piston, but with the one-part piston, cooling oil passes through a separate drilling to the bottom of the big end. Both engines employ restrictors in the big end oil passages incorporating non-return valves.

Despite the three-part large end construction, both engines use horizontally split two-part big end shells. The Vee engine big ends run side by side on the crankpin.

Cylinder liner

The cylinder liner is made of special cast iron and is provided with cooling water passages drilled to a hyperboloid pattern in its thick upper rim. This special feature ensures an intense cooling of the upper liner part and also ensures, by equalising the temperatures of the connected liner rim and the cylinder block, a perfectly circular liner when the engine is in operation (Figure 16.9).

Two holes for cylinder lubrication are drilled in each liner from the bottom, thus avoiding oil pipes through the cooling water spaces. Each

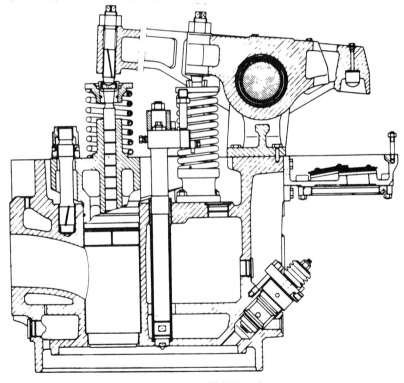

Figure 16.10 Cylinder head and valve gear on TM410 engine

hole feeds one lubricating oil quill arranged half-way along the liner length. A normal 'Assa' cylinder lubricator is provided.

On the TM620 there are four holes per cylinder.

Cylinder head and valves

Both engines use four valve heads, totally enclosed. The cylinder head for the TM410 is shown in Figure 16.10, which is a composite section.

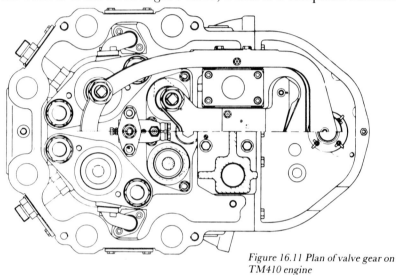

Figure 16.11 Plan of valve gear on TM410 engine

The left half of the section is through the plane of an exhaust valve; the right half is through the centre line.

Figure 16.11 is a composite plan view which shows on the upper side the unusual two-bearing exhaust rocker design. In the case of the TM620 the inlet valve is shorter than the exhaust valve and two fulcrum shafts are used with different lengths of 'Y' shaped rockers to operate each pair of valves. (This feature can be seen in Figure 16.2.)

The cooling water is directed over the flameplate by the intermediate deck, which also serves the function of a strength member so that the flameplate itself can be thinner and thus better able to withstand thermal load.

In both cases the exhaust valves work in water cooled cages and are Stellited on the seating face. SWD find that careful cooling of the seat area, and hence of the sealing face, enables these valves to achieve a life between overhauls on the heaviest fuel not greatly inferior to their life on marine diesel fuel. The inlet valves seat on hardened cast iron inserts mounted directly in the cylinder head.

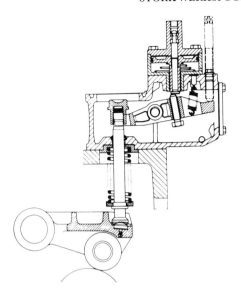

Figure 16.12 *Fuel pump drive and cutout on TM410 engine*

Valve gear

The cam profiles have been designed to avoid rapid acceleration changes, and thus to minimise noise and dynamic stresses. The valve rockers are fitted with needle bearings which require no oil supply; all other contact faces within the cylinder head covers are lubricated by separate impulse oiling equipment. The lubricator for this is driven, in parallel with the cylinder lubricator, from the camshaft gear. The oil used for these purposes is drained separately, to prevent contamination of the crank case oil should a fuel leak occur.

Fuel injection system

One separate injection pump is fitted per cylinder. The system is designed for high injection pressures to obtain good combustion and the high brake mean effective pressures used. The injector is equipped with an easily replaceable nozzle, which is water cooled to prevent carbon deposits when using residual fuels.

The pump on the TM410 is actuated by a spring loaded pushrod which itself follows a lever type cam follower similar to those which actuate the valve pushrods. On the TM620 the pump has an integral roller.

Safety devices

The mechanism for the starting air pilot valves is also used to stop the engine at overspeed. In that event the mechanical overspeed trip

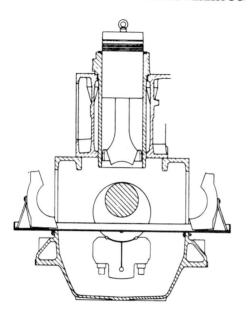

Figure 16.13 Removal of piston and connecting rod on TM410 engine

actuates a pneumatic valve to allow air pressure into an auxiliary cylinder which then keeps every fuel injection pump plunger lifted with the roller free of the cam. Figure 16.12 shows the arrangement used on the TM410 with the starting air pilot valve above right.

Safety devices for low lubricating oil pressure and for low cooling water flow are integrated in the engine systems and are thus independent of the external electrical alarm system. Standard electric alarm equipment, or alarm equipment and other safety functions to the specifications of clients and classification societies, can be installed in separate panels.

Maintenance

Special attention has been paid, in the design stage, to the requirements of quick and easy maintenance. Some of the matters dealt with are mentioned below.

1. Hydraulic tightening for all important bolts and studs.
2. Easy removal of the big end bearing parts to facilitate the withdrawal of a piston by the dismantling of both bearing caps along a horizontal slide through the crankcase openings (Figure 16.13).
3. A gasket-free joint between cylinder head and exhaust pipe.

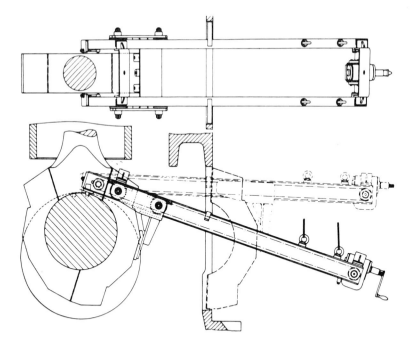

Figure 16.14 Withdrawal of large end cap on TM620 engine

Using these and other measures to facilitate maintenance enables removal times to be achieved, of which the following are typical:

Cylinder head and piston, with connecting rod:	1½ hours
Set of main bearing shells:	1 hour
Valve cage (complete with valve):	20 minutes
Fuel injector:	10 minutes

The weight of TM620 components poses additional problems, as more components exceed what a man can handle unaided. This has been countered, however, by designing maintenance tools. For instance, see Figure 16.14 which shows the tooling developed for withdrawing a large end cap. This tool is suspended on a rail alongside the engine and allows a man to move the caps into or out of the engine by means of a simple handle.

Other examples are a hoisting gear for the hydraulic jacks used to stress all the cylinder heads simultaneously, as well as for hoisting the cylinder head itself (Figure 16.15), and a combined tool for hoisting as

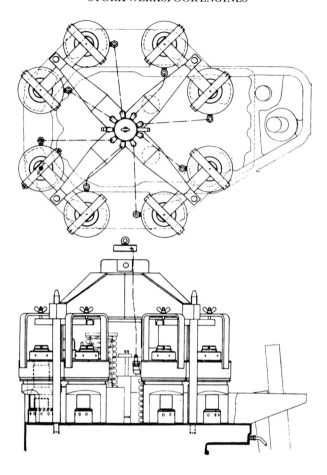

Figure 16.15 Hydraulic tool for cylinder head studs with lifting assembly

well as for hydraulic fitting and extraction, of the cylinder liner. This tool also allows mechanical tilting of the liner, thereby reducing the required headroom.

Acknowledgement

The editors would like to acknowledge the assistance of Stork Werkspoor Diesel in the preparation of this chapter.

D.A.W.

17 Sulzer four-stroke engines

The principal medium-speed engine made by Sulzer Brothers is the Z40, of 40 cm bore × 48 cm stroke.

Sulzer are also responsible for the A25 engine, which will be described in Chapter 23, as well as more than one other medium-speed design which have been involved in projected co-operation arrangements with other manufacturers.

The Z40, which remains wholly a Sulzer concept, and is licensed by them worldwide, originated in the mid-1960s. The initial examples were in two-stroke form, although the basic design and much of the development allowed for these engines to be produced either as two-stroke or as four-stroke engines.

In the 1970s the climate of opinion among medium-speed engine makers and users swung decisively to the four-stroke cycle. Sulzer, despite satisfactory two-stroke experience, decided that market forces made it advisable to follow suit, and all Z40 production for several years (well over 80% of all engines built) has been in the four-stroke form.

In 1980 the Z40 was completely redesigned and in its new form designated ZA40 incorporating several significant changes.

Cylinder ratings of the two forms of the 40/48 theme are given in Table 17.1

Table 17.1 Cylinder ratings of the Z40 and ZA40

Engine	Z40	ZA40
Max speed rev/min	560	560
Cyl output kW	550	640
Mep bar	19.54	22.74
No. of cyls in-line engines	6 and 8	6 and 8
No. of cyls Vee engines	10,12,14,16,18	12,16

In-line engines are designated by the suffix L (e.g. 6ZL40) and Vee engines by the suffix V (e.g. 12ZAV40). It is anticipated that in due course other cylinder combinations will be added to the ZA range. The Vee angle in all cases is 45°. The engines are directly reversible.

274 SULZER FOUR-STROKE ENGINES

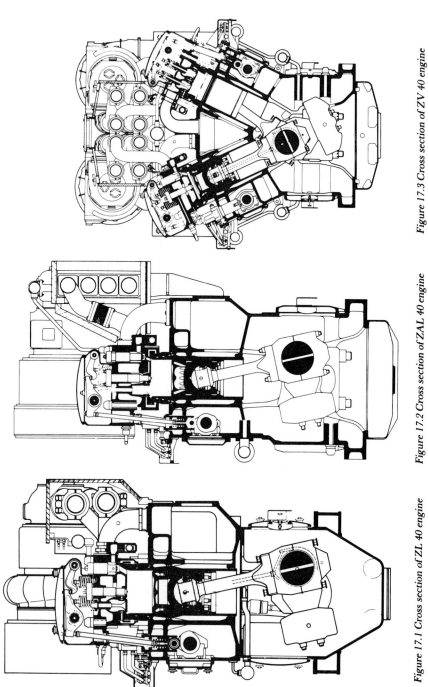

Figure 17.3 Cross section of ZV 40 engine

Figure 17.2 Cross section of ZAL 40 engine

Figure 17.1 Cross section of ZL 40 engine

Crankcase

A fabricated design is used for the ZL engine (Figure 17.4). This is a rigid component carrying the crankshaft in a conventional manner. The housing extends well above the crankshaft centre line to carry the cast iron cylinder block with which it forms a rigid and compact construction.

Figure 17.4 8ZL 40 crankcase and cylinder block

For the ZAL engine an entirely new monobloc casting has been adopted with underslung crankshaft (Figure 17.2). The differences between the two engines can be better appreciated from Figure 17.5, which also shows the tie-bolts, and the main lubricating oil supply.

For the Vee engine crankcase in both the Z and ZA engines, a one-piece cast iron design has been used incorporating an underslung crankshaft. Tie-rods transmit the horizontal forces of the bearing loads from the bearing caps to the frame and provide rigidity (Figure 17.6). For the ZAV engine the main differences stem from the increased crankshaft scantlings necessitated by the uprating.

ZL 40

ZAL 40

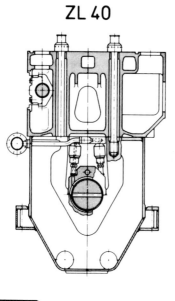
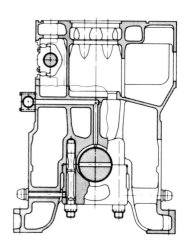

Figure 17.5 Engine casing design

Z 40

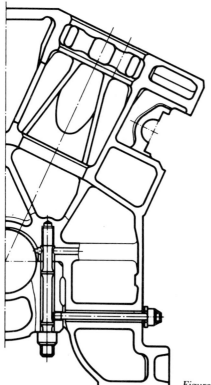
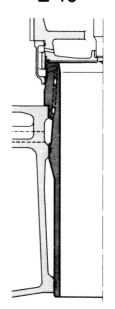

Figure 17.7 Z 40 cylinder liner design

Figure 17.6 ZV 40 crankcase section

Cylinder liner

For the Z range the cylinder liners are of cast iron with a generously dimensioned top part. They are water cooled through cooling bores in the upper part to provide a favourable temperature level. Thermal stresses and deformations are kept low. The middle section of the liner is gallery cooled (Figure 17.7).

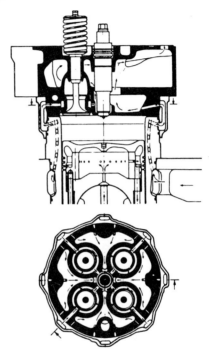

Figure 17.8 Cooling of cylinder liner and cylinder head. Z40 engine

For the ZA design the liners are still bore cooled, but reinforced to take the higher peak pressures. The bores are arranged in such a way that the liner bore surface temperature is high enough to prevent sulphur corrosion without fitting insulating tubes.

Bath nitriding has been specified for better running conditions and reduction of wear of the piston rings. Figures 17.8 and 17.9 indicates the difference between the Z and ZA designs in the combustion chamber region.

Cylinder head

For the Z40 the double bottom water cooled cylinder head is made of special cast iron. It has two inlet and two exhaust valves as well as a centrally arranged fuel valve (Figure 17.8).

On the earliest Z engines, the flow was less precisely controlled by the drillings, which, as Figure 17.8 shows, directed about half of the coolant flow directly to the valve cages and injector tube, while the rest flowed fairly freely across the flameplate constrained by the intermediate deck.

The ZA design, on the other hand, relies exclusively on lateral bores in the very thick flameplate to achieve a very precise control of the cooling flow, as is shown in Figure 17.9. It provides a high mechanical safety and permits ideal cooling conditions for the flameplate and the valve seats. Its higher stiffness also improves the valve sealing behaviour due to reduced distortion of the flameplate and, with it, the valve seat.

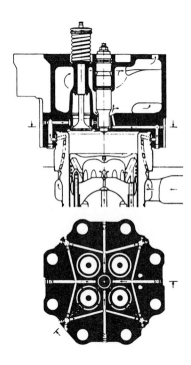

Figure 17.9 Cooling of cylinder liner and cylinder head — ZA 40 engine

The separate water transition jacket on the Z40 head, leading water from cylinder cooling passages to cylinder head, is, in the case of the ZA40, a part of the casting, thus reducing costs, adding to the stiffness of the cylinder head and giving support to the cylinder liners.

The rocker gear is practically identical to the Z40 tandem arrangement, with many interchangeable parts, like rocker arms, thrust pads, etc. The cast rocker gear box is new, as the oil supply bores have been integrated in the casting.

Valves and seats

The Z40 valve arrangement (Figure 17.10) provides the following features:

Identical design of the inlet and exhaust valves.
High valve reliability with heavy fuel operation through the use of specially developed plasma coated valve seats.
Perfect sealing by means of the symmetrical shape of the valve seat insert and the symmetrical flow of the cooling water.
Effective and simple cooling of the flameplate due to radial cooling water passages round the valve seat inserts.
Low pressure losses due to a large cross sectional area available for gas passages.
Double security by means of a two spring support.

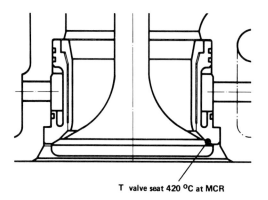

T valve seat 420 °C at MCR

Figure 17.10 Press fitted valve seat insert–Z 40 engine

For the ZA version, the valve seats are similar to those for the Z40 but only the exhaust valve seats are water cooled. The choice for valve design and materials is based on the vast experience gained on Z40 engines in service. The exhaust valves are now made in Nimonic and fitted with rotators, the inlet valves in a ferritic valve steel with Colmonoy 6 or similar type coating.

The dismantling and fitting time of the cylinder head has been vastly reduced by eliminating separate oil pipes, improved connections of exhaust and starting air pipes and a better arrangement of the exhaust pipe insulation panels.

The arrangement is designed to achieve an adequate overhaul life on heavy fuel without the need to resort to valve cages.

Fuel injection system

A helix controlled fuel injection pump unit, which can be cut out individually, is provided for each cylinder. The pump casing has two separate chambers, one connected to the inlet and the other to the spill pipe. Priming is carried out from the upper chamber, and the excess fuel not required for injection is returned to the lower chamber.

The main features of the fuel injection system are:

Fuel leakage into the crankcase is prevented by an oil barrier groove at the lower end of the plunger sleeve.

A high fuel delivery rate coupled with a declining velocity of the fuel cam was introduced to avoid an unnecessarily high injection pressure peak towards the end of injection.

A stepped helix prevents secondary injections and cavitation in the system.

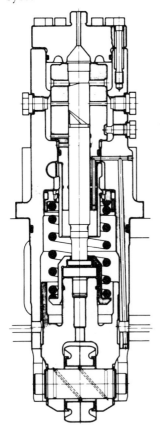

Figure 17.11 Reinforced fuel injection pump

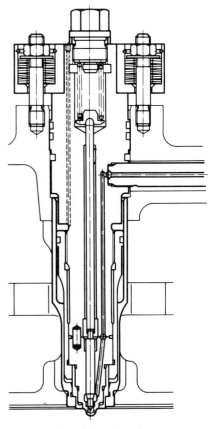

Figure 17.12 Fuel injection valve arrangement

The arrangement of the Z40 fuel pump can be discerned from the right-hand side of Figure 17.13. Later Z40 engines had a reinforced fuel pump of the same general type, and this fuel pump has been adopted for the ZA40 range of engines (Figure 17.11). The injector for all engines is shown in Figure 17.12.

The high-pressure pipe, relief valve and fuel injection valve are as far as possible identical between the Z and ZA engines, but adjustments had, of course, to be made for them to fit the new cylinder head.

The fuel nozzles are water cooled to suit the burning of heavy fuel oil.

Camshaft

The camshafts are carried at the side of the engine in bearings split parallel to the axis of the cylinders. They carry the inlet and exhaust cams, as well as the fuel injection cam for each cylinder. The arrangement is shown in Figure 17.13. The cams are shrunk on hydraulically and can be removed or repositioned at any time using a hydraulic

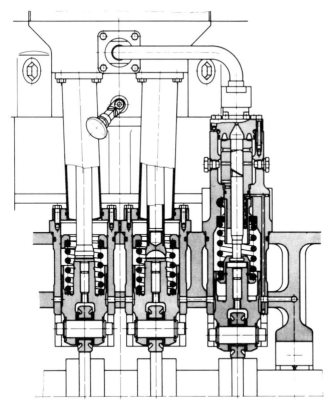

Figure 17.13 Camshaft with valve drive and fuel injection pump arrangement

pump. This design results in a very rigid connection and excludes any risk of the cams becoming loose. Furthermore, it allows very simple adjustment of the timing should this be necessary.

The camshaft and cam arrangement of the non-reversible ZA40 engines will be the same as on the Z40 as described above, with the exception of the fuel cam which will henceforth be of the split design in order to facilitate service exchange, as it is more prone to becoming damaged in service than are the valve cams. However, for easier fitting of the cams in the case of reversible ZA engines, tapered sleeves will be arranged between camshaft and cams (Figure 17.14).

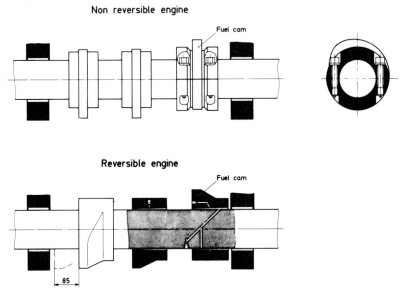

Figure 17.14 ZA 40 camshaft design

Crankshaft

The crankshaft (Figure 17.15) is forged in one piece from alloy steel, and machined all over. The necessary counterweights to obtain favourable bearing loads and vibration free running are bolted on to the crankwebs. It is also drilled to convey the lubricating and cooling oil, which enters at the main bearings and passes through the connecting rod and small end bearing into the piston crown. A vibration damper is normally fitted at the front end of the shaft, where auxiliary drives can also be taken off (cooling water, bilge pumps, etc.). At the rear end of the crankshaft is the flywheel with a barring rim. The last main bearing at the flywheel end is fitted to the outside of the end wall and contains the crankshaft thrust bearing.

Figure 17.15 6Z 40 coupling end

For the non-reversible engine, the lubricating oil, cooling water, and fuel transfer pumps can be fitted to the damper end cover.

The ZA40 crankshaft is very similar to the Z40 design. The main changes are increased pin and journal diameters, each increased by 20 mm.

Bearings

The main bearings are of the tri-metal type and the big end bearings of the flash type. With the in-line Z40 engines, the main bearings can be easily dismantled upwards; whereas with the Vee types, and the in-line ZA40 engines, where the underslung crankshaft arrangement is advantageous, they are removed downwards with equal facility. The bearing cap bolts are tightened hydraulically with a simple device to realise correct pre-loading.

Connecting rod

The connecting rods are drilled to enable lubricating oil to be supplied to the top end bearing and cooling oil to the piston crown. The bottom end of the shank is bolted on to the big end. This conveys the following major advantages:

Lower withdrawal height for the piston.
Piston can be withdrawn without dismantling the big end bearing.
Large pin diameters can be accommodated to obtain low specific bearing loads.

Z 40   ZA 40

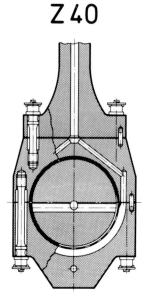
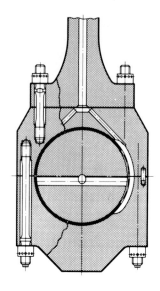

Figure 17.16 Connecting rod big end design

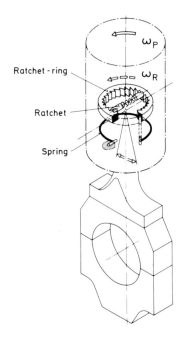
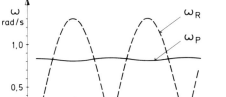

ω_P = Angular velocity of the piston
ω_R = Angular velocity of the ratchet-ring

Figure 17.17 Drive principle of rotating piston

For the ZA40 design the big end had to be reinforced because of the increased crankpin diameter. For increased safety and easier servicing, the connecting rod studs will in future be tightened hydraulically.

The differences in the two versions can be compared in Figure 17.16.

Piston

A noteworthy feature of the Z40 is the rotating piston, and this is retained for the ZA40. The symmetrical piston is cooled with lubricating oil and, besides the reciprocating motion, performs a slow rotating motion around its longitudinal axis. The principle of operation can be followed from Figure 17.17.

For the rotation of the piston, the connecting rod is provided with two pawls positioned slightly out of the centre of the spherical small end bearing. When the connecting rod performs its swinging movement relative to the piston, the pawls impart an intermittent rotating motion to a toothed rim, from which it is transmitted to the piston by means of an annular spring. The flexible connection maintains the forces necessary for rotating the piston at a constant low level. The advantages claimed for this feature are listed below:

At each stroke, a new oil-wetted portion of the piston is always in contact with the pressure side of the liner: this minimises the danger of piston seizures.

The piston rings rotate as well, and so local overheating of the liner resulting from blow-by at the ring gap can be avoided.

Due to the symmetrical design the thermal and mechanical deformation is also symmetrical.

Lubricating oil is distributed uniformly; a favourable factor in respect of the wear rate on cylinder liners and piston rings.

Smaller clearances are possible between piston and liner. This reduces piston slap and consequently lowers the danger of cavitation on the cylinder liner waterside.

The spherical top end bearing eliminates the edge pressure problems associated with gudgeon pin bearings.

The separate cylinder lubrication system, together with the oil scraper placed at the lower end of the piston skirt, permit exact control of the amount of oil reaching the running surface.

Low total oil consumption remains constant over long running periods.

A cross section of an earlier Z40 piston is given in Figure 17.18. In later engines, the diaphragm in the spacer between the piston carrier and the crown, and also the restrictor at the central exit from the piston carrier, had already been omitted.

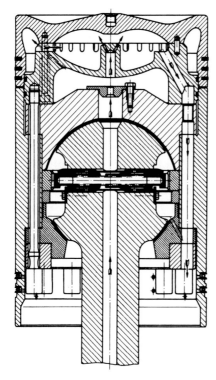

Figure 17.18 Z 40 rotating piston

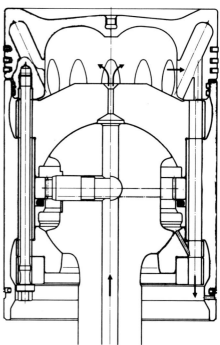

Figure 17.19 ZA 40 rotating piston, bore cooled

SULZER FOUR-STROKE ENGINES 287

The changes introduced for the ZA40 engine are intended primarily to reduce manufacturing costs and to simplify the assembly. The piston crown is bore cooled, which allows the elimination of the spacer ring between crown and the carrier. One, rather than two, oil control rings are used at the bottom of the skirt. Otherwise the principle remains exactly as the Z40. Figure 17.19 shows the ZA piston.

Turbocharging

The pulse turbocharging system is employed. The combustion air delivered by the charger is cooled before entering the combustion chamber.

For the in-line engines, only one turbocharger and one intercooler are required. They can be arranged at either end of the engine (see Figure 17.20).

For the ZV40 engines, two units were arranged at the free end (one on the 10ZV40).

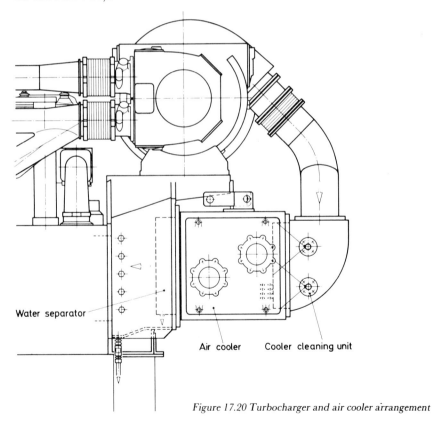

Figure 17.20 Turbocharger and air cooler arrangement

For the ZA engines pulse charging is retained, for its advantages at part loads and in rapid load acceptance. The pressure charger and cooler on the ZAV engine can now be fitted at the drive end if required. To facilitate this the governor has been moved to the side of the crankcase.

The exhaust pipe system has been redesigned completely. Based on service results from the Z40 engines, solid supports of the pipe assemblies have been chosen with additional expansion bellows units in the pipe between them. This system not only reduces manufacturing costs, it also increases reliability and simplifies maintenance considerably.

The insulation of individual pipes has been discarded, and a design of insulating panel introduced which considerably facilitates service removal (Figure 17.21).

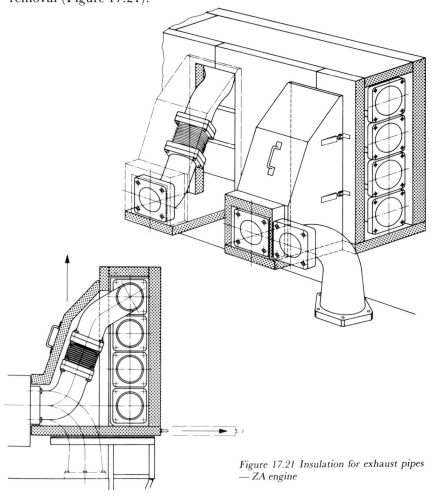

Figure 17.21 Insulation for exhaust pipes — ZA engine

Up to a pressure ratio of 1:3.5, roller bearings will be employed in the turbocharger. Above this, plain bearings connected to the engine lubricating oil system have to be employed. A pressure ratio of 1:3.5 corresponds to a cylinder output of approximately 660 kW at the time of writing, and at sea level.

The latest design of flange-mounted air intercooler for the Z40 engine, together with the facility for fitting a cooler cleaning installation, has also been adopted for the ZA40 engine.

Auxiliary pumps

A versatile range of drives is available for fitting at the free end of both the Z40 and the ZA40 engines. The possibilities for a Vee engine are shown in Figure 17.22. The cylinder lubrication system is shown diagrammatically in Figure 17.23. It is basically the same for all engines.

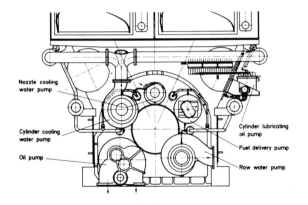

Figure 17.22 Engine pumps fitted on damper end — ZV 40/48 engine

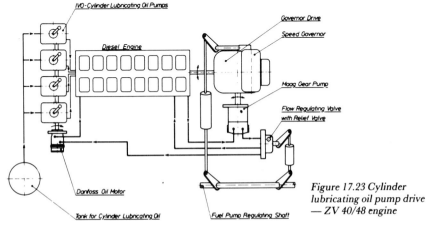

Figure 17.23 Cylinder lubricating oil pump drive — ZV 40/48 engine

Maintenance

Particular attention is given to easy accessibility and the possibility of fast and simple service work with the help of purpose-designed tools.

The weight of all parts has been kept as low as possible so that handling is made easier. Examples are given in Figures 17.24 and 17.25, which show respectively the simultaneous hydraulic tightening of cylinder head studs, and the use of a hydraulic attachment for tightening the main bearing studs.

Figure 17.24 Tightening cylinder heads

Figure 17.25 Tightening main bearing studs ZL engine

The same tools can be used for tightening the horizontal tie-rods for the main bearing caps, the bolts for the crankshaft counterweights, the foundation bolts and the big end bolts.

Acknowledgement

The editors would like to acknowledge the assistance of Sulzer Bros. Ltd in the preparation of this chapter.

D.A.W.

18 GMT engines

GMT, or, to give them their full name, Grandi Motori Trieste, is a company founded in the late 1960s to pool the diesel engine expertise of FIAT, Ansaldo and CRDA, and to rationalise the large Italian diesel engine industry. A large modern factory, incorporating all design and development facilities, was set up in Trieste and a comprehensive range of engines has been introduced covering from high-speed to large-bore engines. Some GMT engines are licence built, but the majority are produced at the Trieste factory.

The current medium-speed engines made by the group are the B420, developed from the A420, and the B550, which are now both established designs.

Originally each of these was offered in six, seven, eight, nine and ten cylinders in-line, (suffix L) as well as twelve, fourteen, sixteen, eighteen and twenty cylinders in Vee form (suffix V). In the current range, the in-line options have been restricted to six, eight and nine cylinders, also five cylinders for the 420. All the Vee options are maintained except for the B420 where the sixteen cylinder is currently the largest. The 420.10 engine is built in-line, but the B550.10 is a Vee engine.

Current rating details are given in Table 18.1.

Table 18.1 General characteristics

Engine type	A420	B420	B550
Cylinder bore (mm)	420	420	550
Piston stroke (mm)	500	500	590
Max speed (rev/min)	514	514	450
Cyl output (kW)	515	566	1030
Mep bar	17.8	19.6	20.5

All engines of both types are direct reversing if required and operate on the four-stroke principle. The Vee angle is 40°. All engines are turbocharged and intercooled. Since both sizes of engine belong to a family, they share many common characteristics. Both will therefore be described together, noting differences between them as appropriate.

Figure 18.1 shows the cross section of an A420L and Figure 18.2 that

GMT ENGINES

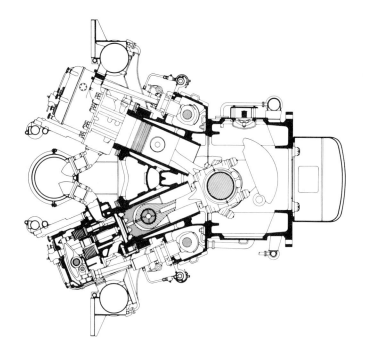

Figure 18.2 Cross section of GMT B420 V engine

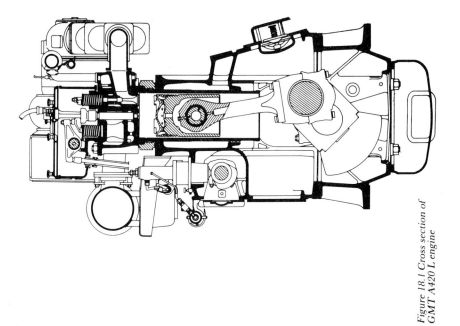

Figure 18.1 Cross section of GMT A420 L engine

of a B420V. The differences to be noted are the much more massive big end in the latter, optional constant pressure exhaust system, and changes in the piston, small end, fail system valve gear and air inlet passages. More use is made of hydraulic tensioning.

The Vee version of the B550 engine is shown in cross section in Figure 18.3.

Although the design of each type is intended to satisfy a variety of applications, particular note has been taken of the requirements of low headroom, simplicity and reliability, not to mention high specific output. The components of the engines are described briefly below.

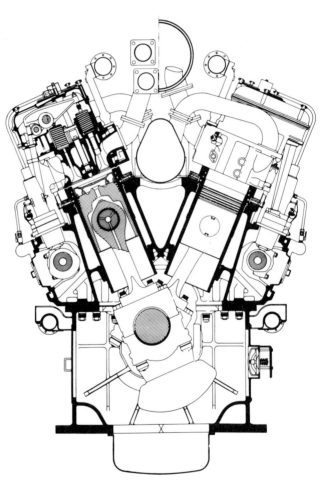

Figure 18.3 Cross section of GMT B550 V engine

Bedplate and frame

The design of the main components of the GMT 4-stroke medium-speed engine takes into account the following fundamental requirements:

To contain the engine internal forces (inertia forces and gas pressures) with a consistent safety margin.
To have maximum rigidity in order to reduce the structure vibrations and the vibration of the elements installed on the engine.
Reduction of weights and overall dimensions within reasonable limits and compatible with the engine arrangement requirements.
Proper choice of the material and methods of construction (iron castings or fabricated steel structures) as a function not only of the above requirements, but of production and transport costs.
Good accessibility to the various parts for easy and correct maintenance.

For such reasons there are some structural differences between 420 and 550 engines and, within each series, between the in-line version engines and the Vee version ones.

A characteristic common to all GMT 4-stroke medium-speed engines is the use of a bearing structure divided horizontally into two parts. The lower part forms the bedplate which supports the crankshaft, an arrangement which assures structural rigidity and greater accessibility for maintenance of main bearings.

The bedplate is made of high resistance cast iron, grey or nodular according to the dimensions, for all the in-line version engines and the 420 Vee engines.

The construction adopted for the B550 Vee engine is a welded assembly of cast steel elements and plate.

The cylinder block assembly for all engines is of high strength cast iron. It houses the liners and carries the camshaft bearings in separate casings. The arrangement of the camshafts offers two advantages:

To isolate each camshaft housing communicating with the cylinder heads through the pushrod sheaths from the crank chamber, in order to prevent possible air or gas escapes from raising the pressure in the main crankcase.
To provide wide doors for the normal inspection and maintenance of camshafts and relative bearings.

When the bedplate is made of cast iron, the cylinder block is rigidly connected to it by tie-rods which pre-load the whole structure in compression. On the other hand, when the bedplate is made of fabricated steel this is not necessary and the two parts are joined by a series of studs.

According to type and dimensions of the engines (diameter, number and arrangement of cylinders) both cast bedplate and frame are divided into sections so as to limit the dimensions of the single parts both from the point of view of casting and of transport or handling.

The position of the join between bedplate and frame is determined mainly as a function of the accessibility to the internal parts and of the construction adopted. Figure 18.4 shows a B550 8L main structure being assembled, and Figure 18.5 that of a B550 16V engine.

Figure 18.4 Assembling the main structure of a B550 8L engine

Figure 18.5 Assembling the main structure of a B550 16V engine

Crankshaft and bearings

The crankshaft is made of high strength alloy steel, forged in one piece. It has non-hardened pins and is provided with counterweights to avoid excessive loads on the main bearings and on the engine main structure.

Main bearings and connecting rod big end bearings are of thin shell, tri-metal type, namely with an anti-friction metal coating on the steel shell further coated with an electrolytic layer of soft material in order to improve running-in of the bearing itself.

The bearings have no central groove in the supporting area, and this increases bearing load capacity at parity of dimensions.

Connecting rod

The three parts of the forged steel connecting rod, that is, the shank and the horizontally split big end, are designed to allow the pistons to be withdrawn without disturbing the big end, and importantly, to reduce the headroom needed for piston withdrawal. Serrated joint faces are not necessary.

Piston

The composite pistons have an aluminium skirt and a steel crown, and embody 'cocktail-shaker' oil cooling. Earlier engines had coil cooled aluminium pistons with Alfin top ring groove inserts.

The piston has four piston rings and one or two scraper rings arranged above the piston pin to control lubrication and oil consumption.

The piston ring in the first groove is provided with a high wear resistant special coating. A chromium plated piston ring is used in the second groove. Grey cast iron piston rings without coatings are sited in the third and fourth groove and in some engines they are provided with special inserts and with electrolytic layers of soft materials to favour their bedding during running-in. Spring-loaded scraper rings are provided, with chromium plated edges.

Figure 18.6 shows the running gear of a B550 engine. As can be seen in Figures 18.1 to 18.3 the shape of the combustion chamber has varied with different stages in the engines' evolution.

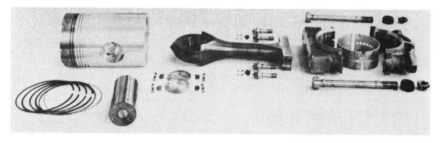

Figure 18.6 Running gear of a B550 engine

Cylinder liners

The cylinder liners are cast in special cast iron highly resistant to wear and mechanical stresses.

The construction of all marine 420 and 550 engines entails a strong backing ring round the upper part of the liner, which also conveys the cooling water up to the liner collar (Figures 18.1 to 18.3). Moreover, the upper inside of the liner is also screened by an annular thermal shield made of high-temperature resisting steel.

With this constructional characteristic the following advantages are obtained:

The presence of the strong backing cylinder allows the thermal deformations of the upper part of the liner to be limited as well as the stresses due to the cylinder inside pressure. Moreover, as such fatigue stresses are superimposed on a pre-compression tensile condition due to strong backing, the behaviour of the cast iron liner is favoured.

The good circulation of the cooling water lowers temperatures in the liner inside wall, thus improving lubricating conditions and, quite important, a distribution of temperatures along the liner itself at a very regular pattern. The presence of the thermal shield helps to obtain these results, particularly to the liner upper part.

The lubrication of the liner inside wall, besides the oil spraying effect coming out of the crank gear, is achieved either by suitably positioned oil jets flowing over the walls themselves or by cylinder lubricators suitably positioned on the liner and fed by volumetric metering pumps mechanically driven by the engine itself.

Cylinder heads and valves

The cylinder heads are made of nodular cast iron. They incorporate two intake and two exhaust valves, the fuel injector and starting and cylinder relief valves.

Figure 18.7 is a view of the B550 cylinder head, and Figure 18.8 a section. The B420 cylinder head is similar. Both include the intermediate deck form of construction. The intake port of the A420 was differently aligned, as can be seen in Figure 18.1. The air intake and exhaust ducts are shaped to minimise pressure loss and promote good air flow.

The exhaust valves are housed in water cooled valve cages to allow the valves to be disassembled without removing the cylinder head and to assure a satisfactory life of the seats.

The water cooling of the valve cage is supplied from an independent circuit, as shown in Figure 18.9 and the supply to each cage can be cut off by a cock mounted on each cylinder head. The cock also acts as an extractor for the whole cock which must be removed before the valve cage is extracted from the cylinder head.

In order to keep constant temperature around the exhaust valve seat, the valves are continuously rotated by a special device, as shown in

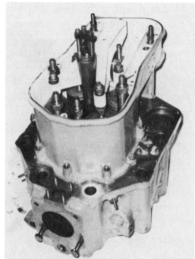

Figure 18.7 General view of a B550 cylinder head

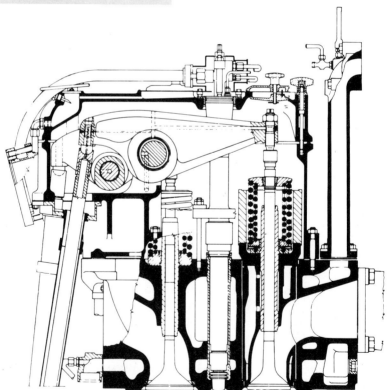

Figure 18.8 Sectional view of B550 cylinder head

Figure 18.9. The exhaust valve seat is Stellite coated and the stem is coated with a special layer of corrosion resisting material.

In case of particularly heavy engine duty, mostly in connection with the type of fuel being used, cooled valves have been developed. These function by letting oil or water circulate inside the valves themselves, which, in such cases, are drilled all along the stem and have a hollow inside the head; the shape of such a hollow has been designed and set up as a best compromise between cooling and mechanical resistance of the valve itself.

The use of water as coolant for this purpose involves some constructional complications mostly directed to avoid any possibility of lubricating oil pollution in the event of water leakages from the valve inlet and outlet circuitry. It is much easier instead to use oil circulation since it is the same oil used for general engine lubrication, coming through holes drilled in the rocker arm.

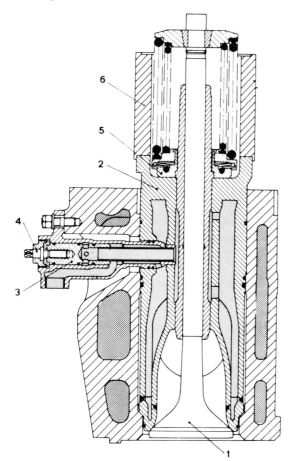

Figure 18.9 Exhaust valve cage showing water cut-off valve
1 Exhaust valve
2 Valve cage
3 Cut-off cock
4 Shutter — extractor
5 Valve rotator
6 Spacer

Timing system

Control of intake and exhaust valves as well as of the injection pumps is accomplished by means of the camshaft, roller cam followers, pushrods and rocker arms.

The camshaft is formed by as many sections as there are cylinders. Every section carries intake, exhaust and injection cams. The whole camshaft is withdrawable sideways after releasing the camshaft bearing keeps which are split parallel to the cylinder axis. The Vee version engines have two camshafts.

In reversible engines the camshaft is equipped with a double series of cams, one for ahead and the other for astern running. In such cases it can be displaced axially in order to bring the ahead or the astern cams under the respective followers. The displacement is controlled by a pneumatic servo-motor and a series of pneumatic devices that, operating in sequence, carry out the reversing manoeuvre in a completely automatic and safe way (Figure 18.10).

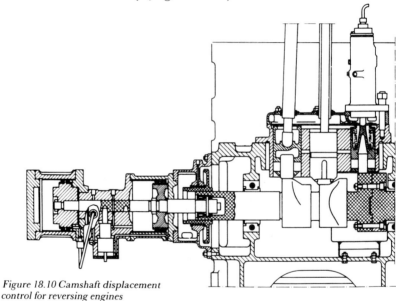

Figure 18.10 Camshaft displacement control for reversing engines

Fuel injection system

The injection system is made up by Bosch-type single pumps with rack control. Each pump is provided with spragging gear so that each cylinder can be isolated without stopping the engine (Figure 18.11, item 26). The fuel valves are provided with water cooling ducts in order to prevent carbon deposits when operating with heavy fuel oil. Figure 18.12 shows (typically) the B550 injector. When the engine is stopped

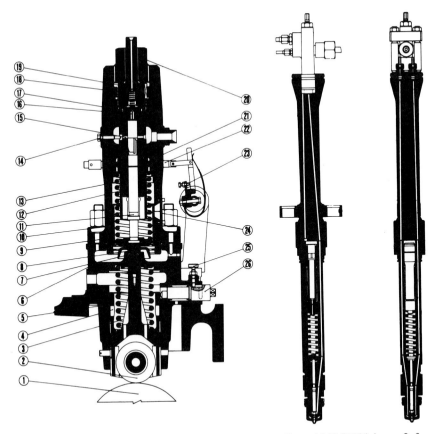

Figure 18.11 B550 fuel pump, mounting and drive

Figure 18.12 B550 injector. Left hand section through fuel gallery, right hand section through cooling circuit

heated fuel can be circulated up to the injector so that it can restart on heavy fuel.

Operation, governing, safety devices

Starting, stopping and, if required, reversing the engine is by a pneumatic system controlled from the engine, or remotely. This system includes some automatic sequences and safeguards applying to the ahead or astern positioning of the camshaft. These block starting and cut-off fuel supply to the injectors if the camshaft itself is not correctly positioned.

Most devices forming the starting and manoeuvring system are assembled on an easily accessible panel. All controls are pneumatic except the fuel control which is mechanical.

The choice and the type of application and of the characteristics of the governor and of its equipment depend on the type of duty the engine must perform and on the particular requirements of the propulsion plant or of the generation system (governor speed droop, load sharing, synchronisation, remote control, automation). The mechanical overspeed governor is separately driven from the main timing gear train.

The overspeed governors operate in different ways, according to whether propulsion or power generation is involved. In the first case the operation is to limit speed with automatic resetting, that is, it falls out as soon as the engine rotational speed returns to the normal value. When there are clutches and driven shaft alternators a suitable sequence of operations is involved to grant the continuity of electrical supply for the ship services.

In the second case, operation always stops the engine, setting the fuel pump racks to zero. The device must be manually reset before restarting the engine itself.

A further safety device is provided in case rack trouble or casual jamming of some fuel control lever leads to overfuelling (particularly dangerous for the power generating unit when the electric machine is de-energised). This device is adjusted to operate at a higher speed value than the previous one and acts on the supercharging air flow cutting off completely the delivery of the turbochargers. This too is a shut-down type device and must be manually reset before starting the engine.

The standard equipment of engine safety devices is completed by a general lubrication oil low-pressure shut-down device that, through a pneumatic servomotor, acts on the fuel cut-off till the engine stops.

Remote control and automation systems

The engines comply with the requirements of the Classification Societies for duty in a 24-hour unmanned engine room. In particular they are ready for application of automatic and remote controls and regulations, and in such cases they do not require inspection or normal manual attention at intervals shorter than 24 hours.

A series of connections has been pre-arranged on the engine for the installation of the detectors necessary to the automatic control system (See Figure 25.1). The engine can thus be easily connected to any automation or remote control system, either pneumatic or electric, for any type of propulsion plant. In particular, such arrangements allow the following functions to be achieved:

1. Combined control of engine revolutions and propeller pitch with automatic load adjustment, or propeller pitch reduction automatically in case of engine overload.

2. Connection between controls and the overload capacity of the engine and/or clutch.
3. Propellor synchronisation in order to minimise vibrations transmitted to the hull.
4. Automatic load sharing among engines mechanically coupled by reduction gear.

Equipment

A principle aim of engine design has been to allow long maintenance intervals, to facilitate maintenance itself, and to cut the time required to carry out the necessary operations. To this end every engine is provided with suitable equipment including, besides the conventional tools and the standard measuring and checking instruments, special tools designed for the correct execution of the most important operations and for the reduction of maintenance times; in particular, hydraulic tools for the controlled tightening of bolts and nuts of the most important parts of the engines (Figure 18.13). Such equipment generally consists of:

1. A special portable high pressure oil pump.
2. A precision pressure gauge to check the oil pressure in the different tightening stages.
3. One or more hydraulic jacks to stretch simultaneously the connecting parts.
4. A spacer to position the jack.

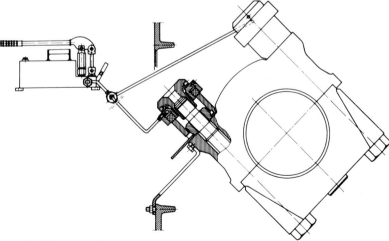

Figure 18.13 Equipment for tightening big end bolts

5. A spanner to tighten or loosen the nut manually.
6. Oil pipe between pump and jack.

In addition, the following tools are supplied:

(a) special tools to expedite such traditional manual operations as, for example, valve seat grinding;
(b) devices to withdraw, assemble, lift and handle various parts;
(c) micrometers, gauges, tachometers and pressure gauges for measurements and checks;
(d) various spanners both for general and specific duty.

Auxiliary equipment

Independent pumps, heat exchangers, filters, cleaners and heaters must be provided for engine services as can be seen in the schemes relating to the various circuits.

Seawater circuit

The seawater circuit is obviously a full-flow open type and is used for cooling all other service fluids (water, oil, supercharging air).

When operating in cold seas it is advisable to recirculate to the pump inlet a part of the outlet water in order to keep water temperature at the prescribed level.

Freshwater circuits

To cool the engine parts exposed to the heat of combustion or exhaust gases fresh water is used, treated with anti-fouling and anti-corrosive additives, and which circulates in closed circuit with a header tank. A distillation plant can be included in the circuit if desired for producing fresh water.

Suitable vents are needed at points in the system where air bubbles might otherwise accumulate.

Fuel valve cooling is carried out by treated fresh water circulation in an independent closed circuit in order to limit water pollution from possible fuel leaks through the fuel valves themselves.

The water outlets from the individual fuel valves flows through separate ducts in order that they may be checked.

When burning heavy fuel exclusively from pier to pier, it is necessary to make provision for heating the water so that the fuel entrained in the high-pressure fuel system may be maintained at a safe viscosity when the engine is standing by.

General lubrication and piston cooling oil circuit

All engine parts, with the exception of the turbochargers, are lubricated by the same oil. Pressure oil is fed from the bus rail to the main bearings, then via the drillings in the crankshaft to the connecting rod, from where it is conveyed to the gudgeon pin for piston cooling and lubrication (Figure 18.14).

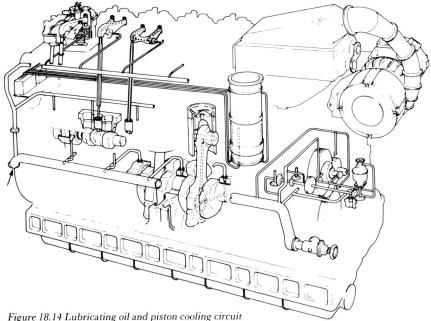

Figure 18.14 Lubricating oil and piston cooling circuit

Separate feeds from the manifold convey lubricating oil to the camshaft, cam followers, timing gears, governor drive, rocker gear, etc. (The cylinder lubrication is arranged separately.)

Oil is drawn from the sump or sump tank by the separately driven pressure pumps and delivered to the cooler via an automatic full-flow self-cleaning filter, filtering down to 20 μm. At the same time, lubricating oil is circulated through centrifugal purifiers at a rate equivalent to about 2.5% of the main engine circuit pump delivery. Given a normally proportioned system, this rate is equivalent to treating the whole of the lubricating oil charge about every six hours of operation.

Fuel circuit

A typical fuel circuit for the GMT range of four-stroke medium-speed engines is shown in Figure 18.15. These engines can burn fuel oil with

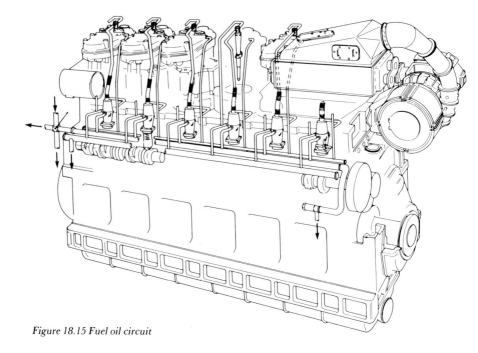

Figure 18.15 Fuel oil circuit

viscosities of the order of 3500 seconds Redwood No. 1 at 100 °F. The fuel must be purified, filtered and heated in order to achieve the correct viscosity at the engine busrail.

The engine will normally be started and stopped on gas oil or diesel oil and a mixing tank is incorporated to allow a gradual transition between the two types of fuel.

However, if the engine is stopped for only a short time, and ambient conditions are not excessively rigorous, the engine may be started and manoeuvred directly with heavy fuel. In such cases, of course, it is necessary when the engine is stopped to maintain heating to the fuel pipes to ensure that the fuel is kept at the correct viscosity.

Starting air circuits

The engines are started by compressed air fed to the cylinders at 30 kg/cm^2. In the Vee engines normally only the cylinders of one bank are used for this purpose.

The starting air distributor is mechanically driven from the camshaft (only one pipe is shown in Figure 18.16 for reasons of clarity) and controls the admission, at the cylinder air start valve, of the starting air itself.

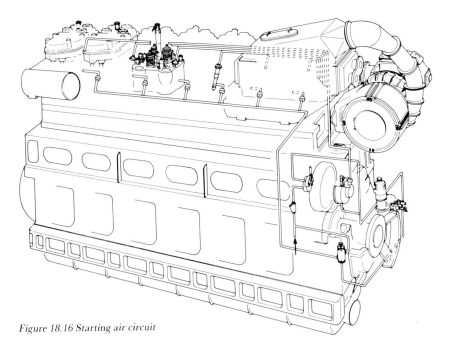

Figure 18.16 Starting air circuit

Acknowledgement

The editors would like to acknowledge the assistance of GMT in the preparation of this chapter.

D.A.W.

19 MaK engines

MaK Maschinenbau, a part of the Krupp group, offer basically five types of four-stroke engine. Of these the three larger sizes fall within the arbitrarily defined medium-speed range. There are, however, subdivisions among them. A mention should be made of the newest MaK engine, the M35, which although running at a higher speed, is of a power within the range of the older engines. It is briefly described at the end of the chapter. All MaK engines are available in directly reversing form.

The nomenclature of MaK has traditionally been based on the stroke, the first two figures normally indicating the stroke in centimetres, while the third indicates the variant of the basic type. There are, however, variants whereby in engines of the same bore, the stroke is shortened but the engine nevertheless retains the same general classification, as can be seen in Table 19.1. The new M35, however, breaks with this tradition, as the stroke is 38 cm and the bore 35 cm.

Table 19.1 Technical data

Engine type	452	453	551	552	601	35
Cylinder bore mm	320	320	450	450	580	350
Piston stroke mm	450	420	550	520	600	380
Speed rev/min	500	650	450	500	425	650/750
Cyl/output kW	220	330	533	610	980	400
Mep bar	14.6	18.1	16.2	17.7	17.4	17.5
No. of cyls	6,8	6,8,9,12,16	6,8,12,16	6,8,9,12,16	6,8,9	6,8
Year of introduction	1965	1971	1967	1971	1971	1981

DESIGN

All MaK engines have a cast iron structure which confers the advantage of good damping properties. The three larger engine families described in this chapter follow traditional design practice with the crankshaft carried directly in the bedplate. Vee angles are 45° in all cases.

The way in which the larger engine types are divided horizontally into sections is a function of the size of the parts to be machined, and it also offers advantages for handling in yards where crane capacity is limited. Vertical sectioning is not necessary even for the longest engines. The evolution of each of these engine ranges during their life to date has featured increases both in the speed of rotation and the degree of turbocharging.

During recent years it has been possible to obtain a considerable increase in the mean piston speed for satisfactory operation so that 9–10 m/sec can be reached today.

Exhaust gas turbocharging today makes brake mean effective pressures of 18–19 bar possible. Further increases, would of course, be possible even with single stage pressure charging. For generator engines the degree of supercharging is necessarily limited by the required capacity to accept the sudden imposition of substantial load.

The pressures towards further uprating will have to be tempered with the requirements of reliability, simple maintenance and low operating costs.

ENGINES WITH A BORE OF 320 MM

This family of engines was the first to be developed of the three medium-speed engines which now make up the MaK range and first appeared as the M451 in 1961.

The M453 is the third generation and, as Table 19.1 shows, the stroke is reduced to 420 mm to permit the higher crankshaft speed.

Figure 19.1 indicates in cross section the layout of the M453 in-line engine, and Figure 19.2 that of the Vee version.

Although MaK were familiar with the arrangement of the engine block with underslung crankshaft and had experience with such design, this engine was built in the classical form with two parts, the bedplate and an engine block, the main bearings being located in the bedplate.

The Vee engine block is also solid and bolted together with the bedplate by tie-rods reaching up to mid-height. The top part of the blocks is not in the tie-rod range and therefore subjected to tensile stresses which are taken by ribs.

Fitting and seating the cylinder liner in the block is very important. Sufficient cooling of the liner top is ensured for the higher outputs as well. Temperatures are low enough so that only small thermal loads are imposed, and overall loading, including combustion, is within a range that is still permissible for centrifugal castings.

The wall thickness of the liner below the top guide was increased from 22 to 26 mm when switching over to the M453 variant. This was not for reasons of strength or deformation, but to make possible the

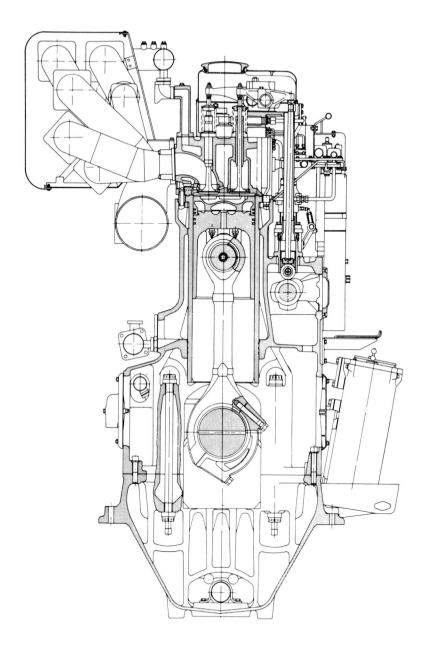

Figure 19.1 Cross section of MaK M453 in-line engine

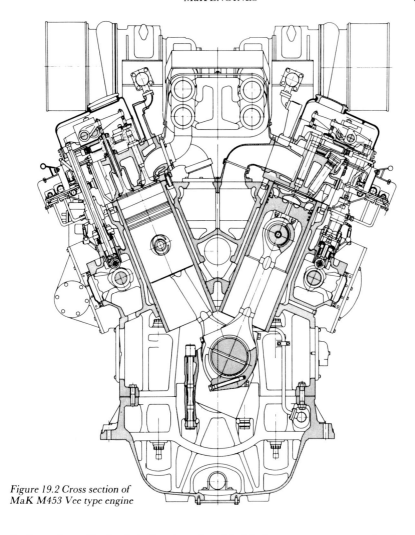

Figure 19.2 Cross section of MaK M453 Vee type engine

fitting of a cylinder lubrication should this become necessary for heavy fuel operation. This requirement has not yet become acute; however, it cannot be excluded in the light of a further fuel deterioration which must be expected.

The liner surface is bath nitrided as for all MaK engines. This provides several advantages:

1. Improved anti-friction properties compared with the bare precision-bored and honed cast iron surface.
2. Lower wear rate also for piston range.
3. Improved resistance against corrosion on the water side.

A new connecting rod had to be taken because of the larger crankpin diameter. Preference was given to the oblique split between the rod and the bearing cap. Due to higher lateral forces resulting from the oblique cut, this was serrated and highly pre-loaded.

The in-line engines were constructed with a further main bearing at the flywheel and inside the crankcase so that overhung arrangement of flywheel plus clutch was possible. This was not necessary for the Vee engines, whose crankshaft is so strong that the lighter flywheels can be overhung without causing excessive crankweb deflections.

The use of the aluminium piston has been limited by the increasing heavy fuel oil operation. This is mainly due to heavy wear in the area above the top piston ring caused by the abrasive heavy fuel residues.

Built-up pistons are more wear resistant, produce smaller clearance variations between warm and cold and therefore less fouling, have smaller wear rates on the hardened groove flanks and have lower piston ring temperatures due to better interior cooling.

For ratings above 275 kW/cyl at 600 rev/min and for heavy fuel above 110 cST/50 °C a composite piston with steel crown is used. Pistons are equipped with three chromium plated compression rings of crowned design and one oil scraper ring.

A four valve vermicular cast iron cylinder head is used. The size of the cylinder head allows the valves to be fitted from the underside without the cages usual for the larger engines. There is no great difference between the time required for changing the head and that for the cages only. From the thermal point of view it is better to have the valve seat insert directly in the flameplate. The form strength is also better than with cages. The valve seats are inserts shrunk into the head and are made of hard and wear-resistant cast iron.

In heavy fuel operation it is very important to keep the temperature of the valve seat insert as low as possible. This is achieved by an annular cooling channel (Figure 19.3) in the bore taking the seat insert in the bottom on the cylinder head. A proper sealing of this channel is ensured by shrink fitting an oversize ring insert into the cylinder head.

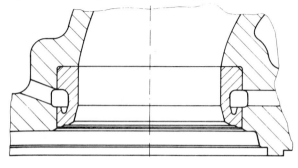

Figure 19.3 M453 cooled exhaust valve seating

With good dissipation of heat from the valve head the valve surface temperature is kept well below the melting point of the sodium and vanadium oxide compounds that are problematic in heavy fuel oil operation. This helps avoid the build-up of deposits on the valve seat which can cause valve burn-out.

Valve rotating devices are naturally installed to keep seat wear to a minimum and maintain the valve's sealing properties. The valve head has been strengthened in order to meet the increased mechanical load on the exhaust valves. It should be mentioned here that it is possible in most cases to convert old type cylinder covers to valve seat cooling. A frequently encountered experience on valve controlled heavy fuel oil engines is stem corrosion on the exhaust valves — this is caused by the accumulation of combustion products containing sulphuric acid. This problem has been solved on all MaK engines by a special seal between the valve stem and valve guide to prevent the ingress of gases, which also provides an exactly defined quantity of lubricating oil for stem lubrication. This prevents deficient lubrication or excessive lubrication accompanied by a build-up of carbon: any acidic combustion products which do get through are neutralised by the alkaline lubricating oil.

The above measures enabled the standard valve stems, coated with Stellite 12, to be retained for heavy fuel oil operation. The available options are Nimonic valves whose use could be justified for very intermittent engine operation (e.g. main propulsion of shorthaul ferry boats) but it must be considered that the price of Nimonic valves is about three times higher than that of the standard ones. Figure 19.4 shows the valve arrangements in more detail.

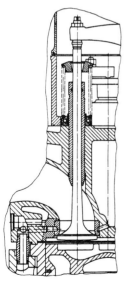

Figure 19.4 M453 exhaust valve arrangement

As far as the rest of the engine details are concerned Figures 19.1 and 19.2 indicate the general layout. Note particularly the generous ribbing surrounding the resilient tie-bolts securing the bedplate to the crankcase, the extra liner register, just below the combustion zone, the push-on connections between the cylinder head air inlet bends and the air chest outlet.

ENGINE WITH A BORE OF 450 MM

This family has been manufactured since 1967. The 551 version is illustrated by the cross sections in Figure 19.5 for the in-line and Figure 19.6 for the Vee version.

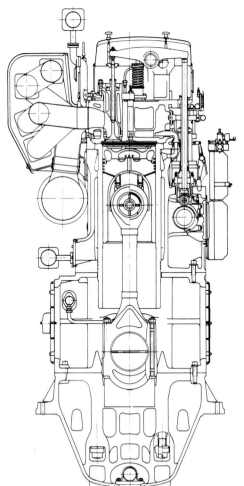

Figure 19.5 Cross section of MaK M551 in-line engine

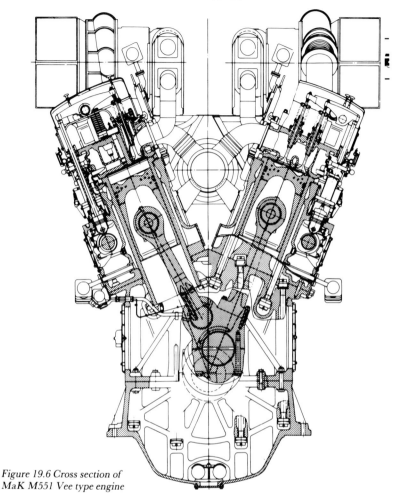

Figure 19.6 Cross section of MaK M551 Vee type engine

For reasons of manufacture, the in-line engine is made in three main parts: bedplate, frame, and cylinder block. The Vee engines have two cylinder blocks on the frame. These main parts are of grey cast iron and secured together by tie-rods. The proven design of arranging the main bearings in the bedplate has also been applied here.

The monoblock bedplate is ribbed according to the direction of the force lines to achieve extreme rigidity of the bearing supports. The frame walls are also very solid for rigidity and noise damping. There is a large clearance above the main bearings for removing the caps. The crankcase doors give good accessibility to the running gear for maintenance and inspection. The tie-rods have been arranged so that the entire engine structure is pre-loaded and no tensile stresses will occur at any point due to combustion loads.

The aluminium piston with a double ring carrier of special cast iron is only used in diesel oil operation with M551 (533 kW/cyl). Built-up pistons are generally used for heavy fuel (See Figure 19.9). The cylinder liners are of the same proven type as for the M453 engines. The bath nitrided working surface, however, is fed with fresh oil from four holes. The oil feed can be adjusted and depends on the engine speed.

This type of cylinder lubrication has proved good, particularly in part load operation with heavy fuel (pier to pier) because of the low wear rates on liners and piston rings.

The cylinder head has two inlet and two exhaust valves, the latter having cooled cages and, in heavy fuel operation, additional cooled seat rings. The exhaust outlet and the air inlet are on the same side, which provides the same good accessibility of the camshaft side as for the M453 engines. The M552 engines are the second generation of the M551 type, from which they were derived in 1972.

Figure 19.7 shows the later in-line engine and Figure 19.8 the Vee engine.

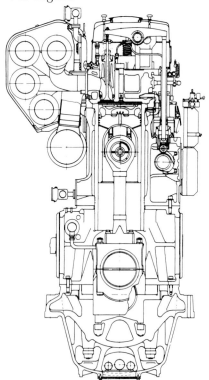

Figure 19.7 Cross section of MaK M552 in-line engine

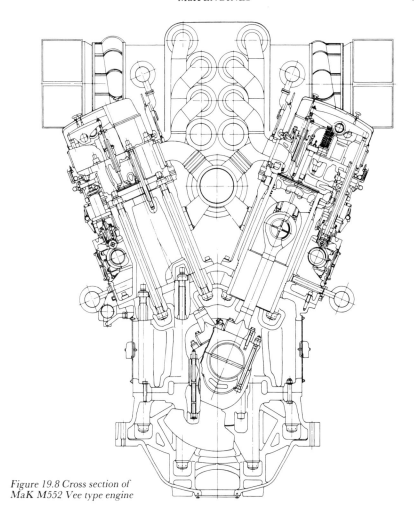

Figure 19.8 Cross section of MaK M552 Vee type engine

The M552 has an output per cylinder of 610 kW as against the 533 kW of the M551.

The M552 engine differs from the M551 only in its strengthened running gear and bedplate, and a piston stroke reduced from 550 to 520 mm. Also, crankshaft and crankpin diameters have been increased and crankwebs strengthened to conform to Classification Society rules and to control torsional vibrations. Owing to the larger crankpin diameter, it was necessary to fit a new connecting rod, which is no longer single-piece but two-piece (Figure 19.10a) divided into a rod section and a bearing section so that the piston and rod may be withdrawn vertically through the liner. A single-piece connecting rod with an inclined bear-

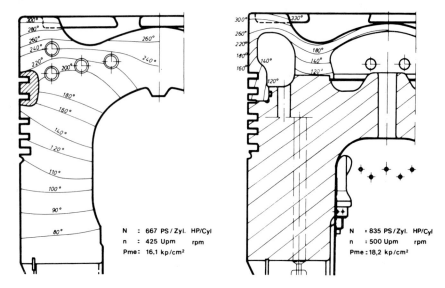

Figure 19.9 Isotherms of M551 (left) and M552 (right) pistons

ing joint was possible, but this design would have meant allowing for greater piston removal clearance above the engine, which is difficult to achieve with the low engine rooms favoured today.

The chosen design, with its horizontal bearing joint and the bolts having threads in the crank bearing section, has been extensively tested and is no longer novel. The transverse measurements of the bedplate and engine housing have not been altered for reasons of production and installation. That is why the piston stroke has been reduced by 30 mm from 550 to 520 mm — in order to maintain the hitherto existing side clearances despite the larger connecting rod bearing.

The loss in stroke volume through shortening the stroke is balanced by the increased speed. Moreover, the strengthened crankshaft with its larger bearing covers led to the previously vertical tie-rods being inclined slightly outwards, but only so much that the engine casing parts, such as cylinder block and crankcase housing, still fit the vertical tie-rod method, because a certain transitional period must be allowed in which the old and the new design are manufactured side by side. Because of the larger big end bearings the articulated connecting rods of the M551 were abandoned and side-by-side rods adopted for the Vee engines. A second crankshaft bearing is placed in the baseplate near to the flywheel in order to dispense with a separate outside pedestal bearing between engine and gearing when there are flexible couplings and clutches, but without allowing crankweb deflections in way of the first cylinder to become excessive.

The next most important part is the engine piston. Because of the higher mean effective pressures and anticipated heavy oil burning, only built-up pistons are used for this engine type. These pistons consist of a forged crown of high-temperature steel and die-cast skirt of special aluminium alloy.

The cooling quantity is such that the ring space is not filled completely, so that the cooling is accomplished with a 'splash' or — more modernly expressed — a 'shaker' effect.

The two upper piston rings are chromium plated on the running surface, thus reducing the wear rate. The cylinder cover as used for the M551 was adopted for the M552 without any alteration. The design strength and the materials have already been improved with the continual development of the M551 to such an extent that the higher outputs are coped with satisfactorily.

Figure 19.10b shows the exhaust valve cage arrangement with seals, rotator and cooling details.

The mean effective pressure now lies at 17.8 bar, which is still at the lower end of the values employed with today's modern engines.

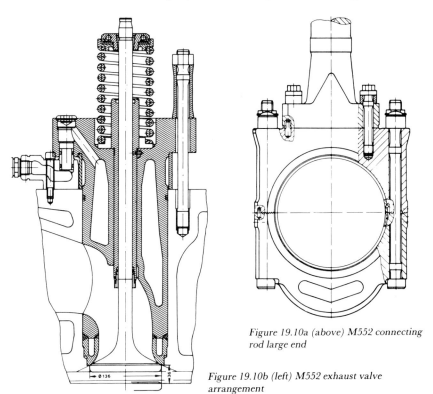

Figure 19.10a (above) M552 connecting rod large end

Figure 19.10b (left) M552 exhaust valve arrangement

ENGINES WITH A BORE OF 580 MM

The M601 engine was designed in 1970. Figure 19.11 shows the M601 engine in cross section. It is to date only made in in-line form. The structure is similar to the well-known types of M551/552. The bedplate, frame and cylinder block are secured together by tie-rods.

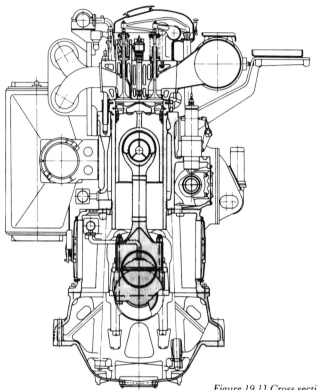

Figure 19.11 Cross section of MaK M601 engine

Particular attention was given to the seating and fixing of the main bearing caps, which are clamped in lateral stops of the bedplate to take the horizontal forces from the bearing. Since a positive fit without clearance needs to be achieved, a round key is provided on both sides of the seating to press the bearing cap outwards against the stops. These keys are also useful aids in assembly as they help correct alignment of the bearing cap after dismantling. The crankshaft bearings are of four-metal type: steel shell, nickel dam, lead–bronze layer and a lead–tin plated overlay. There is an additional end bearing to take the weight of the flywheel and coupling.

The crankshaft, a monobloc forging with large overlap of journals and crankpins, has low transverse stresses. The full power can be taken off the free end of the crankshaft for special drive combinations with dredger pumps or generators utilising the economical, heavy fuel operation of the engine.

The dimensions and wall thickness of the cylinder liners still permit the economical centrifugal casting. Much care has been given to the shape of the collar at the top. It is cooled by a certain number of oblique holes going upwards and connected with radial holes. The outcoming water is collected in a jacket ring, which passes around the cylinder head studs and is bolted on to the cylinder block. Sealing is effected by means of Viton rings.

The inner cooling and sealing arrangement is shown in Figure 19.12. This direct internal collar cooling reduces the temperature at top dead centre of the top piston ring down to 150 °C.

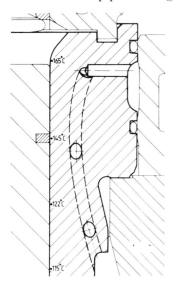

Figure 19.12 Detail showing top of M601 cylinder liner with cooling arrangements and local operating temperatures

A further advantage of the thick-walled cooled collar lies in the low radial thermal expansion and low radial expansion by the gas forces, so that a small clearance between the land above the top piston ring and the cylinder liner can be maintained in order to give better protection to the upper piston rings and the space above the top piston ring.

As in all other MaK engines the cylinder liner's running surface is bath nitrided to provide protection against corrosion, wear, blow-by and formation of burnt spots on the rings. The cylinder liner is lubricated with fresh oil through four outlets like the engines with a bore of 450 mm.

The combustion chamber is made as deep as the crown and edge of the piston permits so that the jets of fuel are sprayed deeply and do not pass beyond the edge of the piston, or onto the wall of the cylinder liner on termination of injection. This feature of MaK engines reduces wear of the piston rings and upper cylinder liner, as such wear would increase considerably if the fuel reached the cylinder liner and destroyed the lubricant layer on it.

It is most likely that the majority of the engines will be running on heavy fuel. MaK have therefore adopted built-up pistons with aluminium skirts and steel crowns. The four compression rings are located in the steel crown. The groove flanks are hardened. The working faces of the rings are oblique by some degrees and have an exact clearance on the rear side.

The working face of the rings is chromium plated. Due to small stroke bore ratio of the engine, almost 'square', the ratio of crankpin diameter/cylinder bore is small, so that the big end can be withdrawn through the liner despite a horizontal split. This results in a simple and light connecting rod without a further split between the big end and the rod body as for the 'marine' big end. The mating faces of the big end bearing are serrated, providing a positive joint.

The cast iron of the cylinder head is of high quality having a ferritic structure with vermicular graphite. This results in good thermal conductivity and high breaking elongation.

In the cylinder head there is an intermediate deck to absorb the gas forces so that the lower deck can be made thinner to limit thermal stresses.

Two inlet and two exhaust valves have been provided for the cylinder head. Only the exhaust valves are caged. Figure 19.13 shows the exhaust valve cage in more detail, and also the hydraulic tightening tool in position. This tool can also be used for withdrawing the cage.

The inlet valves are fitted directly into the cylinder head. Experience has shown that the overhaul intervals can normally coincide with the piston and piston ring inspections. Flow losses are lower without cages and the air flow is noticeably more favourable. In addition, the seat inserts have a smaller deformation margin so that leak problems are considerably smaller than with cages.

The valve cages are of cast steel and secured by three bolts instead of the usual two in order to obtain a more uniform deformation of the cage and load on the joint ring in the head. The joint is relatively high above the valve seat, as compared with the usual arrangement, to keep any deformations in this area away from the valve seat. Below the joint the cage has a slight taper to facilitate removal.

For heavy fuel operation not only the exhaust valves but the inlet valves have rotators in order to increase the life of the valves and seat inserts.

MaK ENGINES

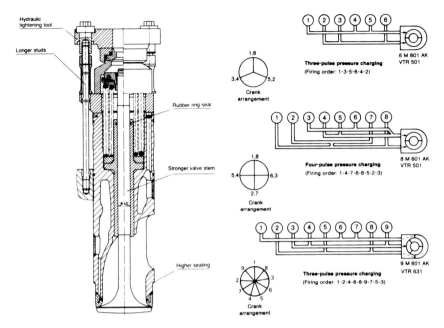

Figure 19.13 M601 exhaust valve

Figure 19.14 Diagrammatic representation of M601 supercharging system

The camshaft is driven via intermediate wheels from the split gear wheel on the crankshaft. The hardened valve cams are shrunk onto the shaft and the fuel cams are adjustable. Reversible engines have twin cams with sliding ramps and the camshaft is hydraulically moved for reversing. Woodward governors are fitted as standard.

The fuel injection pumps and injectors are of L'Orange manufacture and each cylinder has its own pump, which is of the through-flow type.

Pressure changing is by Brown Boveri turbochargers. Impulse pressure charging ensures a favourable efficiency at full and part loads and affords good manoeuvring properties to vessels with fixed pitch propellers.

MaK DIESEL ENGINES M35

The M35 has been developed as a new heavy fuel burning engine which in the MaK engine programme complements the engine series M453 and M551. The main technical data are shown in Table 19.1 (See also Figure 19.15.)

Traditional design was followed with bedplate and entablature of already proven material connected with through-going tie-bolts. This construction ensures rigidity and low noise level. Large inspection

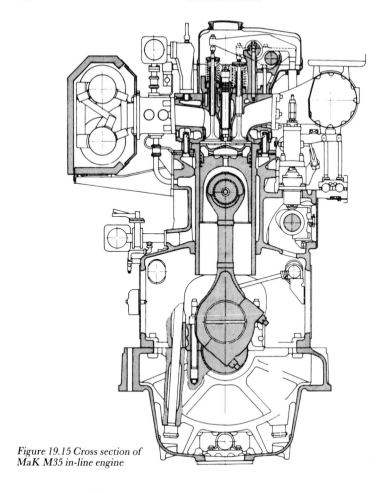

Figure 19.15 Cross section of MaK M35 in-line engine

doors allow for easy access to the crankcase, hydraulic tools being applied for maintenance work on main and big end bearings. The composite pistons and a piston ring configuration in accordance with latest research results meet the high requirements for heavy fuel operation in the future. An obliquely split and serrated big end bearing design has been adopted. The cylinder liners of this, like all MaK engines, are bath nitrided.

The engine has been conceived as a compact design for both main propulsion and auxiliary applications, drawing on previous experiences of diesel engines burning heavy fuel, and already proven manufacturing principles.

The M35 is a range of in-line engines giving easy access to all engine components, simplyfying maintenance. They can be directly reversible.

For operation on fuels with higher viscosity the M35 will be fitted with a separate fresh oil cylinder lubricating system. Two exhaust valves are fitted, in separate cages with optimum valve seat cooling. Cooled intermediate pipe connections are used between cylinder heads and exhaust manifold. Hydraulic tools are used for dismantling and assembling cylinder heads and exhaust valve cages. The design incorporates the pulse charge principle based on BBC's latest turbocharger design.

Acknowledgement

The editors would like to acknowledge the assistance of Krupp-MaK Maschinenbau GmbH and their UK delegate Mr. H. Hoff in the preparation of this chapter.

D.A.W.

20 Deutz engines

Deutz engines are made by Klöckner-Humboldt-Deutz AG and also by licensees.

The Deutz engine which qualifies for this section of the book is the BVM540. In Deutz nomenclature the last two figures indicate the stroke in centimetres.

The first engines of the BVM540 series were introduced in 1968, and had an output per cylinder of 294 kW at 600 rev/min. This has been progressively increased to the current rating which was introduced in 1976, mainly by increases in the brake mean effective pressure. The cylinder output at present is 405 kW either at 600 rev/min for synchronous generation, or at 630 rev/min for marine propulsion. Table 20.1 indicates the salient features of the BVM540 range.

Table 20.1 Technical data

Bore	D	mm	—		370			
Stroke	H	mm	—		400			
Cylinder output	N_{cyl}	HP	—		500			
		(1 kW)			405			
Speed	n	1/min	marine		630			
			stationary		600			
bmep	P_{me}	kp/cm²	marine		18.27 (17.93)			
		(bar)	stationary		19.18 (18.82)			
No. of cyls		—	—	6	8	12	16	
Continuous rating according to DIN 6270	P	HP (1 kW)		3300 (2427)	4400 (3236)	6600 (4854)	8800 (6472)	
Specific fuel consumption	b_e	g/HPh (g/kWh)		152 (206.7)	152 (206.7)	150 (204)	150 (204)	

All these engines are single-acting water cooled four-stroke trunk piston engines, turbocharged and intercooled, and are offered in six and eight cylinder in-line form as well as twelve and sixteen cylinder Vee form, the Vee angle being 48°.

All series 540 engines can be fitted for clockwise or counter clockwise rotation, and marine versions can also be offered in directly reversible form if required.

The engines are suitable for continuous operation from pier to pier on heavy fuel oil if certain safeguards are followed. Figures 20.1 and 20.2 respectively indicate cross sectional arrangements of the in-line and the Vee engines.

The following is an itemised description of the components, with notes of any changes introduced to permit the present rating compared with earlier engines.

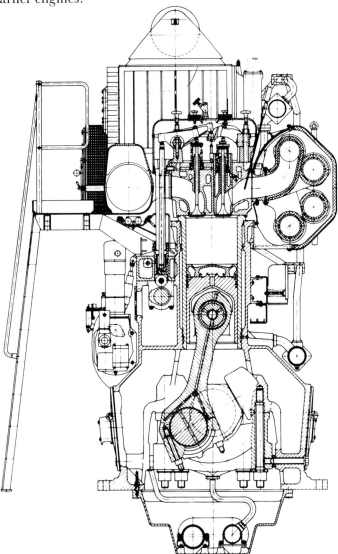

Figure 20.1 Sectional view of in-line engine

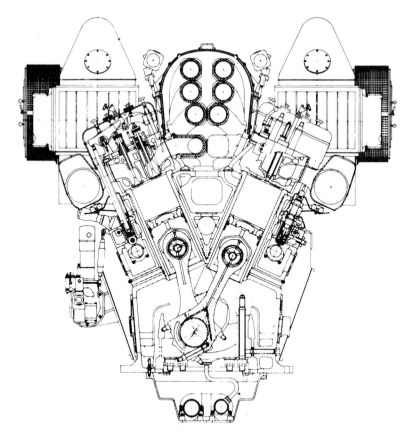

Figure 20.2 Sectional view of Vee type engine

Crankcase

The crankcase is a spheroidal graphite iron casting. The side walls extend well below the crankshaft centre line, thus bracing the crankcase. The crankshaft itself is suspended, that is, the main bearing caps are secured to the crankcase from below. In addition, the bearing caps are braced by two horizontal bolts, and the bearing cap bolts are hydraulically tensioned. To facilitate raising and lowering the main bearing caps, a special removal/replacement tool is provided for this extremely heavy component.

Large inspection holes in the crankcase sides facilitate servicing and enable main bearings and big end bearings to be removed and replaced easily and quickly without the need to dismantle the engine. The lower crankcase, a thin-walled spheroidal graphite iron casting, serves as a dry sump.

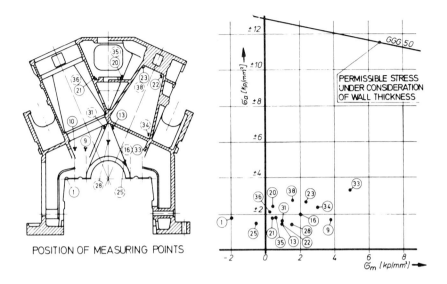

Figure 20.3 Stress analysis of the crankcase of Vee type engine

Tests indicated that no change was necessary to enable the crankcase to meet the newer rating. Figure 20.3 gives an indication of the stress pattern and shows the considerable safety margin.

Crankshaft and bearings

The crankshaft is hammer-forged from high-grade heat-treated steel and has precision-finished journals. The main bearings are lubricated via ports machined in the bearing caps, from the sump oil gallery. The flywheel is secured to the forged crankshaft flange by anti-fatigue bolts, and the camshaft spur gear is fitted between the flywheel flange and the No. 1 main bearing. The vibration damper is located on the crankshaft front end which can take an extension to drive auxiliary equipment or a PTO shaft to suit the overall engine rating. To balance the shaft and relieve dynamic bearing loads, counterweights are fitted on some of the crank webs.

No dimensional changes were found necessary to enable the crankshafts of the earlier series to withstand the higher loadings now current, but the material of the twelve cylinder engine crankshaft had to be raised from 850 N/mm^2 to 950 N/mm^2. Likewise, no change was required in the crankshaft bearing design. The cylinder spacing on both in-line and Vee engines has the same dimension.

Connecting rods and big end bearings

The connecting rods are drop-forged from heat-treated steel alloy, with square section shanks which are fully machined. The big end is split diagonally, the bearing cap is secured to the connecting rod by means of four extra-strong anti-fatigue bolts which are hydraulically tensioned. Vee type engines have their connecting rods arranged a pair per crankpin.

Big end bearings as well as main bearings and small end bushes are of lead–bronze with thin steel backing. Main and big end bearings are protected by a thin coating designed to hold until the unit has been run in. Continuous lube oil supply to big end bearings and gudgeon pin bearings via the main bearings is ensured by galleries and ports machined in both the crankshaft and the connecting rods.

Compared with earlier engines the connecting rod differs by the introduction of a stepped small end whose advantages for carrying higher loads are well understood. The gudgeon pin diameter is not altered. This feature can be appreciated from Figure 20.4, and the general design of the connecting rod from either Figure 20.1 or Figure 20.2.

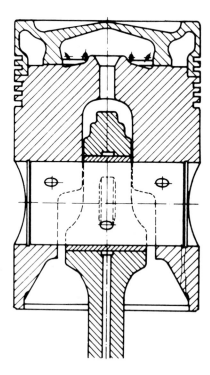

Figure 20.4 Composite piston

Pistons

Earlier engines of the BVM540 series were equipped with a single-piece cast aluminium piston incorporating a cooling coil in the head. With the increase in rating, it has become essential to adopt the composite design having a light metal base and a steel crown, oil being delivered from the gudgeon pin to a cavity immediately below the piston crown, then to the space behind the ring belt. Figure 20.5 indicates the improvement in piston temperature brought about, despite the increase in rating, by the change in construction, and it also indicates the difference in design between the two types. Note that the combustion chamber shape has been altered from a flat bowl to a domed configuration.

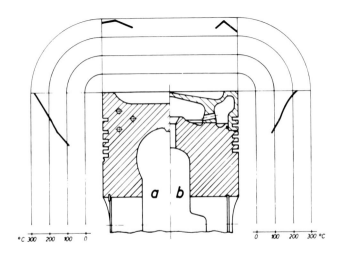

a) 400 HP/CYL. (295 kW/CYL.) 600 1/min   b) 550 HP/CYL. (405 kW/CYL.) 600 1/min
   (ALUMINIUM PISTON   (COMPOSITE TYPE PISTON)
    COOLING PIPE TYPE)

Figure 20.5 Comparative piston temperatures aluminium vs composite

The top piston ring has a plasma-sprayed chrome-molybdenum-carbide running face; the second ring has a chromium plated running face honed to a convex shape. Two further compression rings are carried in the upper ring grooves of the lower part of the piston together with a segmental oil control ring with a helical expander.

Cylinder head and valve gear

Basically the cylinder head design has not changed throughout the life of the BVM540 engine though there have been many detail improvements both to facilitate heavier duty with inferior fuels and to facilitate servicing.

The individual cylinder heads are of spheroidal graphite cast iron and are secured to the crankcase by eight socket head capscrews which are hydraulically tightened. The BVM540 cylinder head has always featured a relatively thin flameplate supported by the air and exhaust gas passage walls. An analysis of the stress condition indicated that no further change was required to carry the present rating (Figure 20.6).

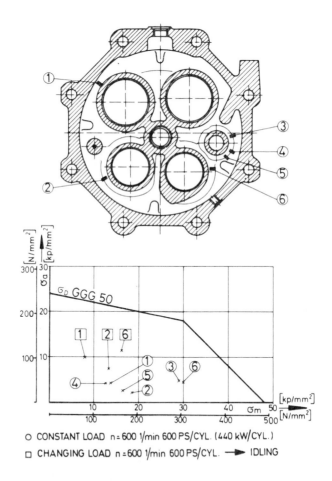

○ CONSTANT LOAD n = 600 1/min 600 PS/CYL. (440 kW/CYL.)
□ CHANGING LOAD n = 600 1/min 600 PS/CYL. ⟶ IDLING

Figure 20.6 Stresses at the cooling water side of flameplate

Each cylinder head has two inlet valves, two outlet valves, an air starting valve, a safety valve and an indicator cock. The exhaust valves are caged, as is normal for an engine of this type and, as can be seen from Figure 20.6, they are asymmetrically disposed on the flameplate. As a result the injector too is disposed slightly towards the air side of the cylinder. This is also shown in Figure 20.7.

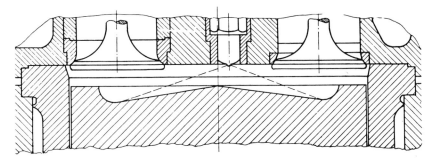

Figure 20.7 Combustion chamber

For reasons of high thermal and corrosion resistance the two exhaust valves are made from Nimonic 80A; this has applied since the earliest engines in the series. Both inlet and exhaust valves are all fitted with valve rotators normally of Rotocap design. For heavy fuel operation, however, the Rotocaps on the exhaust valves are replaced by a Deutz design automatic valve rotator. Both inlet and outlet valves now have removable stem guides, which are made of a material featuring high thermal resistance. The inlets seat directly on hardened seat inserts in the cylinder head itself and are not caged.

The exhaust valve cage design has been altered in conjunction with the introduction of the present cylinder rating. The changes, which can be appreciated from Figure 20.8, are designed partly to facilitate the higher rating and partly to reduce the time and effort needed to exchange the valve cage. The improved shape of the port will be noted from Figure 20.8.

Valve seat cooling has also been introduced to keep the seat temperature below 500 °C. In addition, a wear resistant seat is bonded to the nodular cast iron exhaust valve cage by high temperature vacuum brazing.

As regards the simpler overhauling arrangements, which enable a valve cage to be removed in 20 minutes, these depend on two features:

1. The exhaust valve cage can be removed without the cooling water having to be drained from the cylinder head. It is only necessary to take off the rocker chamber cover held by four wingscrews and

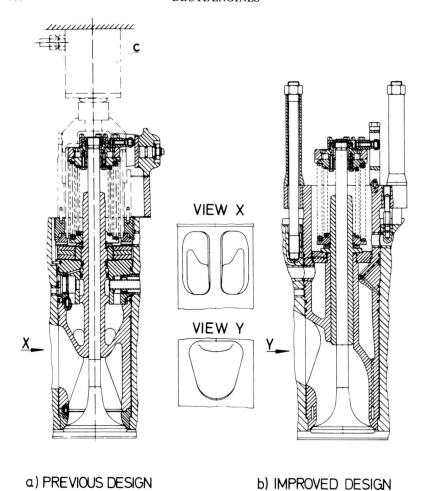

Figure 20.8 Exhaust valve cage

then to separate the water spaces of the exhaust valve cage from those of the cylinder head by means of a slide valve using a hand lever (Figure 20.9). After reassembling the exhaust valve cage the connection is restored by returning the valve to its original position, using the same lever. If this is forgotten, a boss on the rocker chamber cover will automatically open up the passage (Figure 20.10). Furthermore, this boss locks the slide valve in position when the rocker chamber cover is mounted and thus protects against unintentional actuation of the lever.

2. Instead of the former method of fastening the exhaust valve cage in the cylinder head (Figure 20.8, left-hand), with a hydraulic

DEUTZ ENGINES

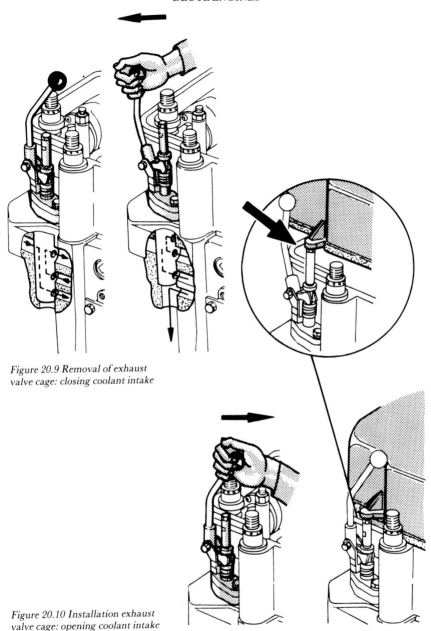

Figure 20.9 Removal of exhaust valve cage: closing coolant intake

Figure 20.10 Installation exhaust valve cage: opening coolant intake

clamping tool (C) via a disc spring, the new exhaust valve cage design is fastened with two resilient bolts only without requiring any special device (see Figure 20.8b).

The rocker arrangement can be appreciated from the sectional views of the engine and also from the plan view (Figure 20.11). Note the completely segregated fuel connections. The rocker arms are lubricated via separate ducts connected to the engine lube oil circuit. The valve gear compartment features an oil-tight light metal cover having inspection windows to facilitate valve rotator inspection.

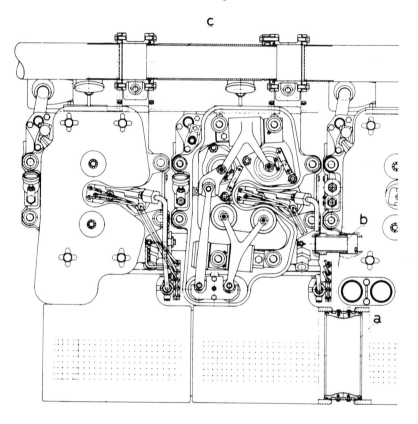

Figure 20.11 Cylinder head plan showing valve gear and plug in connections for charge air (a) starting air (b) and cooling water (c)

Compared with earliest engines, disassembly and assembly of the cylinder head has been simplified to such an extent that it takes at most only one hour to remove. Reassembly is simplified by the fact that the charge air pipe is divided into individual sections as previously (Figure 20.11) with plug-in connections (a), the starting air pipe which now runs from cylinder head to cylinder head has also movable connecting plug pieces (b), and the water manifold now consists of universally jointed and identical sections. Note that the exhaust thermometers are fitted into the exhaust duct of the cylinder head, and independent of the exhaust pipes, outside the exhaust cladding.

Cylinder liners

The water cooled cylinder liners which are centrifugal castings, are supported at the joints by the crankcase. Each cylinder liner and head assembly is secured to the crankcase by hydraulically tensioned socket head cap screws so as to ensure a gas-tight and water-proof joint. The water jacket is sealed off the crankcase by two seals located at the lower end of the liner, and the seals are arranged to form a compartment between the water jacket and the crankcase from where any water that may leak through the seal, can be drained off, thus facilitating leakage checks.

Camshaft and gear assembly

In-line engines have one camshaft; Vee type engines have two — one on each cylinder bank. The camshaft itself is fabricated from individual, hardened cams and journals as required for the number of cylinders the engine has. The camshaft is suspended from below the housing that accommodates the cam follower guides and the individual fuel injection pumps. The camshaft housings are installed from above in the camshaft recess.

Reversible engines have inlet and outlet cams designed to facilitate clockwise and counter-clockwise rotation of the crankshaft, and in addition are fitted with an injection control cam. To reverse, the cams are moved in an axial direction by a pneumatic plunger. The camshafts are gear-driven from the flywheel end of the crankshaft by gears that have been hardened and polished.

Fuel injection pump

All BVM540 engines have Deutz helical-groove-plunger type fuel injection pumps. Fuel metering is controlled by rods that are mounted in the camshaft housings, which synchronise the plungers of the fuel injection pumps. Plungers are fitted with a barrier to prevent fuel from contaminating the engine lube oil system. The fuel lift pump feeding the fuel into the injection pumps through the filters, is installed separately from the engine and driven by motor.

The testing carried out to introduce the current rating did not indicate any need for change in the injection system from that of earlier engines.

Speed control

The engine speed is controlled by a Woodward hydraulic governor gear-driven from the flywheel end of the engine. There is an option of several types of governor for different duty cycles, designed for fixed or variable speed control.

Turbochargers

The in-line models and the twelve cylinder Vee type model have one BBC turbosupercharger, the sixteen cylinder engine has two. These can be mounted on the flywheel side or on the damper side optionally and depending on the engine mounting system. The hot air is cooled by water circulating through the intercoolers fitted on the engine. The exhaust gas ducts made from a material of high thermal resistance, are insulated and covered by a contact protection plate. The duct sections are connected by corrugated pipe connectors. On Vee type engines, the hot air ducts are accommodated in the cavity formed by the two cylinder banks.

Whereas the same general turbocharger type is used as on earlier models a higher output version is specified for the current rating in order to maintain the excess air ratio despite the higher boost pressure.

Lubricating oil system

All engine bearings are serviced by the pressurised engine lube oil system. The turbosuperchargers have their own lube oil system. A dry sump is supplied as standard, and there is an option of flat and high inclination types. The lube oil cooler, which is installed separately from the engine, absorbs heat from the lube oil, passing it to the circulating cooling water.

An engine-mounted 40 micron screen filter has been provided for engine mounting. To enable this to be cleaned simply and without shutting the engine down, a four-chamber design has been selected with a number of filter cartridges being arranged in each chamber. The lubricating oil flows through three chambers and the fourth serves as a stand-by unit. For cleaning purposes one contaminated chamber is shut off and automatically substituted by the stand-by unit. After removal of the filter cover the contaminated filter cartridges of this filter chamber are cleaned by a compressed air gun inserted into each cartridge. The filter cover can only be refitted with the cock closed.

Cooling system

The engine and the hot air ducts are cooled by water circulation, the coolant prepared to specification to contain the correct amount of anti-corrosive. By special order, two rotary pumps can be fitted on the pump housing. These are linked to and driven by the engine and can be provided with a cut-in/cut-out facility to serve the system whenever necessary. So the coolant is circulated either by these pumps or one of the pumps is circulating the coolant while the other one is moving raw water. Recooling of the circulating coolant takes place in a separately installed heat exchanger. The lube oil cooler and the injection valve coolant cooler are linked to the cooling system. Where a transmission oil cooler is fitted, this unit can be linked to the raw water system.

Thus a single-circuit cooling system has been provided where the engine jackets as well as the charge air coolers and oil heat exchangers are cooled by fresh water in a single circuit with one fresh water pump. This arrangement enables the seawater heat exchanger to be mounted away from the engine, generously dimensioned and easily cleanable. It also means that the air and oil coolers on the engine do not require cleaning except at very rare intervals. Figure 20.13 shows the cooling circuit diagrammatically.

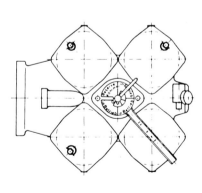
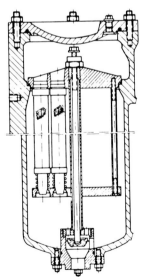

Figure 10.12 Lube oil micronic filter

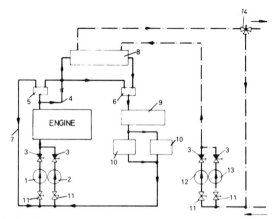

Figure 20.13 Single-circuit cooling system
1 Fresh water pump
2 Auxiliary fresh-water pump
3 Relief valve
4 By-pass line
5,6 Thermostats
7 By-pass line
8 Water cooler
9 Oil cooler
10 Charge air cooler
11 Shut-off valve
12 Raw water pump
13 Raw water auxiliary pump
14 Proportionery valve

The high temperature thermostat (5) is designed to keep the engine outlet temperature constant at approximately 80 °C. Water from the cooler (8) flows first through the lubricating oil heat exchanger and enters the charge coolers (10) at a temperature of approximately 45 °C virtually independent of engine load conditions. This ensures that there is no risk of the charge air temperature dropping below 45 °C and this excludes the risk of condensate deposits in the air-intake system even with high relative humidity and charge air pressure.

The system ensures that even under full load the charge air temperature does not exceed 60 °C, which was the basis on which the engine development was carried out.

Starter system

All 540 engines are started by air admitted to the cylinder, the valves being operated by a timed control supply of starting air.

Acknowledgement

The editors would like to acknowledge the assistance of Deutz Engines Ltd. in the preparation of this chapter.

D.A.W.

21 Mirrlees Blackstone engines

Mirrlees Blackstone Engines are part of the range of Hawker-Siddeley Diesels Ltd.

The medium-speed designs come from the Mirrlees factory, which traces its diesel associations back to the first British licence from Dr Diesel at the end of the last century.

The current engine, the K Major, started life in the 1950s as the 'K', one of the first large medium-speed diesels in the world. Subsequent development and uprating in the light of very extensive field experience has led by stages to the present K Major. The K Major is a four-stroke, direct injection, turbocharged and intercooled oil engine built in-line and in 45° Vee form. It has been progressively developed over the years and appropriate redesigns undertaken to exploit fully the technological advances in engine design and related fields. It is directly reversible for marine applications. Figure 21.1 shows the K Major Mark 2 in-line engine and Figure 21.2 the Vee engine, both in cross section.

The principal characteristics of the engine are given in Table 21.1.

Table 21.1 Technical data K Major engines

Engine type	Mk2	Mk3
Cylinder bore mm	381	400
Piston stroke mm	457	457
Cylinder configuration in-line	3,5,6,7,8,9	6,8,9
Cylinder configuration Vee engines	12,14,16	12,16
Bmep bar	19	19
Speed rev/min	600	600
Piston speed m/s	9.14	9.14
Cyl output kW	492	545
Compression ratio	11.35	10.67

Full power can be taken off either end of any engine in the range. Very significant increases in specific power output have been achieved without prejudice to high standards of reliability and durability and within the known critical parameters to ensure satisfactory operation on residual fuels. Many thousands of hours of operating experience, test data and development knowledge have culminated in the Mark 3 engine, the latest in the K Major series.

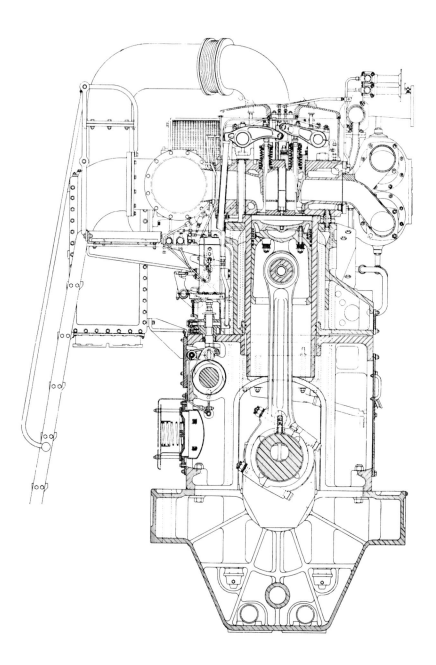

Figure 21.1 Cross section of Mirrlees Blackstone K Major Mk 2 in-line engine

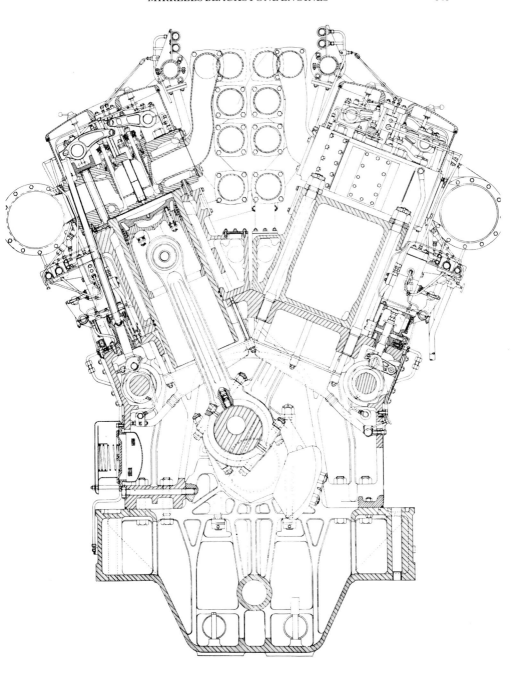

Figure 21.2 Cross section of KV Major Mk 2 in-line engine

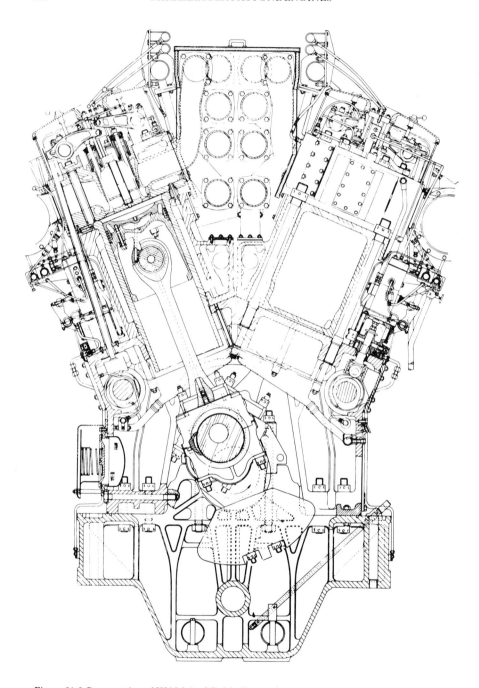

Figure 21.3 Cross section of KV Major Mk 3 in-line engine

The Mk3 version is shown in cross sectional view in Figure 21.3.

While those new features which were developed of necessity for the Mark 3 remain confined to the Mark 3, and those which have been proved successful on the Mark 2 are retained on the latest Mark 2, some standardisation in the future is not to be ruled out.

The K Major is now available in two type designations, the Mark 2 with a bore of 381 mm, and the Mark 3 with a bore of 400 mm, both having a common stroke of 457 mm. For the purposes of this chapter the design and construction features described will be based on those of the latest Mark 3 design. For reference purposes comparative technical data for both Mark 2 and Mark 3 are listed in Table 21.1.

K Major engines are rated in accordance with BS5514, which is fully in accord with ISO 3046 and meets with the requirements of the major Classification Societies. They cover continuous shaft ratings from 2240 kW (3000 bhp) to 8710 kW (11 680 bhp) at speeds up to 600 rev/min.

In-line engines are built in units having six, eight and nine cylinders, while Vee form engines have twelve and sixteen cylinders.

The stroke-to-bore ratio is 1.14:1 and the mean piston speed at the maximum continuous operating speed of 600 rev/min is 9.14 m/s. The brake mean effective pressure at the maximum continuous rating is 18.96 bar and typical fuel consumption at this rating, dependent upon the heat value of the fuel, is of the order of 204 g/kWh.

The Mark 3 is specially designed to burn residual fuel reliably and efficiently and in the following description reference is made to those features which contribute to this heavy fuel burning ability. The company's experience in this specialised field is outstanding and includes familiarity with operation on catalytically cracked fuels and on fuels having high levels of vanadium or Conradson carbon — all of which are likely to become more common as fuel quality deteriorates in the next decade.

The K Major engine is of cast iron construction and the rigidity of the design ensures that stress levels are well within acceptable limits, at the current ratings. Strain gauge measurements and mathematical models have confirmed this and show that factors of safety are generous.

The frame

The column, which forms the upper part of the crankcase, consists of a single cast iron box section structure which houses the camshaft — or two camshafts in the case of the Vee form engine. The column extends below the level of the crankshaft centre line, and large inspection doors are provided on both sides to permit easy access to the running gear. The doors on one side carry Biceri type crankcase pressure relief valves

Figure 21.4 Column and cylinder block for Mirrlees Blackstone Vee engine

and incorporate approved flame traps, in accordance with the recommendations of the Classifications Societies. Figure 21.4 shows a sixteen cylinder Mark 2 column with one cylinder block in position.

On earlier K Major engines centrifugal filters were fitted to the column adjacent to each main bearing, but these are no longer necessary where the full oil treatment advised for operation on heavy fuel is applied. The cylinder block is a single-piece iron casting for each bank of cylinders. It is mounted on the column and secured by through bolts which extend into deeply tapped bosses formed in the top deck of the column, thereby relieving tensile stresses in the cylinder block.

On the in-line engines through-bolts of high tensile steel pass from the top of the cylinder blocks to lower locations athwart the main bearings. The block forms the water jacket for the cylinder liners. The fresh water used for cooling enters at the lowest point and the outlets to the cylinder heads are at the highest point, to ensure freedom from vapour pockets.

Cylinder liners

The individual cylinder liners are centrifugally cast in special close-grained iron. The liner is located and secured by a flange at the upper end, while the lower end is fitted with two synthetic rubber rings to permit downward expansion relative to the cylinder block. It has not so far been found necessary to depart from the simple liner flange design.

The piston

The piston is of two-part construction. The alloy steel crown incorporates an inner load carrying boss so that the combustion loads are not transmitted through the outer wall which carries the compression rings. The outer wall can then be a relatively thin section which both reduces the heat flow path from the crown to the rings and assists efficient oil cooling of the ring belt. Under overload conditions, the measured top ring groove temperature is only of the order of 180 °C.

The crown is attached to the piston skirt by four high tensile steel studs (Figure 21.5). Lubricating oil to cool the piston crown is fed from the big end bearing via a longitudinal drilling in the connecting rod and through drillings in the gudgeon pin to the annular chamber behind the compression rings. The oil then passes through a transfer drilling to the

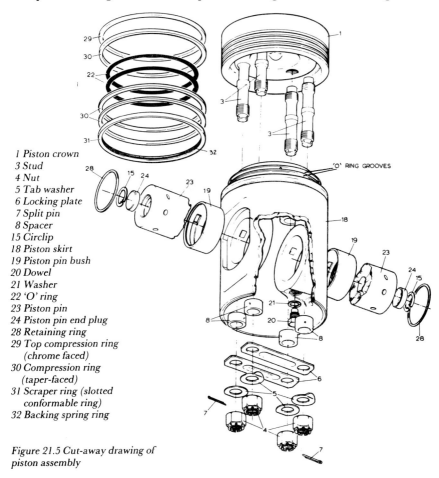

1 Piston crown
3 Stud
4 Nut
5 Tab washer
6 Locking plate
7 Split pin
8 Spacer
15 Circlip
18 Piston skirt
19 Piston pin bush
20 Dowel
21 Washer
22 'O' ring
23 Piston pin
24 Piston pin end plug
28 Retaining ring
29 Top compression ring
   (chrome faced)
30 Compression ring
   (taper-faced)
31 Scraper ring (slotted
   conformable ring)
32 Backing spring ring

Figure 21.5 Cut-away drawing of piston assembly

central chamber below the piston crown and returns from there through another drilling, to the crankcase (Figure 21.6).

The piston skirt is an alloy iron casting which carries a conformable oil control ring in a groove above the gudgeon pin. Mark 2 engines have two oil control rings. Drillings are provided to convey lubricating oil from the gudgeon pin bosses to the piston crown. These are drilled at an angle to the vertical axis of the piston so that they break into the bores of the gudgeon pin bosses at a point displaced about 30° from the vertical axis. This can be seen in Figure 21.3 and also in Figure 21.5. This feature removes the drilling from the most highly stressed region of the boss and so transfers the stress-concentrating effect of the hole to an area of lower stress, thereby improving resistance to fatigue.

The fatigue strength of the gudgeon pin bosses is further improved by steel bushes which are frozen into the bores for the gudgeon pin. These

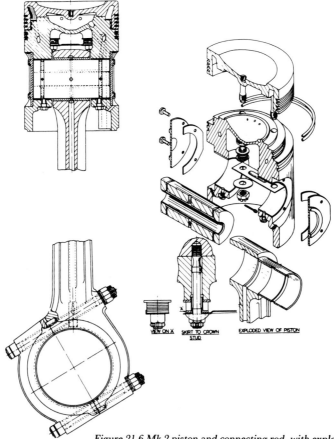

Figure 21.6 *Mk 2 piston and connecting rod, with exploded view of piston*

bushes are lined with phosphor-bronze to provide a durable bearing surface for the gudgeon pin which is fully floating and retained by circlips. This allows a greater part of the piston's width to be used for pin support.

The Mark 2 retains for the present the older design's gudgeon endplates (Figure 21.6).

Cylinder heads

The individual cylinder heads are cast in special close-grained iron and they incorporate an internal intermediate deck in order to intensify the cooling of the combustion face and other vital surfaces. The cooling water is conveyed from the cylinder block to the head by ferrule type connections.

The cylinder head to cylinder liner gas seal is a solid copper ring and each cylinder head is secured to the cylinder block by eight alloy steel studs. The two inlet valves operate in renewable guides and they have renewable seat inserts as a standard feature. Valve rotators are also incorporated and the valves have a special valve stem coating to minimise wear and corrosion. To ensure satisfactory operation when burning heavy residual fuels and to facilitate maintenance, the two exhaust valves are arranged in detachable water cooled cages. These can be discerned in Figure 21.3.

Heavy and residual fuels usually contain a number of combustible trace elements and these often include sodium and vanadium salts. The vanadium compounds, which are formed during the combustion of the fuel, have a relatively low melting point and unless measures are taken to limit the temperature of the exhaust valve, these compounds are deposited on the sealing face of the valve. They prevent the valve from seating effectively and promote destruction of the sealing face by local guttering or burning.

Normally, heat flows from the head of the exhaust valve by two paths. The first is through the seat on the valve head to the cage or cylinder head. The second is through the valve stem and its guide to the cage or cylinder head. The deposition of vanadium compounds on the seating surface of the valve reduces the heat flow through this path, resulting in an increase in valve temperature which promotes rapid deterioration of the valve. It is therefore necessary to ensure that the temperature of the seating surface of the valve does not exceed about 550 °C.

The K Major exhaust valve assembly is illustrated in Figure 21.7. The valve cage is cooled by water which is fed into the top of the cage and conveyed by a drilling to an annulus formed near the seat. The

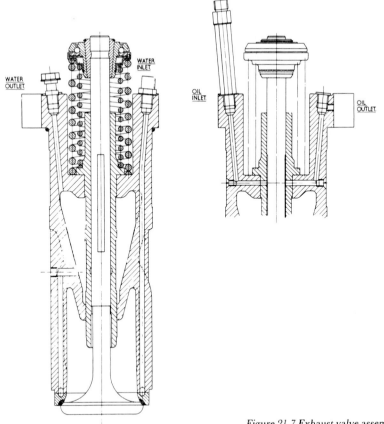

Figure 21.7 Exhaust valve assembly

water flows around the annulus and circulates around the valve guide, to convey heat from the valve stem, before passing out of the top of the cage.

The valve cage is a 3% chrome-molybdenum steel casting which has the upper part of the water cooling annulus machined in its lower face. The valve seat is a wrought steel ring of the same material as the cage, with the lower part of the water cooling annulus machined in its upper face: these two parts are permanently joined together by electron-beam welding. Both the valve facing and seat are armoured with A60 high nickel alloy.

Heat transfer from the stem of the valve to the water cooled valve guide is much improved by a reduction in the running clearance between these parts; and by employing positive lubrication of the valve stem and its guide this clearance has been reduced by some 75% relative to that which is typical for an uncooled assembly. The valve stem temperature has been reduced significantly and the build-up of hard carbon deposits — which may lead to valve sticking — has been virtually eliminated.

The positive, intermittent supply of lubricant to the exhaust valve stem and guide is achieved by feeding lubricating oil into drillings which connect with shallow local flats machined on opposite sides of the valve stem. When the valve is in the open position, lubricant flows down the flat on one side, around an annular groove formed in the bore of the guide and up the flat on the opposite side of the stem. The pressure of the lubricating oil supply to the exhaust valve is regulated by a pressure-reducing valve which responds to the charging air pressure in the engine inlet manifold. The pressure of the lubricating oil supply is proportional to the charging air pressure, and since this and the exhaust pulse pressure are directly related, the lubricating oil pressure is varied to suit the exhaust pressure. By this means, over-oiling at low engine loads and oil starvation at higher loads, are avoided.

In addition to the water cooling of the exhaust valve cage and controlled pressure lubrication of the valve stem, a valve rotating device is incorporated in the upper spring plate of the assembly. By slowly rotating the valve, a more uniform temperature is obtained around the periphery of the valve and its seat, and the rotation also assists the positive lubrication of the valve stem.

Valve gear

The two exhaust valves are operated from a single pushrod through separate valve levers on a common shaft, the motion being transferred from the pushrod by two linked levers. The two inlet valves are operated directly by a single pushrod through individual levers on a common shaft.

Camshaft

A large-diameter camshaft is provided for each bank of cylinders, and separate inlet, fuel and exhaust cams for each cylinder are mechanically attached to this. The inlet and exhaust cams are of polynomial form, to minimise shock loadings on the valve gear. The fuel cams have a falling velocity injection characteristic. The plunger type roller cam followers are pressure lubricated and the camshaft runs in renewable shell type bearings. It is driven from the flywheel end of the crankshaft by a compound train of hardened alloy steel gears.

Crankshaft

The crankshaft is fully machined from a high-grade chrome-molybdenum steel forging and balance weights are attached to the crank webs. Large diameter main bearing journals and crankpins which overlap, and webs of generous proportions, result in a particularly rigid crankshaft which is well in excess of Classification Society requirements for the maximum engine output. Drillings are provided to convey lubricating oil from the main bearings to the big end bearings. A Geislinger type damper is fitted at the forward end of all current engines. Earlier engines used viscous fluid dampers.

Bearings

The crankshaft runs in steel-backed, thin shell bearings. They are lined with a lead–bronze material together with a lead–tin–copper overlay. Service experience confirms that this combination provides a low wear rate with good tolerance to dirt contamination.

Bedplate

The main bearings are carried in the bedplate which is an iron casting of deep section with integral longitudinal girder members to provide a rigid support for the crankshaft. A lubricating oil gallery extends the length of the bedplate and vertical drillings convey the oil to the main bearings. The main bearing caps are located transversely by abutments and they are each secured by four high tensile steel bolts. This arrangement can be seen in Figure 21.8 and in the cross sections.

Figure 21.8 Bedplate of Mk 2 in-line engine

Connecting rods

The Mark 3 connecting rods are of three-piece construction with palm ended rod and large end bearing housing with part grooved overlay plated lead–bronze lined bearing shells. Oilways are provided in the connecting rod to allow lubrication of the piston pin bearing together with a copious flow of cooling oil to the pistons (Figure 21.9).

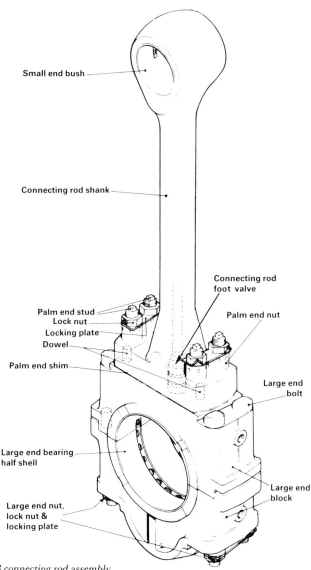

Figure 21.9 Mk 3 connecting rod assembly

Mark 2 engines retain connecting rods of the automotive type machined from nickel alloy steel stampings. The large end bearing housing is split obliquely to facilitate removal of the piston and connecting rod through the cylinder. The large end bearing cap is secured by four alloy steel bolts, and the large end bearings are thin steel shells lined with aluminium tin. The small end bearing bush is of aluminium alloy. Figure 21.6 shows the piston and connecting rod assembly of earlier Mark 2 engines.

Lubricating oil system

For Mark 3 and current Mark 2 engines, a wet sump lubricating system is normally employed. Alternatively, engines can be arranged for gravity drain dry sump operation. Oil is circulated by one engine driven pump on the in-line engines and by twin pumps on the Vee form engines. They are mounted externally at the free end of the engine. Lubricating oil is delivered via a full flow fine filter and an oil cooler, to the crankshaft and camshaft bearings, gudgeon pins and all other important running gear: this system also provides the oil for piston cooling.

Figure 21.10 Mirlees Blackstone K8 Major direct reversing marine engine

A useful facility provided on all K Major engines for some time now is the means to monitor the piston cooling. Figures 21.1, 21.2 and 21.3 all show a 'D' shaped oil catcher tray attached to the bottom of each liner to one side of the connecting rod. The spent oil discharge from the piston crown is directed to that side of the rod and thus falls mostly into the catcher. From there it is led to a tundish just inside the crankcase door before returning to the sump.

A Rototherm thermometer carried in the crankcase door is oriented to record the temperature of this oil in the tundish.

Water cooling

A closed circuit water cooling system with a shell and tube heat exchanger is usually employed and the fresh water and raw water pumps are separately driven by electric motors.

The fresh water pump may be used to aftercool the engine after it has been shut down. A separate treated water cooling system is provided to remove heat from the fuel injector nozzles and the exhaust valve cages.

Fuel injection

Individual fuel injection pumps are provided for each cylinder. A metered quantity of fuel is delivered to a low inertia type injector having a multi-hole nozzle which has provision for water cooling. The high-pressure fuel pipe and its connections, between the pump and the injector, are arranged outside the valve gear cover to minimise the risk of fuel contamination of the lubricating oil due to leakage from these sources. The high-pressure pipes can be sheathed to comply with the requirements for unattended engine operation.

Turbocharging

One turbocharger is fitted to each bank of cylinders and it normally draws air through an intake filter and a silencer. The air is delivered from the turbocharger to the engine inlet manifold, via an intercooler which may be either of the water cooled or air cooled type.

Controls

A pneumatic control system is employed which provides remote push-button starting and stopping. Protection devices designed to shut down the engine in the event of low engine oil pressure, and engine overspeed, are fitted as standard in addition to the mechanical hydraulic overspeed trip on the engine which returns the fuel injection pump control shaft to the 'no fuel position': a mechanical spragging device renders the fuel pumps inoperative.

Starting is by compressed air which is admitted to the engine cylinder through non-return valves. The admission of the air is timed by the fuel pump tappets.

Maintenance

A range of maintenance tools is available to facilitate all the routine overhauling tasks.

For the K Major Mk 3 the opportunity has been taken to introduce hydraulic tightening/untightening equipment for all the main threaded fasteners on the engine frame/bearings. The corresponding change in the design of the nuts can be clearly discerned in Figure 21.3 compared with Figure 21.1 and 21.2.

Acknowledgement

The editors would like to acknowledge the assistance of Mirrlees Blackstone (Stockport) Ltd. in the preparation of this chapter.

D.A.W.

22 The rest of the field

The engines described in this chapter are of 250 kW/cylinder or over, running at or below 600 rev/min and not covered in the previous eight chapters. The philosophy adopted by the editors in describing medium-speed engines is that the makes which have been numerically most successful, or are longest established and still in production, warrant a chapter each.

Since the range of makes is very extensive, and many makes offer more than one design, and since their fortunes may change for technical or other reasons, it may be that during the life of this edition an engine or make described in this chapter is more qualified for a more extensive description in a separate chapter on its own.

All that one can say is that the relative success of the principal representatives will be reviewed when the seventh edition is prepared, as it was before this edition. In the meantime, this chapter will highlight the principal features of the most significant engines not otherwise considered, and the status of some of the others. It needs also to be said that notwithstanding the power and speed limits stated in the first lines of this chapter, development is tending to produce engines which, while following a 600 rev/min philosophy, have a capacity for higher speeds, even up to 750 rev/min.

There are also engines which run very slowly, either because they are two-stroke or because they are built with very long strokes, and are able to operate as direct drive units. They will be considered in this chapter to the extent that their build philosophy and power capability suggest that this is more appropriate.

JAPANESE DESIGNS

The Japanese marine diesel industry is complex, vigorous, and includes some practice which differs from that current in other countries with marine diesel tradition.

For instance, many Japanese makers find that their traditional local markets prefer very low speeds (200–300 rev/min) and will accept very long stroke engines in consequence. A range of powers is often provided by a series of sizes all with six cylinders only. At the time of writing, such engines remain of traditional construction without enclosure for cylinder heads and valve gear. Most Japanese engine builders catering for international owners have built designs licensed from the major European builders. However, several have produced designs which follow modern European practice, and some of these designs have created considerable interest.

At the same time a not insignificant number of international owners have displayed a growing willingness to accept the uncomplicated verities of some of the Japanese traditional designs. A selection of Japanese designs is described.

Hanshin

The Hanshin Diesel Works includes in its range two relevant six cylinder units, the 6LUS40 and the 6LUS54. Both are uprated versions of earlier engines and are four-stroke. The last two figures in the type code correspond to the bore in centimetres.

Table 22.1

Type	6LUS40	6LUS54
Bore mm	400	540
Stroke mm	640	850
Cyl output mcr kW	373	680
Speed rev/min	300	230
Bmep at mcr bar	18.4	18.2

The engines are very similar and the 6LUS40 is illustrated in the cross section in Figure 22.1. They are simple and rugged, characterised by heavy cast iron sections and conventional running gear.

They feature two-part 'cocktail shaker' oil cooled pistons. The three-part frame is of cast iron, held together with long tie-rods. The crankshaft is carried in saddle bearings. The camshaft is mounted low in the frame in a separate trough which includes the lower half bearings, the split coinciding with the cylinder jacket to crankcase joint face.

The larger engine has four caged valves, these for each function being actuated by a simple rocker via a bridge piece, air and exhaust manifolds being on the same side of the engine. The 6LUS40 is simpler as the air manifold is on the opposite side. Both engines are designed to burn heavy fuel and are directly reversible.

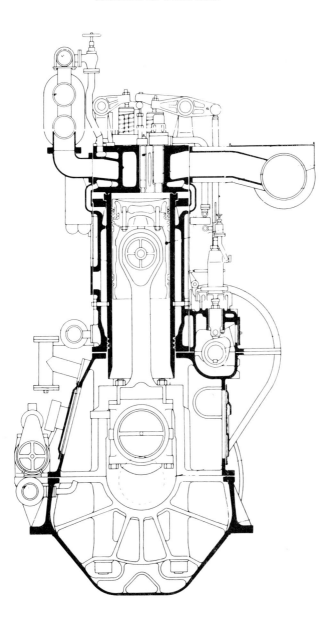

Figure 22.1 Cross section of Hanshin 6LUS40 engine showing its heavy cast-iron construction and conventional running gear

Akasaka

Akasaka Diesels follows a similar philosophy to Hanshin, the leading particulars of their current five models (all produced in six cylinder form only) being listed in Table 22.2. These engines are of extreme stroke/bore ratio and are developed particularly for slow speed and economy.

Table 22.2

Type	A28(R)	A31(R)	A34	A37	A41
Bore mm	280	310	340	370	410
Stroke mm	550	600	660	720	800
Cyl output mcr kW	186	223	275	323	409
Speed rev/min	320	290	270	250	230
Bmep at mcr bar	20.4	20.3	20.1	19.8	20.1

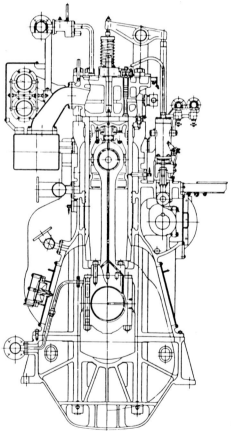

Figure 22.2 Cross section of the Akasaka A31 engine

While the two smaller engines do not really qualify for this chapter they are included to illustrate the family concept. The suffix 'R' indicates that they have reversing gear of the multi-plate hydraulic clutch type.

This range of engines, although largely new, is developed from older designs of similar philosophy. All are four-stroke and burn heavy fuel. About 70% of full output may be taken from the forward end of the crankshaft.

A cross section of one engine, the A31, is shown in Figure 22.2. The conventional structure is evident with, in this case, two valve heads. The frame is in two parts held together with long tie-bolts.

The large end is of three-part construction; the piston is composite, with 'cocktail shaker' oil cooling. The cylinder head has an intermediate deck, and the camshaft is mounted at mid-height, removable sideways.

Mitsubishi

Mitsubishi Heavy Industries Ltd. are heavily involved with licence building medium- and slow-speed engines, but have a prolific range of their own designs from outboard sizes to the very largest crosshead types.

In the medium-speed field they offer a compact trunk piston two-stroke engine, at present in Vee form only, the UEV42/56C. The leading particulars are given in Table 22.3 and a simplified cross section in Figure 22.3.

Table 22.3 UEV42/56C engine

Bore mm	420
Stroke mm	560
Cyl output mcr kW	485
Speed rev/min	380
Bmep at mcr bar	9.75
Nos. of cyls	12,18

Mitsubishi use a welded steel one-piece bedplate and frame structure. The crankshaft is threaded through large circular openings in the transverse diaphragms, and the lower main bearing housing has a corresponding semi-circular profile. The lines extend deeply into the crankcase and are cooled by scavenge air below the air inlet part belt. The three-part big ends run side by side.

There are three exhaust valves in the head, all driven by the same rocker lever. The valve gear is fully enclosed.

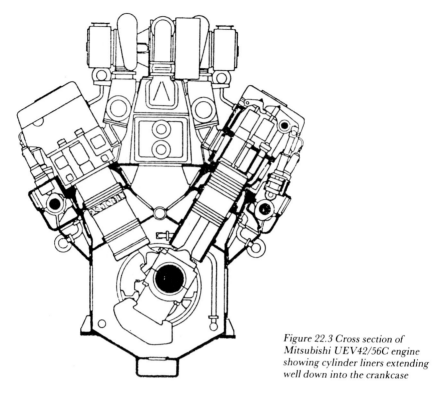

Figure 22.3 Cross section of Mitsubishi UEV42/56C engine showing cylinder liners extending well down into the crankcase

Mitsui

Mitsui Engineering and Shipbuilding Co. Ltd. are, like Mitsubishi, also major licence builders of established European designs. Their venture into designing and building on their own account was much more recent, however, and was, for a large medium-speed design, competitive with the leading European types. This emerged as the 60M, a four-stroke direct reversing engine of 600 mm bore and capable of 1120 kW/cyl; but by a bitter misfortune its launch coincided with the 1973 oil crisis and a critical reduction in its intended market.

Mitsui's reaction to this setback was to adapt their test experience to a similar engine of smaller size which has won some significant orders. This is the 42M, whose leading particulars are given in Table 22.4. The engine is four-stroke and directly reversing.

The frames of each variant are in two parts, as shown in Figure 22.4: a welded steel crankcase structure made from cast sections and plate, and a cast iron cylinder housing held together with tie-bolts. The underslung crankshaft bearings are secured by vertical and lateral bolts. All are hydraulically tightened. The design is very stress

Table 22.4

Type	L42M	V42M
Arrangement	in-line	45° Vee
No. of cyls	6,7,8,9	10,12,14,16,18
Bore mm	420	420
Stroke mm	450	450
Cyl output mcr kW	558	558
Bmep at mcr bar	20.2	20.2
Speed rev/min	530	530

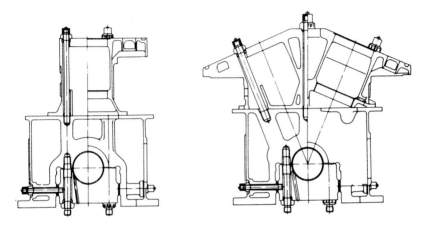

Figure 22.4 Engine frame structure of the Mitsui 42M engine in Vee and in-line cylinder form

conscious: for instance, Mitsui control the lateral fit between bearing keep and the frame by means of tapered adjusting liners before tightening, to avoid stressing the frame unduly when tightening.

For the crankshaft bearings Mitsui have been able to keep within the stress capacity of thin white-metal (on a lead–bronze backing in the case of the crankpins) with an overlay in each case. While this is a function of the control of maximum bearing pressures, it also means that damage to a bearing does not necessarily lead to shaft damage.

The combustion area and cylinder head can be appreciated from Figure 22.5. The composite piston, bore cooled liner, three deck head, caged exhaust valve and rigid high-pressure fuel line, can all be discerned, as well as the camshaft bearing arrangement.

In addition to the normal four compression rings in the crown (the second is Ferrox-filled to improve running-in) there are two conformable SOC rings at the top of the cast iron piston skirt, plus two broad lead–bronze bands above the gudgeon and one below, also to assist bedding-in.

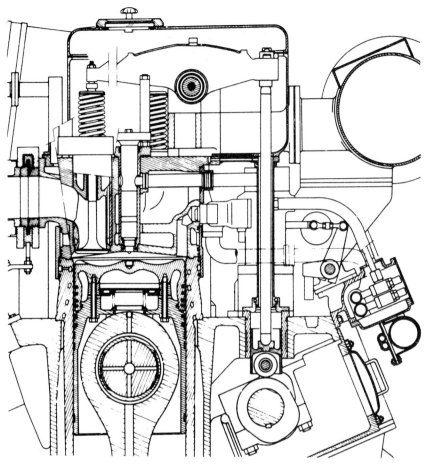

Figure 22.5 Cross section through cylinder head and piston of 42M engine. A pair of inlet and exhaust valves are provided for each cylinder

After careful consideration of alternatives against the requirement of rigidity, the design chosen for the connecting rod is a three-part marine type but with serrations at the big end joint (Figure 22.6).

The exhaust valve cage can be removed easily by taking out the push-rod and lifting the rocker clear.

Valve rotators are used on the exhausts, and guide vanes are formed in the inlet passages of the head to encourage swirl in the incoming air. To reduce the water loading of charge air in humid ambients, Mitsui include a mist catcher after the intercooler to precipitate it.

The injector sleeve is provided with small water passages at the seat. These communicate with the cylinder head water passages and allow a convection flow which gives adequate cooling to the nozzle for opera-

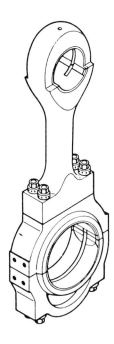

Figure 22.6 Connecting rod of Mitsui M42 engine

tion on all fuels, without the complication of water cooling the nozzle as such.

Much thought has been given to service and maintenance problems, and a full range of tools has been designed to facilitate removal and refitting of parts.

OTHER JAPANESE TYPES

Niigata are substantial manufacturers though not primarily for the marine market. Their designs include some of high power and of more European than traditional Japanese appearance. These include the 40CX of 400 mm bore × 520 mm stroke giving 405 kW/cyl at 450 rev/min, made in six or eight cylinders in-line. They also make a series of long stroke sixes for traditional Japanese marine markets. Fuji Diesel make a variety of engines: these include the S40 of 400 mm bore made in long stroke (620 mm) and shorter stroke (580 mm) versions designated C and B respectively.

Daihatsu, while best known for their higher speed engines, briefly described in the next chapter, make the DS32 of 320 mm bore × 380 mm stroke in six to sixteen cylinder combinations. Cylinder output is 294 kW at 600 rev/min, the bmep being 19.2 bar.

EUROPEAN DESIGNS

Most European designs have been covered in chapters dedicated to them, but some of those left out deserve mention.

MWM

Moteren Werke Mannheim have extended their range to higher powers with the introduction of the TBD510 in-line and TBD511 engines. As MWM's output tends more to high speed it is not altogether surprising that they have adopted a speed of 750 rev/min at mcr. It is included, nonetheless, because it is a high-output engine and intended to be competitive with engines classified as medium speed, and is likely to be sold at lower speeds. (See Table 22.5 and Figure 22.7.) The new range succeeds the TBD501 which was a longer stroke engine running at 574 rev/min.

Table 22.5

Type	TBD510	TBD511
Arrangement	in-line	Vee
No. of cyls	6,8	12,16
Bore mm	330	330
Stroke mm	360	360
Cyl output mcr kW	350	350
Speed rev/min	750	750
Bmep at mcr bar	17.7	17.8

The frame is of SG iron with underslung crankshaft, the bearing keeps being secured by vertical and lateral bolts.

Composite pistons are used, and the cylinder heads incorporate a double deck to control flameplate distortion and stress. Exhaust valves are caged despite the relatively small size, particularly for operation on heavy fuel.

The cylinder liner is not bore cooled, but to segregate thermal stresses it is supported at a level as far below the top as possible. Water circulates up to the gasket to cool the combustion zone.

The camshaft is withdrawn sideways. Connecting rods are of automotive pattern and run side by side on the Vee engine crankpins.

Wichmann

At one time there were many two-stroke engine types in service, and one of the most popular was the Polar loop-scavenged type which ceased manufacture when Nohab introduced the four-stroke F20 range.

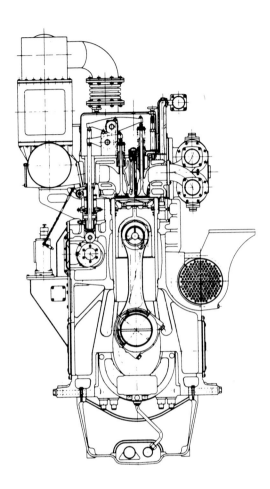

Figure 22.7 Cross section of MWM TBD510/511 engine

An engine which has filled the void left by the passage of the Polar two-stroke, and with which it shares some affinity, is the Wichmann AXA loop-scavenged engine, made by Wichmann Motorfabrik in Norway. Figure 22.8 shows its cross section and main features. Its cylinder dimensions are 300 mm bore × 450 mm stroke and it is made in four, five, six and seven cylinder units. It runs at 375 rev/min to produce 249 kW/cyl at a bmep of 12.56 bar.

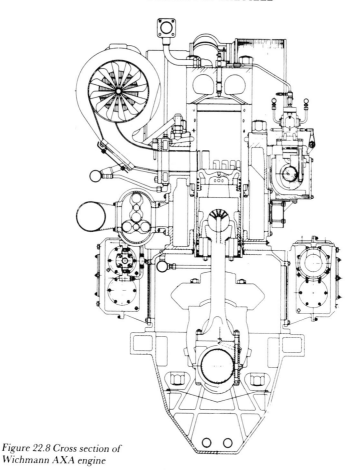

Figure 22.8 Cross section of Wichmann AXA engine

Allen

W. H. Allen will be mentioned in the high-speed section but the 370 engine just meets the power criterion for inclusion in this section. Figure 22.9 and Table 22.6 give the principal data.

Table 22.6 Allen 370 engine

Bore mm	325
Stroke mm	370
Cylinder	6–9
Cyl output mcr kW	244
Speed rev/min	600
Bmep at mcr bar	15.8

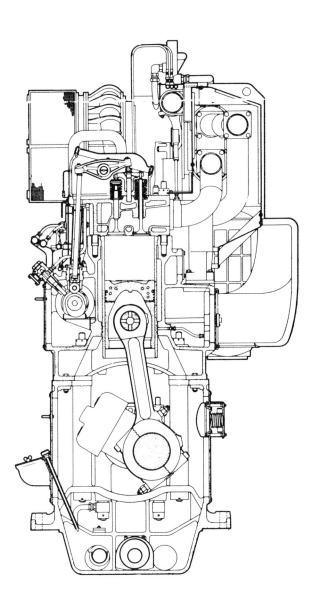

Figure 22.9 Cross section of Allen 370 engine

The cast iron frame is in two parts held together with tie-bolts, and the crankshaft is carried in the deep bedplate.

A three-deck head is used but cages are not necessary. The connecting rods are split diagonally at the big end. The piston is an aluminium component, coil cooled.

The 370 range is developed from the S37 range of the same cylinder dimensions which has been supplied over many years for propulsion of coastal and special purpose vessels.

Ruston

The original Lincoln company, now part of the GEC Diesels group, introduced the AT engine of 318 mm × 367 mm bore and stroke in the 1960s and many of these are in marine service. A successor was introduced in 1980, with the bore increased to 350 mm and designated AT350. The engine is made in six, eight and nine cylinders and produces 373 kW per cylinder at 600 rev/min.

Bolnes

While Smit-Bolnes engines are no longer in production at the time of writing, there are many examples still in service of this unusual crosshead two-stroke engine. The crosshead took the form of a lower piston. Uniflow scavenge was employed with a single valve in the head. (See also the Bolnes design described in Chapter 23.)

B & W

While the smaller B & W four-stroke engines are very much in being, and are described in Chapter 23, the larger engines have ceased production since B & W pooled their resources with MAN.

US DESIGNS ENTERPRISE

Enterprise

The Enterprise Division of Transamerica De Laval is one of two US makers offering engines in this class, and the only one with any marine experience. Figure 22.10 shows a typical cross section and the main features. Bore and stroke are 432 mm × 533 mm. Engines are built in-line and in Vee form up to twenty cylinders of up to 504 kW/cyl at 450 rev/min. The dominance of the US market by volume produced

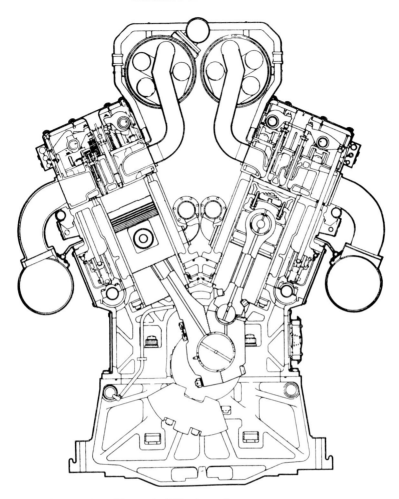

Figure 22.10 Cross section of Enterprise RV series engine
(Courtesy of Diesel Engine and Gas Turbine Progress)

locomotive designs has tended to restrict the demand there for engines of slow- or medium-speed construction, but the Enterprise is popular among those who prefer such construction.

D.A.W.

23 High-speed engines and auxiliaries

PRINCIPLES

Much of the discussion in Chapter 13 applies also to the engines considered here.

It is perhaps necessary to stress that *high speed* in the chapter heading reflects the current marine convention whereby *medium speed* covers the range up to 600 rev/min only, so that at present everything above that speed must, by default, be high speed.

Of course, the marine convention tends to close the lists altogether at 1000 rev/min, but it has to be recognised that, in the non-marine fields, the division between medium and high speed now occurs somewhere around 1000–1200 rev/min.

Also, whatever inclinations or preferences each of us has for generously rated machines running at modest speed — and despite instances of unhappy experience befalling some examples of more daring specification — the marine engineer will always be governed by two basic tenets:

> The machinery must work as intended, reliably, and go on doing so for the life of the vessel.
>
> The machinery chosen must also give the lowest overall cost.

Cost is not solely a question of fuel, lubricants and maintenance. All owners to some degree have to recognise the constraints of capital cost, especially when interest rates rise well into double figures.

Whether capital is borrowed on the financial market or taken from reserves, it has to be serviced. Over the life of the vessel the sum of interest charges and operating costs may indeed be lower with a low initial cost but high maintenance cost design provided it is reliable. If the owner uses a grant to build he can obviously get more ship if he lowers his standards. Perhaps this might be the only way the vessel will be authorised. The equation of costs is very much influenced by whether the vessel will trade on the owner's account or on time charter, or whether it is to operate on bare-boat charter.

It is therefore inevitable that the lower initial cost of high-speed engines, plus their advantages in space and weight to the vessel as a whole, will always be tempting. If, in addition, the owner is able to find adequate reassurance about the time between overhauls, the reliability, the fuel and maintenance costs, he may decide to join those who have reasonably painlessly overcome their misgivings about speed. It is impossible to generalise in this context, because success depends not so much on the characteristics of the design as on the maker's diligence in rigorously eliminating the cause of every fault reported especially with regard to detail.

More particularly in high-speed rather than in medium-speed engines it may be possible to assess the prowess of an engine in other duties and environments. One needs to be cautious, as maintenance conditions may be significantly different — as Chapter 13 points out. Nevertheless, one cannot ignore success in a locomotive application, particularly as this field usually involves heavy duty, limited periodic maintenance and long overhaul life. Naturally no responsible maker entering the marine field will — or dare — take the marine environment for granted. Adaptations from any other application to marine service must be just as thoroughly thought out and proved as it would be for a specifically marine engine.

Need and capability in the high-speed propulsion category have been successfully married in a substantial number of instances, particularly in the power range extending to 3000–4000 kW. Such engines are particularly suitable for tugs, ferries, trawlers, coasters, etc., but have been applied to propel vessels up to 9000 tons dw on a single screw.

AUXILIARIES

Whatever an owner's feelings for such engines as main propulsion units, he seldom chooses engines running below 720 rev/min for auxiliary power, except for very large installations such as fully refrigerated vessels.

The years 1950 to 1970 saw a considerable increase in the electrical loading of deep sea cargo vessels due to the adoption of electric deck machinery, higher standards of crew accommodation, greater mechanisation, automation, etc. These have caused the typical sea load to rise to 200–400 kW, and the harbour load when working cargo to nearer 500 kW.

Prophecy is always risky and the rise in the 1970s has been small but it does not seem that there are any factors waiting to force a resumption of the increase. Refrigerated vessels, or those requiring climatisation for soft fruit cargoes, usually have an unbalanced requirement on the two legs of their round voyage, and this will usually entail the installation of more, as well sometimes, of larger units — even up to 1.5 MW.

Given the universal use of ac machinery, auxiliaries are obliged to operate at one of the appropriate synchronous speeds:

600, 720, 900, 1200 or 1800 rev/min for a 60 cycle distribution system;
600, 750, 1000 or 1500 rev/min for a 50 Hz system.

Marine practice is to have three generators in a general cargo vessel, one to carry the normal sea or harbour load, one on standby and one available for maintenance. Refrigerated vessels usually carry four sets and use two on the inward leg of the voyage; but the solution chosen must reflect the needs of each case.

At the time of writing the most popular speed is 720 or 750 rev/min for the 500 kW size, rising to 1200 or 1500 rev/min for the 200 kW size which suffices for coasting vessels. In fact it is becoming very difficult to find makers offering machinery for such powers running at lower speeds, because of the limited demand. This applies not only to the engines but to the electrical machines. There is a growing tendency to take advantage of cp propellers by using the main engine, or one of them, to provide the sea load. This saves fuel, noise, and wear on the auxiliaries; but it remains necessary to transfer the load back to an auxiliary generator when coming into port.

The one-generator-on-the-board philosophy, particularly if driven by the main engine, means that an automatic start facility is essential for the stand-by generator, so that in the event of the running set failing power is not lost.

Since a turbocharger takes time to accelerate, no diesel engine can accept instantaneously more than a load equivalent to a bmep of 10 bar or so. If the owner wishes to use an engine rated at more than this, the automatic start must include selective reconnection to limit the bmep equivalent of the initial load to the prescribed 10 bar (or less if it is still required to be smoke free when starting). If the engine on stand-by is continuously primed, or time-cycle-primed, so that the lubrication system is always charged, it can be started and generating in 8–10 seconds, after the failure signal. Otherwise it will take at least 13–15 seconds.

Until the late 1970s it was held that it was pointless to use heavy fuel in auxiliaries — and it is still problematical in the main at speeds over 1000 rev/min. With the worsening fuel situation, and since the propulsion versions of many auxiliary engines are being adapted for fuels up to 1000–1500 secs Redwood No. 1, there is a tendency for auxiliaries now to be operated on such fuel.

HIGH-SPEED ENGINES

The following pages contain a brief résumé of some of the engines available under the category of high-speed engines. The field is vast, with dozens of makes and many models. It will be useful, therefore, to group them on a geographical basis, which will allow some generalisations to be made. As far as is possible emphasis will be put on the more significant — whether because of novelty or numbers used — and on the lower speeds, because they are used for propulsion. Consideration of the smallest and highest revving engines belongs properly to other more specialised volumes. Unless otherwise stated, engines discussed in this chapter are not direct reversing. Where reference is made to 'heavy fuel' it should be taken to mean a viscosity rating of 1500 secs Redwood No. 1 at 37 °C unless a lower value is indicated.

European designs

There are over 40 separate engine designs available from over two dozen makes in at least 12 countries. Several are licensed worldwide. Many examples of about 240–260 mm bore are developed from earlier designs with maximum speeds of up to 1000 rev/min. Some engines approaching this size are completely new with envisaged speeds of 1100–1200 rev/min, though their initial experience is at lower speeds. Just under half retain the conventional arrangement of crankshaft carried in a bedplate. The rest including most, but certainly not all, of the newer designs have adopted the underslung arrangement.

Examples are available of engines running at speeds up to 1500 rev/min generally with bore sizes of 180–200 mm. Some engines are two-stroke and run at much lower speeds, and some are borderline — between 'medium' and 'high' speed philosophy.

Bedplate mounted crankshaft, up to 1000 rev/min (four-stroke)

Engines in this group mostly have bores of 240–260 mm. Table 23.1 sets out some features of a representative selection.

Much the most successful of this group in terms of horsepower sold is the Ruston RK, though it is probably equalled in numbers by the Mirrlees Blackstone E. The Ruston RK traces its development back to 1942 when the then English Electric Company redesigned an earlier machine of the same cylinder size. It was designed with naval and locomotive applications in mind. Development has been continuous and it is, at the time of writing, the mainstay of Ruston Diesels Ltd., a constituent of GEC Diesels. Many thousands have gone into locomo-

Table 23.1

Make	ABC	Allen		Crepelle	Mirrlees Blackstone	Normo	Ruston		SWD
Type	D2	S12	S37	PSN3	EMk2	LDM	AP230	RKC	F240
Country of origin	Belgium	UK		France	UK	Norway	UK		Holland
Bore mm	256	241	325	260	222	250	230	254	240
Stroke mm	310	305	370	280	292	300	273	305	260
Cyl output mcr kW	223	139	191	188	117	127	179	195	165
Speed rev/min	1000	750	600	1000	1000	750	1000	1000	1000
Bmep at mcr bar	16.9	16.4	12.7	14.3	14.1	14.0	19.0	15.5	16.9
No. of cyls in-line	6	4,6,8	—	4,6,8	6,8	6,8,9	4,6	6	6,8,9
No. of cyls Vee	—	—	12,16	12,16	—	—	—	8,12,16	—

tives and large numbers are found in all types of marine service.

The development has included all the usual features associated with uprating, but it has some notable features which can be discerned in Figure 23.2. These include the ingenious and unique inclined-bolt big end which affords, despite a large crankpin diameter to cylinder bore ratio, a rigid housing with a simple horizontal split, while the rod can still be withdrawn through the cylinder. Although uprating eventually makes a composite piston inevitable, Ruston have been able to keep a one-piece aluminium alloy piston with a salt-cored inclined elliptical oil cooling cavity despite the permitted bmep at lower speeds than 1000

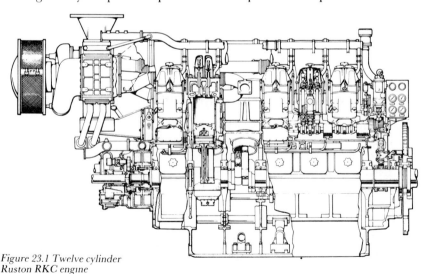

Figure 23.1 Twelve cylinder Ruston RKC engine

rev/min rising almost in inverse proportion to the speed. For instance, at 720 rev/min the bmep is 18.9 bar.

The RK engine has the joint face on each side of the main bearing cut at an angle of 15° to the horizontal to guarantee that the keep fits tightly against the abutment. Except for the cylinder head studs all major fastenings are bolts or stud bolts. All major oil supply lines are in the form of pipes rather than drillings, the makers preferring in this way to guarantee their cleanliness.

The RK camshaft runs in tunnel bearings and is withdrawable in two cylinder sections. The Vee engines have phasing gears; the six in-line has chain drive. The eight Vee now has a short simplified Lanchester balancing system. Other details can be seen in Figure 23.1. In the 1960s the RK engine was one of the first of its size to be developed to burn heavy fuel where Nimonic exhaust valves are fitted.

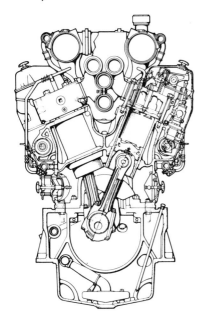

Figure 23.2 Cross section through Ruston RKC Vee engine

The other Ruston engine listed — the AP230 — is a recent development of the AP2 mainly by increasing the bore from 203 mm. This engine was intended, and has been very popular, for auxiliary generation. The four-cylinder version is limited to 500 kW at 720/750 rev/min precisely for this market (Figure 23.3). The cams are in clusters clamped to a conventionally mounted camshaft via cotters.

The Mirrlees Blackstone E Mark 2 is a development of the original Blackstone E which has sold well since the late 1940s, and is still avail-

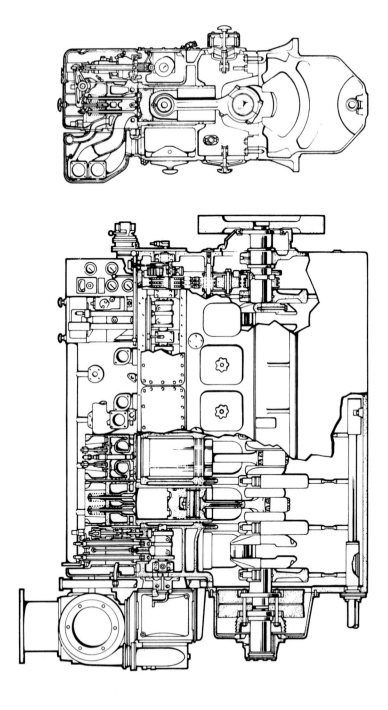

Figure 23.3 Sectioned elevation and end elevation of Ruston 6AP230

able. The older E is made in four cylinder form as well as six and eight, and it is also available in twin bank form with twelve or sixteen cylinders. The twin bank engines comprise two crankcases cum cylinder housings of standard pattern (but one the mirror image of the other) bolted back-to-back and onto a common bedplate holding the two crankshafts. These were meshed to a common output shaft between them by gearing aft of the flywheels.

Compared with the E, the E Mark 2 has been uprated about 15% and redesigned above the crankshaft (Figure 23.4). Internally, an automotive type rod (diagonal split) has replaced the conventionally split big end, and the head now has four valves. Externally, the characteristically neat, all enclosed arrangement of fuel pumps, tappets and camshaft has given way to externally mounted fuel pumps set at an angle so that the HP fuel lines can be led outboard of the air manifold to the injector which is external to the head. The rocker enclosures are individual and the fuel pump is now at the side of them rather than between them.

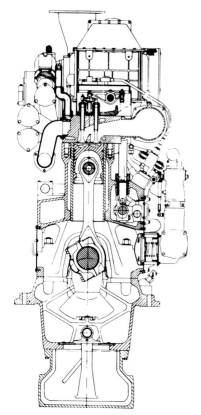

Figure 23.4 Mirrlees Blackstone ESL Mk 2

W. H. Allen have been making the S12 and S37 for many years but on a smaller scale than the engines described above. Their uprating policy has always been cautious, but they have been very successful in the marine field. The S37 is very like the 370 described in Chapter 22, and is the basis from which the 370 was developed as an uprated, in-line, version of the S37 which is sold mainly as a Vee. Like the 370, the S37 has a very deep bedplate cam crankcase, the joint with the cylinder housing being well above the crankshaft.

The S12 is an engine of very similar philosophy to the Ruston RK except that it retains a conventional big end, and the fuel pump is mounted to one side of the valve pushrods and is not enclosed. An unusual feature is that the rockers are splayed and the valve stems for each function are inclined to the vertical. That is to say, the combustion chamber has a slightly penthouse form, which leaves more room for the injector. Allen have offered the S37 to burn heavy fuel for many years and this applies now to the S12 also.

ABC, the Anglo-Belgian Company, have been making the DX engine for many years as a choice of cylinder combinations in in-line

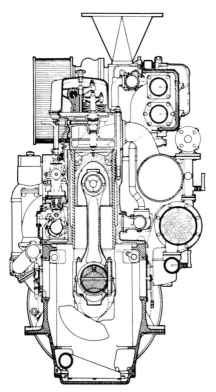

Figure 23.5 ABC DZ engine

form largely for marine purposes. The DZ (Figure 23.5) draws heavily on this experience as well as established technology, to produce an engine of about double the cylinder output of the DX. Note the deep bedplate, the conventional big end, the welded two-part piston, and the HP fuel connection taken through the head to the injector.

SWD's F240 engine is a successful surviving design of the original Kromhout Motoren Fabriek which was taken over by SWD when the Dutch diesel engine industry was rationalised: it has been further developed by SWD. Figure 23.6, shows in cross section the principal features: note the short stroke and relatively low camshaft. Pistons can be either one-piece aluminium as shown, cast iron or composite, depending on application. Air and exhaust manifolds are on the same side of the engine away from the fuel lines. The engine normally burns Class BI fuel, but worse fuels can be burnt by agreement with SWD.

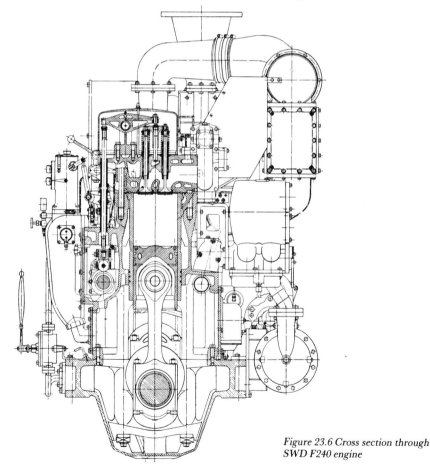

Figure 23.6 Cross section through SWD F240 engine

Normo is the name used by Bergen Mekaniske Verksteder to designate Bergen engines intended for marine propulsion. Bergen have made the LDM for some years, and its features are discernible in Figure 23.7. It has a very deep fabricated bedplate cam crankcase, while the cylinder housing remains cast iron. Air and exhaust manifolds are on opposite sides and the two valves for each function are in parallel. The normal fuel is diesel but worse fuel can be used if required.

Crepelle et Cie have made diesel engines for a long time but the PSN3 is a relatively new design (Figure 23.8). It has a very deep cast bedplate-cum-crankcase held to the cast cylinder housing by tie-bolts. Note the through-flow heads, cooled one-piece pistons, external HP fuel system. Long stroke versions of the in-line engines are available for lower speeds.

Mention should also be made of the VD26/20 engines of Schwermaschinenbau Karl Liebknecht in the German Democratic Republic. These are conventional six or eight cylinder in-line machines running up to 1000 rev/min. Bore and stroke are 200 mm × 260 mm. Output is 110 kW/cyl.

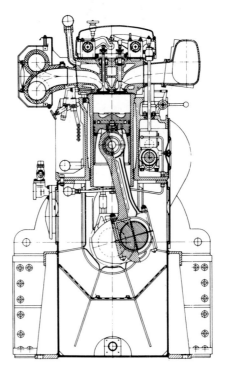

Figure 23.7 Cross section through Normo LDM engine

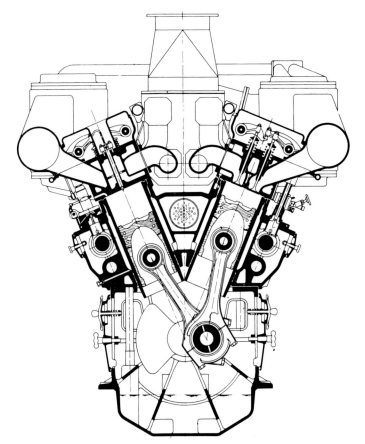

Figure 23.8 Cross section through Crepelle PSN3 engine

Engines with underslung crankshafts, up to 1000 rev/min (four-stroke)

All of the engines in this group are relatively new, that is to say, they were first produced in the 1960s or 1970s, or are fairly drastic redesigns of earlier engines.

We will consider a selection of them in two groups, alphabetically chosen, in Tables 23.2 and 23.3.

B & W Alpha are best known for their integrated propulsion concept, whereby one factory produces the engine, gearbox, tailshaft and propeller. The engines were originally two-stroke, but the T23 (in-line) and V23 (Vee) four-stroke range has been established since the mid-1960s, and the corresponding S28/U28 since the early 1970s.

The two engines differ very little except in size, so Figures 23.9 and

Table 23.2

Make	B and W Alpha		Deutz	GMT	MaK	MAN		MWM
Type	T/V23	T/U28	BVM628	B230	M332	L/V20/27	ASL/ASV 25/30	440/441
Country of origin	Denmark		West Germany	Italy		West Germany		
Bore mm	225	280	240	230	240	200	250	230
Stroke mm	300	320	280	270	330	270	300	270
Cyl output mcr kW	114	195	184	186	160	100	220	128
Speed rev/min	825	775	1000	1200	750	1000	1000	1000
Bmep at mcr bar	13.9	15.3	17.4	16.7	17.3	14.15	17.9	14.0
No. of cyls in-line	5,6,7,8	5,6,7,8	6,8,9	4,6	6,8	4,5,6,7,8,9	6,8,9	6,8
No. of cyls Vee	8,10,12,14,16,18	8,10,12,14,16,18	12,16	8,10,12,14,16,18,20	12	12,14,16,18	12,16,18	12,16

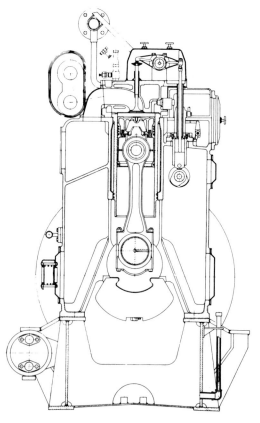

Figure 23.9 Burmeister and Wain 225/300 and 280/320 in-line engine typical cross section

23.10 show the T23 and V23 respectively as representations also of the S28/V28. Note the air chamber in the frame. The fuel pumps and HP lines are enclosed but segregated from the lube oil spaces. The pistons are of cast iron. The camshaft is formed in sections, is driven by phasing gears, and is tunnelled in the cylinder block. The fuel pump is outside the valve pushrods on the 23; between them on the 28.

In more recent T23 engines the connecting rod, shown with a conventionally split big end has been replaced by the automotive type shown in Figure 23.10 on the V23, and similar to that used on the

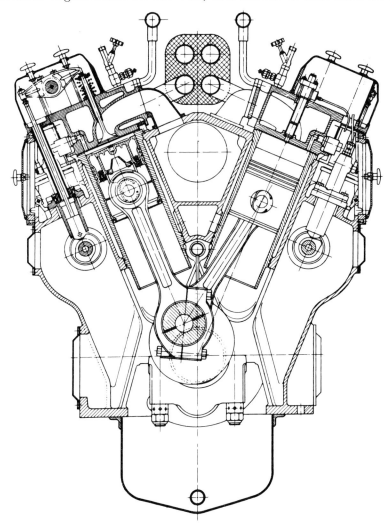

Figure 23.10 Burmeister and Wain Vee type engine

S28/V28. The later S28/V28 engines have four valve heads with an intermediate deck. The T23 is normally coupled to the gearbox through the free end (i.e. non-flywheel end) of the crankshaft. The 28 size engines will burn heavy fuel; the 23 is restricted to fuel of 200 secs Redwood No. 1.

The Alpha system provides either fixed pitch or cp propellers each with the appropriate gearbox. The cp servomotor is mounted within the gearbox output wheel. The gearboxes have hardened helical teeth, sleeve bearings and Michell type thrust bearings. Of the combinations shown in Table 23.2 the 8T23, the 5S28 and the 8U and 10U28 are not offered as propulsion sets. As alternator sets the 8V23 is not offered.

Deutz Engines have, in 1981, replaced the 528 series by the 628, which has been developed from it. The 528 engine had a smaller bore and a two valve head. The 628 is intended to accept heavy fuel up to 2000 secs Redwood No. 1. Figure 23.11 shows the general features of the in-line version. The Vee engines have a Vee angle of 48°. Note the one-piece cast SG iron crankcase and cylinder housing, which carry the combustion loadings directly from cylinder headstuds to main bearing

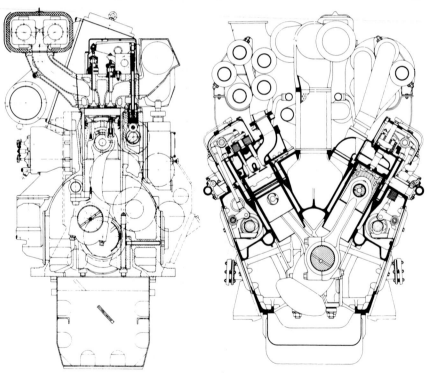

Figure 23.11 Deutz BVM 628 engine

Figure 23.12 Cross section through GMT B230 engine

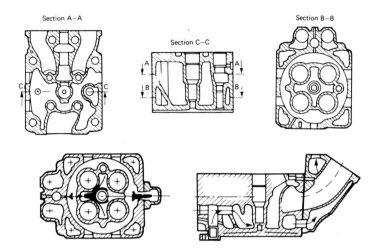

Figure 23.13 BVM 628 cylinder head showing coolant flow

studs. Note also the high transverse bolt securing the main bearing keep close to the joint face.

The exhaust piping is fully enclosed by insulated covers. Turbocharging on six and twelve cylinder engines is on the normal pulse system: for the eight, nine and sixteen cylinder engines a single-entry multi-pulse system is used.

Deutz have developed an unusual solution to the problem of flameplate distortion/support: a conical plate supports a reduced flameplate at about two-thirds of the bore radius. This is shown in Figure 23.13 which also indicates how water is directed first to the centre of the flameplate. Nimonic 80A exhaust valves are standard, and armoured valve seats are considered unnecessary. All valves have rotators.

Deutz, like many German makers, prefer camshaft pumps. The long injection lines and soft characteristics associated with these, especially with large engines, are, however, difficult to reconcile with the crisp injection needed at high output. Their solution to shorten the lines is to group either three or four adjacent cylinders at a time into smaller block pumps positioned opposite the centre of the group of cylinders. The HP connection is taken through the head to the injector so that the HP line is little, if any, longer than with individual pumps.

GMT 230 engines are also offered in a long stroke version (310 mm). There is also a very similar series of 210 mm bore. Figure 23.12 shows, typically, the Vee connection. The usual features are visible, but note the annular cavity cooling the monobloc aluminium piston, the lever type cam followers and tunnel type camshaft in segregated compartment.

The frame is cast iron. Heavy fuel can be used. The engine is included in this group despite its 1200 rev/min capability because its philosophy is similar to the rest of the group.

Mention should be made also of another Italian maker, Franco Tosi, who produces the QT320 SSM of conventional underslung design in six to sixteen cylinders, 279 kW/cyl at 730 rev/min.

MaK's M332 series is a development from the M282 of the same bore but which has a shorter stroke (280 mm) and runs at up to 1000 rev/min to produce the same cylinder power. There is also a derated version of the M332, namely the M331 with two valve heads and uncooled pistons producing 125 kW/cyl. The M332 is shown in Figure 23.14.

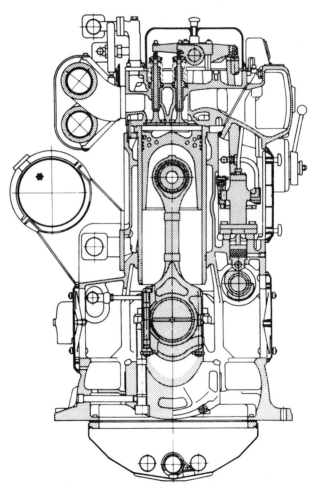

Figure 23.14 Cross section through MaK M332AK engine

The crankcase/cylinder block is a single iron casting and auxiliaries are driven from the flywheel end. For heavy fuel, and for higher ratings than 160 kW/cyl, composite pistons are preferred, in which case there are two chromium plated top rings carried in the crown section, a plain ring and conformable SOC (slotted oil control) ring in the aluminium skirt. Note the inclined air manifold/cylinder head joint for ease of assembly.

MAN's 20/27 engine range is relatively new and the in-line version is shown in Figure 23.15. The engine is unusual in having a very high camshaft where cams act directly on the valve rockers. To keep the engine compact the rocker casing is integral with the head and the fuel pump is ingeniously suspended below the camshaft. Two valve heads are used: when heavy fuel is burnt these are fitted with rotators and a two-part piston is used. On the in-line engines only, the charge-air cooler is integrated into the frame. These engines are constant pressure turbocharged. All auxiliaries can be driven from the camshaft phasing gears.

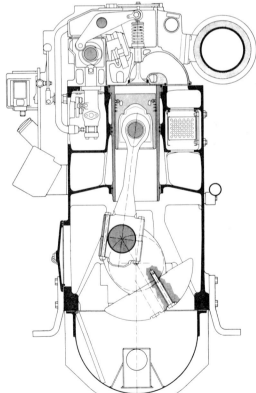

Figure 23.15 Cross section through MAN L20/27 engine

The 25/30 engine is the fruit of the co-operation agreement signed late in the 1960s between Sulzer and MAN and is almost identical with Sulzer's own AS25/30, even to the exterior detail which is more reminiscent of the latter company's style.

The MAN version has some differences, such as the use of composite pistons. The MAN Vee is shown in Figure 23.16, the Sulzer in-line version is shown in Figure 23.20. Note the high flange fuel pump and lever follower. Other details are mentioned below. The MAN and Sulzer ranges include some differences in the cylinder combinations offered.

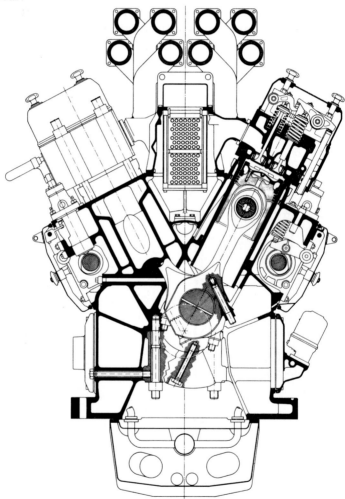

Figure 23.16 Cross section through MAN ASV 25/30 engine

HIGH-SPEED ENGINES AND AUXILIARIES

The MWM 440 and 441 engines are respectively the in-line and the Vee versions. These engines are at the present time the largest in a fairly extensive family using camshaft fuel pumps, and have always tended to prefer compactness to low speed. Their construction follows accepted good practice.

Table 23.3

Make	Mirrlees Blackstone	Nohab Polar	SACM	SEMT	Sulzer		Wärtsilä		
Type	MB275	F20	AG0	PA6	AL20/24	AS25/30	Vasa 22	Vasa 24	Vasa 32
Country of origin	UK	Sweden	France		Switzerland		Finland		
Bore mm	275	250	240	280	200	250	220	240	320
Stroke mm	305	300	220	280	240	300	240	310	350
Cyl output mcr kW	288	164	180	298	102	200	146	156	309
Speed rev/min	1000	825	1300	1050	1000	1000	1200	750	750
Bmep at mcr bar	19.1	16.3	16.7	19.1	16.39	16.29	15.8	15.7	17.6
No. of cyls in-line	6,8	4,6	6,8	—	6,8	6,8	4,6,8	4,5,6,8	4,6,8,9
No. of cyls Vee	12,16	8,12,16	8,12 16,20	12,16,18	—	10,12 16,18	12,16	—	12,16 18

The MB275 engine was developed from new by Mirrlees Blackstone in the late 1970s. Figure 23.17 shows both the in-line and the Vee versions.

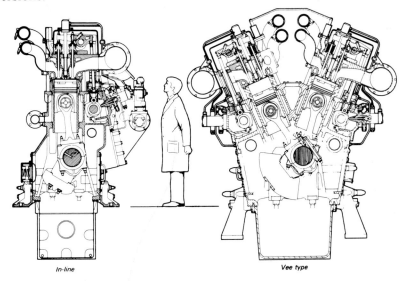

Figure 23.17 Mirrlees Blackstone MB275 in-line and vee engines

The engine is capable of burning heavy fuel. Note the three-part connecting rod, and massive lateral bolting of the main bearing keep.

The piston is of composite construction (steel and aluminium) carrying all the rings in the crown. The camshaft is assembled in a separate housing as a unit and side loaded onto the engine. The housing carries the pumps directly. A spragging gear is incorporated to shut the engine down in the event of overspeed. Full power can be taken from either end of the engine.

Although shown as Swedish and still built there, Nohab engines have become part of Wärtsilä. Nohab's F20 has been manufactured since the mid-1960s and follows fairly standard practice for this type of engine. Figure 23.18 indicates its main features. It has been joined by an uprated version of the same size, the F30, running at up to 1000 rev/min. The F20 has a single-piece cooled aluminium piston and a relatively high camshaft. It will burn heavy fuel up to 1000 secs Redwood No. 1.

SACM (Société des Ateliers et de Constructeurs de Mulhouse) is included despite its high speed, because it follows the construction philosophy of the group. Its marine applications include a number of high performance and special purpose craft.

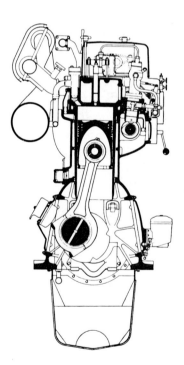

Figure 23.18 Cross section through Nohab F20 engine

HIGH-SPEED ENGINES AND AUXILIARIES

SEMT (Pielstick) introduced the PA6 in the late 1960s and from the outset the emphasis was on compactness, mainly for locomotive application; but this feature is also valuable to the naval architect. Accordingly, the connecting rods are shorter than the classical 4 × the throw. A 60° Vee angle was chosen to be able to pack more equipment within the banks.

All PA6s are Vee engines. The section in Figure 23.19 shows the features which we have now come to expect as normal in a highly rated engine. The phasing gears are at what is normally the non-drive end, though power can be taken from either end. The engine has been developed to burn the full range of heavy fuel, and the nozzle tip is then water cooled. The engine is widely licensed round the world.

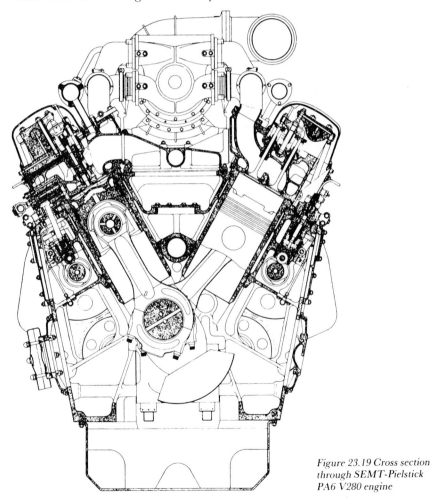

Figure 23.19 Cross section through SEMT-Pielstick PA6 V280 engine

Sulzer's AL20/24 and AS25/30 engines are so very similar that the constructional features of both can be appreciated from Figure 23.20 which shows the ASL25/30. As mentioned previously MAN were associated with Sulzer when the 25/30 engine was put on the market, and both companies manufacture (and licence) it; though as Tables 23.2 and 23.3 show only MAN offer a nine in-line, and only Sulzer a ten in-line. Note in Figure 23.20 the HP fuel connection and the three-deck head, which has now been superseded by a bore-cooled version like the ZA40: also the camshaft is withdrawable sideways and the charge cooler integral with the frame.

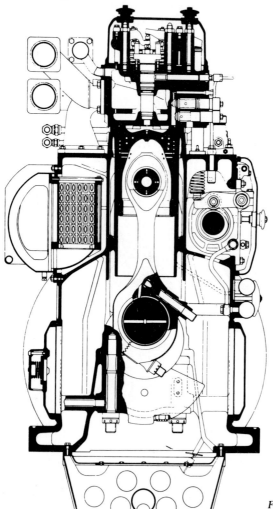

Figure 23.20 Cross section through Sulzer ASL 25/30

On the Vee engine the charge cooler is mounted along the Vee. The AS25/30 engines will burn heavy fuel.

Wärtsilä have made the 14TS engine (actually a bedplate type engine) for many years before it was uprated to become the Vasa 24. It was principally and originally intended as an auxiliary engine, but it has since been applied to a wider range of duties. The Vasa 22 joined it as a new engine in 1970 and is illustrated in Figure 23.21. It represents good modern practice, and brackets the Vasa 24 in the range of outputs offered. Since the 1970s this has been joined by another new engine of the same philosophy but double the cylinder power — the Vasa 32.

It is not possible to mention all engines which qualify for inclusion in these sections, nor can we anticipate new designs, nor changes in the relative fortunes of those included and those not included. We should note, however, the 240CO engine made by Cockerill in six to sixteen cylinders delivering 202 kW at 1000 rev/min, and Franco Tosi who offer the QT320SS and the Q12BSS giving 272 kW/cyl at 730 rev/min and 162 kW/cyl at 1085 rev/min respectively.

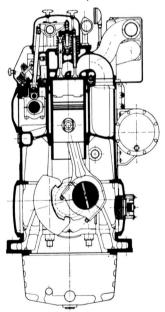

Figure 23.21 Cross section through Wärtsilä VASA 22 engine

OTHER FOUR-STROKE ENGINES

All the engines described up to now have used direct injection, but there are some whose makers find it advantageous to offer engines with indirect injection of which the most refined system was the Ricardo swirl chamber. (See also Chapter 5.)

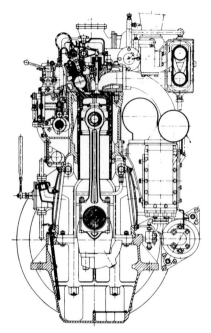

Figure 23.22 Cross section through SWD DRO210K engine

SWD are one manufacturer who include a pre-chamber engine in their range. This is the DRO210K, shown in Figure 23.22. The Ricardo head and individual pump can be clearly seen. An extra advantage claimed for this system is that the relatively clean combustion obtained enables the engine to cope better with prolonged light load running. Bore and stroke (relatively long) are respectively 210 × 300 mm and maximum speed 900 rev/min.

Paxman, part of the GEC Diesels Group in the UK, still make the RPH engine which, in its V6 and V8 forms, has been fitted in many SD14 ships. It runs at 1200 or 1500 rev/min and is of 178 mm bore. Paxman's latest designs, however, are all high performance DI engines primarily designed for naval craft and other duties demanding high power/weight ratios.

SEMT Pielstick at 1500 rev/min also favour indirect injection. Figure 23.23 shows the PA4–185 which runs at up to 1500 rev/min and is capable of accepting heavy fuel. The pre-chamber can be seen at left. Bore and stroke are 185 × 210 mm and it is made in in-line and Vee form.

SACM in France also offer an engine, the MGO, of 195 mm bore × 180 mm stroke, running at this kind of speed.

MTU, Motoren and Turbinen Union, are probably the largest firm dedicated to high performance high-speed engines. Typically their

HIGH-SPEED ENGINES AND AUXILIARIES

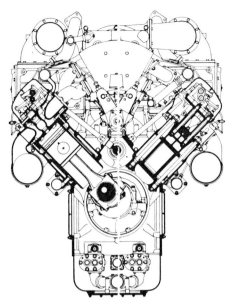

Figure 23.23 Cross section through Pielstick PA4-185 engine

V652TB range is illustrated in Figure 23.24. Bore and stroke are 190 × 300 mm and speed is 1380 rev/min. Indirect injection is used. On the other hand, other MTU models — such as the 956 — use direct injection. Despite their naval and high performance association, MTU have engined many smaller mercantile vessels and provided generating sets for larger ones.

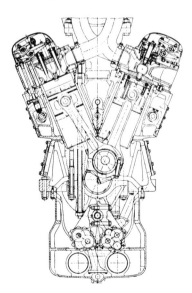

Figure 23.24 Cross section through MTU V652TB engine

MWM provide several ranges of engine in the higher speed category.

Mirrlees Blackstone have introduced the MB190, which in many ways is a higher speed version of the MB275 on a smaller scale.

Hedemora Verkstäder make an engine of similar philosophy to the Pielstick PA4 but use a tunnel crankcase and individual injection pumps.

TWO-STROKE ENGINES

There are now only a handful of European two-strokes in the marine field. The Wichmann AXA engine has been mentioned in Chapter 22. The same firm produce a small version of this engine and in Vee form, the VX, in six to sixteen cylinders. Bore and stroke are 280 mm × 360 mm, and the VX produces 225 kW/cyl at 600 rev/min.

Holland is very much associated with two-stroke engines, and one of the amalgamations of the 1960s brought together Brons and De Industrie. Brons have for many years produced the GV, a uniflow engine of 220 mm bore × 380 mm stroke, made in six to sixteen cylinders in Vee form. It runs at 375 rev/min with an output of 92 kW/cyl. It has been joined by the turbodiesel of the same size and configuration but developing 184 kW/cyl at 600 rev/min.

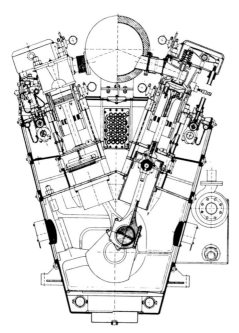

Figure 23.25 Cross section through Bolnes VDNL 150/600 engine

A higher output two-stroke which is no longer made, the Smit Bolnes engine was mentioned in Chapter 22. It had much in common with the Bolnes VDNL150/600, (another crosshead design) which is still made by Bolnes Motorenfabriek. The cross section of this engine (Figure 23.25) very much resembles a scaled down version of the larger engine. The Bolnes VDNL150/600 has a bore and stroke of 190 mm × 350 mm and generates 110 kW/cyl at 600 rev/min. It is made in versions of three to twenty cylinders.

JAPANESE DESIGNS

Many designs exist in the Japanese diesel engine industry, but those of greatest interest in the context of this chapter are the ones most likely to be seen in international shipping.

By far the most prolific is Daihatsu, whose engines are used for propulsion in smaller craft, and for many years as auxiliaries in the majority of the ocean going ships built in Japan. They are also licence-built in other countries.

As mentioned in Chapter 22, Japanese designers either follow the traditional slow-speed long stroke and relatively simple philosophy preferred by their Far Eastern markets, or they follow the pattern which has come to be accepted by their Western customers. But even in the latter case the approach tends to follow well-established practice. The resulting machines are therefore more notable for their dependability than their originality.

Daihatsu engines follow the 'Western' philosophy. A typical cross section of a Daihatsu engine, the larger DS28, is shown in Figure 23.26 and illustrates the point. The engine is of 280 mm bore × 340 mm stroke (i.e. very close to the 1.2 ratio of the engines in the group of the European engines described previously). The crankshaft is carried in a deep bedplate — note the extensive ribbing. Resilient studs hold bedplate and cylinder block (both cast iron) together. The construction features can be seen to include the measures usually appropriate to carry a bmep of 19.9 bar at 720 rev/min, which gives an output of 248 kW/cyl. At the time of writing the engine is available only as a six in-line, and will accept fuel of up to 600 secs Redwood No. 1.

Daihatsu also make the DS26 of 260 mm bore in various combinations, of which the six cylinder version, with the stroke reduced from 320 to 300 mm, and speed increased from 720 to 900 rev/min, delivers up to 196 kW per cylinder.

There is also the smaller DS22 of 220 mm × 280 mm bore and stroke made in various cylinder combinations. The six cylinder again has been

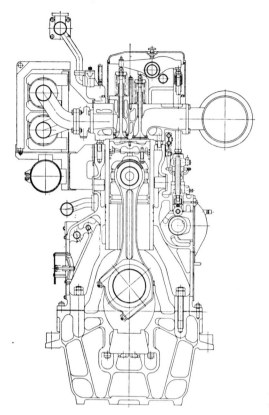

Figure 23.26 Cross section through Daihatsu DS-28 engine

developed to higher speeds (with a reduced stroke of 250 mm). It produces 147 kW per cylinder at either 900 or 1200 rev/min. Daihatsu tend towards a range based mainly on six cylinder engines.

Yammar are the world's largest producer of non-automotive diesel engines in terms of number, but their output is largely of units of 5–10 hp. Their output of larger engines is still significant enough for them to be a major supplier of auxiliaries in Japanese-built ships. Most Yammar 'large' engines are built in six cylinder form and, like Daihatsu, in what a Western trained engineer would regard as good traditional practice. The largest is the ZL, made in six and eight cylinder configurations which produces 221 kW/cyl at 750 rev/min. Bore and stroke are 280 × 340 mm. There are several other sizes: the GL at 240 × 290 mm, the T220KL at 220 × 280 mm, and so on.

Niigata and Fuji also produce engines of similar types covering a wide range of powers; and so do Mitsubishi. See, for instance, Figure 23.27, which indicates the Niigata 28BX. It has a bore and stroke of 280 mm × 320 mm and is produced in six to eighteen cylinders rated at 208

HIGH-SPEED ENGINES AND AUXILIARIES

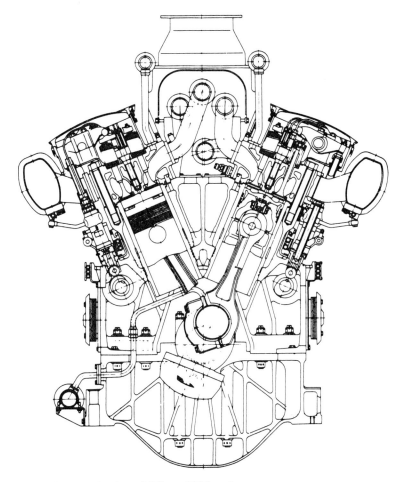

Figure 23.27 Cross section through Niigata 28BX engine

kW/cyl at 720 rev/min. There are also the traditional Japanese designs like Hanshin; and Akasaka provide engines in this category. See, for example, Figure 22.2 or (typically) Figure 22.1.

US DESIGNS

Finally, American high-speed engines need to be considered. The American diesel engine industry is dominated by automotive sizes on the one hand and locomotive sizes on the other. The size of the domestic markets for these have led to such huge production volumes that there is a very considerably cost advantage to non-automotive or non-

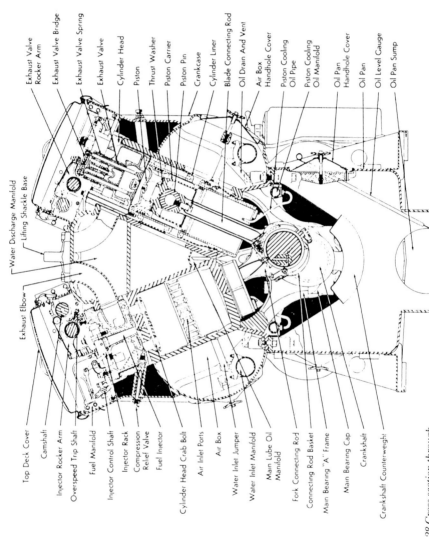

Figure 23.28 Cross section through Niigata 28BX engine

locomotive users to adapt one of these units to their purpose rather than to seek the kind of purpose-built unit their non-American counterpart would. Because of their widespread use in the US there are too many skilled engineers well versed in handling such engines. Nor can it be denied that even when such a daunting concept as the highly rated two-stroke has been produced in something like half a million cylinders for some 50 years — more than half of them for North American railroads — it is bound to have reached a high level of reliability. In fact the engine in question, currently the 645 but formerly the 567, made by the Electromotive Division of General Motors, is numerically the most successful non-automotive diesel ever. There are licensed builders outside America.

The numbers denote the swept volume of a cylinder in cubic inches and the change represented an increase in bore size from $8\frac{1}{2}$in to $9\frac{1}{16}$in (215.7 mm to 230.2 mm) while the stroke remained at 10in (254 mm). This change was made in the late 1960s.

Letters after the type number denote the status of the build, currently 'E'. Otherwise, apart from the addition of turbocharging, this engine has only changed in relatively detailed regards since its inception. It is produced in eight to twenty cylinders in A5 Vee form only. See Figure 23.28, which shows the constructional features.

At the time of writing this engine produces 148 kW/cyl at up to 1000 rev/min. The welded frame is remarkably simple as Figure 23.28 shows. It is made up from simple plate sections into an elliptical form braced by the cylinder housings. A cast steel frame suspended from each pair of cylinder housings carries the main bearings and the firing forces.

The liner and jacket form a self-contained assembly and the cylinder head is bolted to it. The head and liner together are secured to the cylinder housings in the frame by four crab-bolts each of which clamps on two cylinder heads. For servicing it is possible to withdraw as a single unit the entire cylinder set of parts from valves to big end.

Each head has two pairs of exhaust valves and a pump injector each operated by identical rockers bearing directly on a high level camshaft. The piston comprises a carrier and an external oil cooled head and skirt member which is free to rotate about its axis. The large end is closed by two relatively light half straps, since in a two-stroke the load is normally only on the rod half. The frame doubles as air chest, water outlet manifold and lube oil manifold and encloses the water inlet manifold.

At the time of writing the preferred fuel is gas oil. Servicing requirements are not onerous, but as with many designs developed for the US domestic market, they tend to be, if anything, more dependent on repair by replacement than their contemporaries, and the achievement of correct standards. Given these, the 567 and 645 have amply demonstrated their ability to work hard with little further attention.

The next engines to EMD in terms of volume are much less numerous, and the first of these is the Alco. Alco Inc. derives from the acronym of the former American Locomotive Company, itself an amalgam of former steam loco builders.

After several changes of ownership, Alco is now part of GEC Diesels (UK). Alco's engine since the 1950s is the model 251. A four-stroke, the basic concept was first produced in the late 1930s and has been extensively developed.

The forerunner of the 251 was the 244 of the same dimensions, which appeared in the early 1940s. Latterly the 251 has been rated at 191 kW/cyl at up to 1200 rev/min. The bore and stroke are 9in × 10½in (229 × 267 mm). Figure 23.29 indicates the 251 engine in its 45° Vee form (twelve, sixteen or eighteen cylinders) but it is also produced as a six in-line. It is licensed to be built in several countries. Note the externally mounted fuel pump actuated by a rocker, also the HP fuel pipe arrangement, the underslung crankshaft, the serrated main bearing joint, the two-part welded aluminium piston and (for the sixteen and eighteen cylinder engines) the single exhaust manifold.

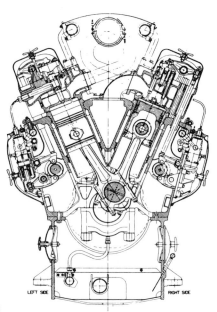

Figure 23.29 Cross section of Alco engine (from 'Locomotive Cyclopaedia)

Since the inclusion of Alco in the GEC group a new engine, based on joint development with Ruston, is being introduced. This is the 270 of 270 mm bore × 305 mm stroke, producing 241 kW/cyl at 1000 rev/min. It is offered as a six or eight cylinder in-line and as a Vee in twelve, sixteen and eighteen cylinders.

Other engines in this class include the General Electric 7FDL, a four-stroke of the same cylinder dimensions as Alco, but differing in many important respects, as Figure 23.30 shows. Note the articulated rod and the large unitary cylinder and head housing which, as with EMD, permits the exchange of a complete line set of parts as a unit. Current rating is 186 kW/cyl at 1050 rev/min. The FDL engine is produced in eight, twelve and sixteen cylinder Vee form.

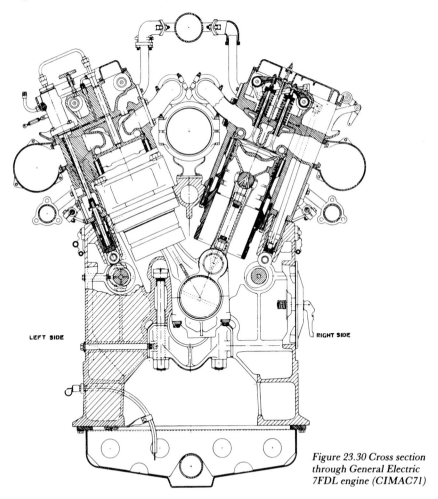

Figure 23.30 Cross section through General Electric 7FDL engine (CIMAC71)

Fairbanks-Morse, part of Colt Industries and a licensee of SEMT, produce a two-stroke opposed-piston engine, the 38TD8½, the last numbers being the bore in inches, that is, 206 mm. The stroke of each piston is 254 mm, and these engines produce 246 kW/cyl at 900 crankshaft rev/min. Six, nine and twelve cylinder versions are currently produced. The crankshafts are dephased to secure optimum port timing and since the exhaust leads, it produces most of the power. To reduce loads on the phasing gears, therefore, the 'air' crank is placed at the top of the cylinder block, and drives the mechanical blower.

Waukesha are a licensee of Sulzer for the A25 engine, but also make their well-established VHP engine which delivers 97 kW/cyl at 1200 rev/min. It is made as a six in-line or a 12 Vee.

HIGHER SPEED ENGINES

For the reasons mentioned at the beginning of the US section the US marine industry, and the industries of countries dependent on the US, make extensive use of faster running engines of which the principal exponents are Caterpillar and Detroit Diesel, whose power range extends to 850 kW at 1200 rev/min, and Cummins whose largest power exceeds 1 MW at 1950 rev/min.

It should be noted that the US designs mentioned are not the only examples: Rolls-Royce and Dorman in the UK; Poyand and Baudouin in France; MTU, Deutz and MWM in Germany are all well established; and there are many others.

It is not considered appropriate to describe engines of this kind in this volume as there are other titles devoted to them, and the philosophy of their design is rather different from the other engines we have considered. Suffice it to say that the engines follow to a greater or less degree automotive practice in construction and in their reliance on regular service attention rather than continual monitoring. Because of their production volume and use in many applications, service will usually be widely available at ports of call.

D.A.W.

24 Fuels and fuel chemistry

Today fuel consitutes not only the single highest cost factor in operating a ship but also the source of the most potent operating problems. The reason for this is that new refining techniques, introduced as a result of political developments in the Middle East in 1973–1974, have meant that fluid catalytic cracking and vis breaking have produced a more concentrated residual fuel of very poor quality. This residual fuel is the heavy fuel oil traditionally supplied to ships as bunkers and used in the majority of motor ships of a reasonable size for the main engine. The high cost of even these poor quality residual fuels means that owners generally have no alternative but to burn them, though some still prefer to use even more expensive intermediate grades produced as a result of mixing residual fuel oil with distillate.

The problems referred to in the opening paragraph are the effects on the engine in terms of wear and tear and corrosion resulting from harmful components in the fuel. It is the duty of the ship's engineer to be aware of these harmful constituents, their effect on the operation of his engines and the remedies offered to counter the harmful properties.

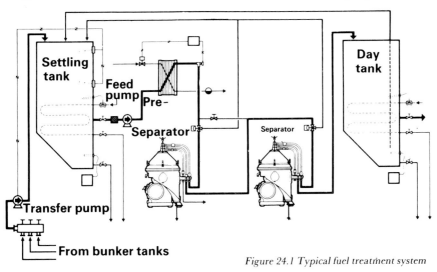

Figure 24.1 Typical fuel treatment system

Table 24.1 Effects of heavy fuels

Properties	Present H.O.	Future H.O.	Effect on engine
Viscosity (Red 1 at 37°C Heating temp.) pumping	3500 50	5200 65	Increased fuel heating required
centrifuging	95	98	
injection	110–120	115–130	
Density at 15°C	0.98	0.99	Water elimination becomes more difficult
Pour point °C	30	30	
Noxious elements			Fouling risk of components
Carbon residue %	6–12	15–22	Increased combustion delay
Asphaltenes %	4–8	10–13	Hard asphaltene producing hard particles. Soft asphaltene giving sticky deposits at low output. Increased combustion delay with defective combustion and pressure gradient increases
Cetane number	30–55	25–40	High pressure gradients and starting problems
Sulphur %	2–4	5	Wear of components due to corrosion below dew point of sulphuric acid (about 150°C)
Vanadium ppm	100–400	120–500	Burning of exhaust valves at about 500°C.
Sodium ppm	18–25	35–80	Lower temp. in case of high Na. content
Silicon and aluminium (CCF slurries)			Wear of liners, piston grooves, rings, fuel pump and injectors

PROBLEMS WITH HEAVY FUELS

The problems of present and future heavy residual fuels are mainly categorised as:

1. Storage and handling.
2. Combustion quality and burnability.
3. Contaminants, resulting in corrosion and/or damage to engine components, for example, burnt out exhaust valves.

Storage problems

The problems of storage in tanks of bunker fuel is one of a build-up in sludge leading to problems in handling. The reason for the increase in sludge build-up is because heavy fuels are generally blended from a cracked heavy residual using a lighter cutter stock resulting in a problem of incompatibility. This occurs when the asphaltene or high molecular weight compound suspended in the fuel is precipitated by the addition of the cutter stock or other dilutents. The sludge which settles in the bunker tanks or finds its way to the fuel lines tends to overload the fuel separators with a resultant loss of burnable fuel, and perhaps problems with fuel injectors and wear of the engine through abrasive particles.

To minimise the problems of sludging, the ship operator has a number of options. He may ask the fuel supplier to perform stability checks on the fuel that he is providing. Bunkers of different origins should be kept segregated wherever possible and water contamination kept to a minimum. Proper operation of the settling tanks and fuel treatment plant is essential to prevent sludge from entering the engine itself. A detergent-type chemical additive can be used to reduce the formation of sludge in the bunker tanks.

Water in the fuel

Water has always been a problem because it finds its way into the fuel during transport and storage on the ship. Free water can seriously damage fuel injection equipment, cause poor combustion and lead to excessive cylinder liner wear. If it happens to be seawater, it contains sodium which will contribute to corrosion when combined with vanadium and sulphur during combustion.

Water can normally be removed from the bunkers by proper operation of separators and properly designed settling and daily service tanks. However, where the specific gravity of the fuel is the same or greater than the water, removal of the water is difficult — or indeed not possible — and for this reason the maximum specific gravity of fuel supplied for ship's bunkers has generally been set at 0.99.

Burnability

The problems that are related to poor or incomplete combustion are many and complex and can vary with individual engines and even cylinders. The most significant problem, however, is the fouling of fuel injectors, exhaust ports and passages and the turbocharger gas side due to failure to burn the fuel completely. Fuels that are blends of cracked residual are much higher in aromatics and have a high carbon to hydrogen ratio, which means they do not burn as well.

Other problems arising from these heavier fuels are engine knocking, after burning, uneven burning, variation in ignition delay, and a steeper ignition pressure gradient. These factors contribute to increased fatigue of engine components, excessive thermal loading, increased exhaust emissions, and critical piston ring and liner wear. The long-term effects on the engine are a significant increase in fuel consumption and component damage. The greatest fouling and deposit build-up will occur when the engine is operated at reduced or very low loads.

The fuel qualities used to indicate a fuel's burnability are: Conradson carbon residue; asphaltenes; cetane value; carbon to hydrogen ratio.

To ensure proper fuel atomisation, effective use of the centrifugal separators, settling tanks and filters is essential; and the correct fuel viscosity must be maintained by heating, adequate injection pressure and correct injection timing. The maintenance of correct running temperatures according to the engine manufacturer's recommendations is also important, particularly at low loads. Additives which employ a reactive combustion catalyst can also be used to reduce the products of incomplete combustion.

High-temperature corrosion

Vanadium is the major fuel constituent influencing high-temperature corrosion. It cannot be removed in the pre-treatment process and it combines with sodium and sulphur during the combustion process to form eutectic compounds with melting points as low as 530 °C. Such molten compounds are very corrosive and attack the protective oxide layers on steel, exposing it to corrosion.

Exhaust valves and piston crowns are very susceptible to high-temperature corrosion. One severe form is where mineral ash deposits form on valve seats, which, with constant pounding, causes dents leading to a small channel through which the hot gases can pass. The compounds become heated and then attack the metal of the valve seat.

As well as their capacity for corrosion, vanadium, sulphur and sodium deposit out during combustion to foul the engine components and, being abrasive, lead to increased liner and ring wear. The main defence against high temperature corrosion has been to reduce the running temperatures of engine components, particularly exhaust valves, to levels below that at which the vanadium compounds are melted. Intensively cooled cylinder covers, liners, and valves, as well as rotators fitted to valves, have considerably reduced these problems. Special corrosion resistant coatings such as Stellite and plasma coatings have been applied to valves.

Low-temperature corrosion

Sulphur is generally the cause of low-temperature corrosion. In the combustion process the sulphur in the fuel combines with oxygen to form sulphur dioxide (SO_2). Some of the sulphur dioxide further combines to form sulphur trioxide (SO_3). The sulphur trioxide formed during combustion reacts with moisture to form sulphuric acid vapours, and where the metal temperatures are below the acid dew point (160 °C) the vapours condense as sulphuric acid, resulting in corrosion.

The obvious method of reducing this problem is to maintain temperatures in the engine above the acid dew point through good distribution and control of the cooling water. There is always the danger that an increase in temperatures to avoid low temperature corrosion may lead to increased high temperature corrosion. Attack on cylinder liners and piston rings as a result of high sulphur content fuels has been effectively reduced by controlled temperature of the cylinder liner walls and alkaline cylinder lubricating oils.

Abrasive impurities

The normal abrasive impurities in fuel are ash and sediment compounds. Solid metals such as sodium, nickel, vanadium, calcium and silica can result in significant wear to fuel injection equipment, cylinder liners, piston rings and ring grooves. However, a new contaminant is the metallic catalyst fines composed of very hard and abrasive alumina and silica particles which are a cause for much concern. These particles carry over in the catalytic cracking refinery process and remain suspended in the residual bottom fuel for extended periods. It has been known for brand new fuel pumps to be worn out in a matter of days, to the point where an engine fails to start through insufficient injection pressure, as a result of catalyst fines in the fuel.

The only effective method of combating abrasive particles is correct fuel pre-treatment. Separator manufacturers recommend that the separators should be operated in series (a purifier followed by a clarifier) at throughputs as low as 20% of the rated value.

PROPERTIES OF FUEL OIL

The quality of a fuel oil is generally determined by a number of specific parameters or proportions of metals or impurities in a given sample of the particular fuel. Such parameters are: viscosity; specific gravity;

flash point; Conradson carbon; asphaltenes content; sulphur content; water content; vanadium content; sodium content, and so on. Two parameters of traditional importance have been the calorific value and viscosity. Viscosity, once the best pointer to a fuel's quality or degree of heaviness, is now considered as being only partially major quality criterion because of the possible effects of constituents of a fuel.

Calorific value

The calorific value or heat of combustion of a fuel oil is a measure of the amount of heat released during complete combustion of a unit mass of the fuel, expressed kJ/kg. Calorific value is usually determined by a calorimeter but a theoretical value can be calculated from the following:

Calorific value in kcal/kg of fuel
$$= \frac{8\ 100\ C + 34\ 000\ (H - \frac{O}{8})}{100}$$

where C, H and O are percentages of these elements in one kilogram of fuel. Carbon yields 8080 kcal/kg when completely burned, hydrogen 34 000; the oxygen is assumed to be already attached to its proportion of hydrogen — that is, an amount of hydrogen equal to one-eighth the weight of the oxygen is nullified. The sulphur compounds are assumed to have their combustion heat nullified by the oxy-nitrogen ones.

Another variant of the formula is:

Calorific value in kcal/kg of fuel
$$= 7500\ C + 33\ 800\ (H - \frac{O}{8})$$

C, H, O being fractions of a kilogram per kilogram of fuel.

To convert the kcal/kg values to SI units as widely used:

1 kcal/kg = 4.187 kJ/kg

Example: A fuel of 10 500 kcal/kg in SI units

10 500 kcal/kg = 10 500 × 4.187 kJ/kg
= 44 000 kJ/kg

The calorific value as determined by a bomb calorimeter is the gross or 'higher' value which includes the latent heat of water vapour formed by the combustion of the hydroden. The net or 'lower' calorific value is that obtained from subtracting this latent heat. The difference between the gross and net values is usually about 600–700 kcal/kg, depending upon the hydrogen percentage.

A formula that can be used to calculate the net value is:

$$CVn = CVg - 25\,(f+w)\ kJ/kg$$

where: CVn = net calorific value in kJ/kg
CVg = gross calorific value in kJ/kg
f = water content of the fuel in percentage by weight
w = water percentage by weight generated by combustion of the hydrogen in the fuel.

Typical gross calorific values for different fuels are:

diesel oil   10 750 kcal/kg
gas oil   10 900 kcal/kg
boiler fuel   10 300 kcal/kg

When test data is not available the specific gravity of a fuel will provide an approximate guide to its calorific value:

Specific gravity at 15 °C	= 0.85	0.87	0.91	9.93
Gross CV kcal/kg	= 10 900	10 800	10 700	10 500

Viscosity

The viscosity of a fuel is its resistance to flow and is a measure of the work done in moving a given mass of the fuel. Viscosity decreases rapidly with an increase in temperature, and thus, in order to handle today's heavy fuels of high viscosity, heating is necessary to thin the oil.

The viscosity value of an oil has no significance unless it is stated at a given temperature. Viscosity is usually measured in seconds Redwood or degrees Engler from measurement using standard apparatus in which a given quantity of the oil is run through a standard orifice at a given temperature. For example, 200 seconds Redwood No. 1 at 37 °C is the time for 50 ml of the oil to run through a standard orifice at 37 °C.

Degrees Engler, on the other hand, is a relative measurement. For example, for an oil of viscosity 8 degrees Engler at 50 °C it means that the run through time is eight times the run through time of the same volume of water at 20 °C. Viscosity of lubricating oils are usually expressed in Centistokes.

416 FUELS AND FUEL CHEMISTRY

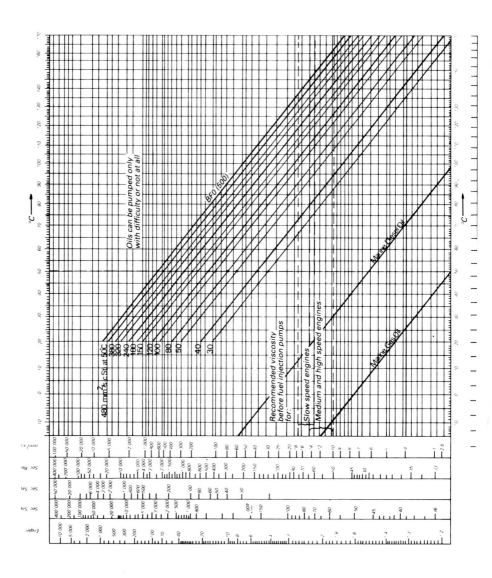

Figure 24.2 A typical temperature/viscosity chart for heavy fuels

Cetane number

The cetane number of a fuel is a measure of the ignition quality of the oil under the conditions in a diesel engine. The higher the cetane number, the shorter the time between fuel injection and rapid pressure rise. A more usable pointer of ignition quality is the diesel index, expressed as:

$$\text{diesel index number} = \frac{G \times A}{100}$$

where: G = specific gravity at 60 °F on the API scale
A = aniline point in °F, which is the lowest temperature at which equal parts by volume of freshly distilled aniline and the fuel oil are fully miscible

Conradson carbon value

This is the measure of the percentage of carbon residue after evaporation of the fuel in a closed space under control. The Conradson or coke value is a measure of the carbon-forming propensity and thus an indication of the tendency to deposit carbon on fuel injection nozzles. The Ramsbottom method has largely replaced the Conradson method of carbon residue testing, but it gives roughly the same results.

Ash content

The ash content is a measure of inorganic impurities in the fuel. These typically are sand, nickel, aluminium, silicon, sodium and vanadium. The most troublesome are sodium and vanadium, which form a mixture of sodium sulphate and vanadium pentoxide, which melt and adhere to engine components, particularly exhaust valves.

Sulphur content

This has no influence on combustion but high sulphur levels can be dangerous because of acid formation, mentioned earlier in this chapter. In recent years there has been a tendency to equate sulphur content with cylinder liner wear, but opinions differ on this matter.

Water content

This is the amount of water in a given sample of the oil and is usually determined by centrifuging or distillation.

Cloud point

The cloud point of an oil is the temperature at which crystallisation of paraffin wax begins to be observed when the oil is being cooled down.

Pour point

This is the lowest temperature at which an oil remains fluid and thus is important to know for handling on board purposes. An alternative is the solidifying point or the highest temperature at which the oil remains solid. It usually lies some 3 °C below the pour point.

Flash point

The flash point is defined as the lowest temperature at which an oil gives off combustible vapours, or the point at which air/oil vapour mixture can be ignited by a flame or spark.

Specific gravity

This is normally expressed in kg/m^3 or g/cm^3 at 15 °C. As the density of the fuel depends upon the density of the individual components, fuels can have identical densities but widely varying individual component densities. Apart from being an indicator of the 'heaviness' of a fuel, when measured by a hydrometer the specific gravity can be used to calculate the quantity of fuel by weight in a tank of given dimensions.

Typical values for standard fuels are given in Table 24.2.

Table 24.2 Typical value for standard fuels

Fuel	$SG (g/cm^2)$	Flash point (°C)	Lower CV (kJ/kg)
Gas oil	0.82–0.86	65–85	44 000–45 000
Diesel oil	0.85	65	44 000
Heavy fuel	0.9–0.99	65	40 000–42 000

(200 secs Redwood No. 1 – 3500 secs Redwood No. 1)

The following is a fuel oil specification of a leading slow-speed diesel engine manufacturer. These properties are considered the worst in each case that can be burnt in the particular engine.

Maximum viscosity	6000 secs Redwood No. 1
Specific gravity (15 °C)	0.990
Flash point	60 °C
Maximum Conradson carbon	18% by weight
Maximum asphaltenes	14% by weight
Maximum sulphur	5% by weight
Maximum water	1.0% by weight
Maximum ash	0.2% by weight
Maximum vanadium	500 ppm
Maximum sodium	30% of vanadium content

Combustion equations

Table 24.3 gives a list of the symbols designating the atoms and molecules of the elements and compounds in liquid fuels. For the atomic weights given the suffix denotes the number of atoms accepted as constituting one molecule of the substance.

Table 24.4 summarises the equations of the chemical reactions taking place during combustion. Nitrogen compounds are ignored. The weight of oxygen needed is evaluated, from the appropriate equation, for every element in the fuel, and the results are added together. The chemically correct quantity of air required to burn exactly the fuel to produce carbon dioxide and water vapour and unburnt nitrogen is approximately 15:1, termed the Stoichiometric Ratio. The atmosphere may be assumed to contain 23.15% of oxygen by weight and 20.96% by volume. At standard atmospheric pressure, 1 kg of air occupies a volume of 0.83 m^3 at 20 °C and 0.77 m^3 at 0 °C. An average diesel fuel should comprise in percentages, carbon 86–87; hydrogen 11.0–13.5; sulphur 0.5–2; oxygen and nitrogen 0.5–1.0. The greatest energy output in the combustion of fuels is from the hydrogen content and fuels having a high hydrogen: carbon ratio generally liberate greater quantities of heat. Typical examples are given in Table 24.7.

Table 24.3 Elements and compounds in liquid fuels and products of combustion

Name	Nature	Atomic symbol	Atomic weight	Molecular symbol	Molecular weight
Carbon	Element	C	12.00	C	12.00
Hydrogen	Element	H	1.008	H_2	2.016
Oxygen	Element	O	16.00	O_2	32.00
Nitrogen	Element	N	14.01	N_2	28.02
Sulphur	Element	S	32.07	S_2	64.14
Carbon monoxide	Compound	—	—	CO	28
Carbon dioxide	Compound	—	—	CO_2	44
Sulphur dioxide	Compound	—	—	SO_2	64.07
Sulphur trioxide	Compound	—	—	SO_3	80.07
Sulphurous acid	Compound	—	—	H_2SO_3	82.086
Sulphuric acid	Compound	—	—	H_2SO_4	98.086
Water	Compound	—	—	H_2O	18.016

Table 24.4 Combustion equations

Nature of reaction	Equation	Gravimetric meaning
Carbon burned to carbon dioxide	$C + O_2 = CO_2$	12 kg C + 32 kg O = 44 kg CO_2
Carbon burned to carbon monoxide	$2C + O_2 = 2(CO)$	24 kg C + 32 kg O = 56 kg CO
Carbon monoxide burned to carbon dioxide	$2(CO) + O_2 = 2(CO_2)$	56 kg CO + 32 kg O = 88 kg CO_2
Hydrogen oxidised to steam	$2H_2 + O_2 = 2(H_2O)$	4 kg H + 32 kg O = 36 kg steam (or water)
Sulphur burned to sulphur dioxide	$S_2 + 2O_2 = 2(SO_2)$	64 kg S + 64 kg O = 128 kg SO_2
Sulphur dioxide burned to sulphurous acid	$SO_2 + H_2O = H_2SO_3$	64 kg SO_2 + 18 kg H_2O = 82 kg H_2SO_3
Sulphur dioxide burned to sulphur trioxide	$O_2 + 2(SO_2) = 2(SO_3)$	32 kg O + 128 kg SO_2 = 160 kg SO_3
Sulphur trioxide and water to form sulphuric acid	$SO_3 + H_2O = H_2SO_4$	80 kg SO_3 + 18 kg H_2O = 98 kg H_2SO_4

Table 24.5 Volumetric meaning of equations

Nature of reaction	Volumetric result
Carbon burned to carbon dioxide	1 vol C + 1 vol O = 1 vol CO_2
Carbon burned to carbon monoxide	2 vols C + 1 vol O = 2 vols CO
Carbon monoxide burned to carbon dioxide	2 vols CO + 1 vol O = 2 vols CO_2
Hydrogen oxidised to steam	2 vols H + 1 vol O = 2 vols H_2O (steam not water)
Sulphur burned to sulphur dioxide	1 vol S + 2 vols O = 2 vols SO_2
Sulphur dioxide burned to sulphurous acid	—
Sulphur dioxide burned to sulphur trioxide	—
Sulphur trioxide and water to form sulphuric acid	—

Table 24.6 Heat evolved by combustion

Nature of reaction	Thermo-chemical equation	Heat evolved per kg of element burned kcal/kg
Carbon burned to carbon dioxide	$C + O_2 = CO_2 + 96\,900$	8 080
Carbon burned to carbon monoxide	$2C + O_2 = 2(CO) + 58\,900$	2 450
Carbon monoxide burned to carbon dioxide	$2(CO) + O_2 = 2(CO_2) + 315\,000$	5 630 (per kg CO burned)
Hydrogen oxidised to steam	$2H_2 + O_2 = 2(H_2O) + 136\,000$	33 900 (a) 29 000 (b)
Sulphur burned to sulphur dioxide	$S_2 + 2O_2 = 2(SO_2) + 144\,000$	2 260
Sulphur dioxide burned to sulphurous acid	—	—
Sulphur dioxide burned to sulphur trioxide	—	—
Sulphur trioxide and water to form sulphuric acid	—	—

Notes. (a) Value if the latent heat of the steam formed is to be *included*.
(b) Value if the latent heat of the steam formed is to be *excluded*.

Table 24.7

Fuel	H:C ratio	Calorific value kJ/kg
Methane	4:1	55 500
Ethane	3:1	51 900
Propane	2.7:1	50 400
Kerosene	1.9:1	43 300
Heavy fuel	1.5:1	42 500
Benzene	1:1	42 300
Coal	0.8:1	33 800

C.T.W.

25 Operation, monitoring and maintenance

While the practices evolved over three generations for direct drive engines and for two generations for medium-speed engines betray substantial differences in important detail, there are basic principles common to both. The operators of either require to achieve:

1. Reliability: there should not be unplanned stops.
2. Long periods between overhauls.
3. Optimum performance to meet the designed duty.
4. Minimum annual cost in fuel, attention and parts.

To achieve these objectives it is obvious, though it should not be taken for granted, that the ship's engineering officers and engine room staff must not only be competent but well versed in the characteristics of the engine in question. This should start with attendance at the maker's works for test bed trials, and also the training courses that most manufacturers provide.

Every manufacturer will provide facilities and recommendations for taking such observations of the engine's behaviour as he considers helpful in maintaining its full efficiency. See, for instance, Figure 25.1 which shows typically for a GMT medium-speed engine the measuring points for data logging, automation and protection. He will also provide for guidance a schedule of maintenance attention recommending intervals at which all necessary servicing work is undertaken. This will be based on his general experience with similar designs of engine, perhaps on long experience of large numbers of engines of the same type, but it will inevitably be pessimistic, so that operators can form a judgement in safety about the ideal intervals for their own machinery. This will be influenced by the owner's own views, by the operating conditions in which the engine must work, and by the vessel's trading pattern: no owner wishes to take his vessel off hire when it is most busy.

As the trend against 24-hour manning continues, more reliance is inevitably placed on condition monitoring plus alarms to replace the eyes and ears of the watch keeping engineers, and on data logging to achieve the continuous monitoring of the engine's behaviour. The crucial alarms which are invariably provided are for low lubricating oil

OPERATION, MONITORING AND MAINTENANCE

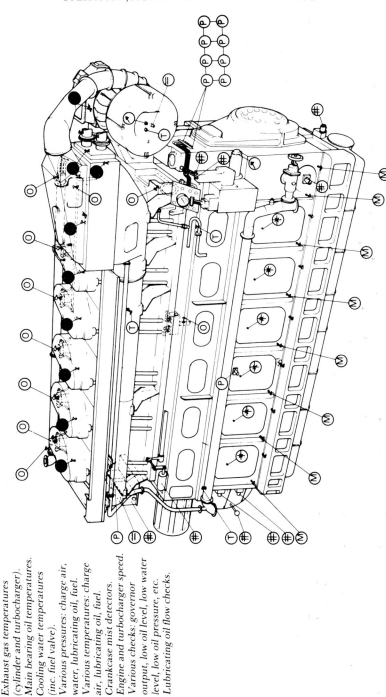

Figure 25.1 *Measurement points for critical data and automation (six cylinder GMT engine)*

○ Cylinder pressures.
● Exhaust gas temperatures (cylinder and turbocharger).
M Main bearing oil temperatures.
O Cooling water temperatures (inc. fuel valve).
P Various pressures: charge air, water, lubricating oil, fuel.
T Various temperatures: charge air, lubricating oil, fuel.
✱ Crankcase mist detectors.
↟ Engine and turbocharger speed.
Various checks: governor output, low oil level, low water level, low oil pressure, etc.
= Lubricating oil flow checks.

pressure, low cooling water level, high cooling water temperature, but many others can be added if required. Some, particularly those listed above, are backed up by a shut down capability, although such protection must involve a trade-off between damaging the machinery and hazarding the vessel.

It must be said, however, that notwithstanding the inevitability — not to mention the ingenuity, and in several cases the noteworthy dependability of data logging or condition monitoring installations — the absence of skilled ears and eyes in the engine room is on occasion sorely missed. Fortunately most engine builders have accepted the challenge of unmanned operation, and have assiduously studied the evidence of operators' experience (and of failures) to learn the lessons necessary to improve reliability and time between overhauls (TBO).

Particularly with the significant inroads into marine service made by medium- and high- speed engines, engines have been introduced which in other environments are normally unmanned and all but devoid of instrumentation. They depend in such cases almost totally on the 'hours elapsed' recommendations of the manufacturer for carrying out servicing, plus one or two crucial alarms. The kind of engine used as auxiliary power in coasters, for instance, may have not even any provision for such things as cylinder readings.

Finally, a word about values. Every reputable maker endeavours to put his designs on the market in a state which represents the optimum compromise between all his design objectives and the constraints of material and component capability. In older times one might draw conclusions from, say, the exhaust temperature achieved in Engine A versus that achieved in Engine B, and say that one was more highly stressed than the other. Such comparisons today may have to be tempered to recognise that one maker may have adopted a different scavenge ratio, or a different valve material, or even just a different arrangement of instrumentation. That is not to discourage the right to an explanation, but the ultimate criterion is still the reliability, and the annual cost as demonstrated by experience.

The procedures and precautions needed to keep machinery in good operational condition fall into three basic categories:

1. Operation
2. Monitoring
3. Maintenance

OPERATION

Maintenance has to include the way in which the machinery is used. A disciplined routine will ensure that nothing is overlooked, and that the machinery is not exposed to unnecessary risk or to unnecessarily rough handling.

The following notes were prepared for one of the earliest editions of this book. They were conceived for direct drive two-stroke engines, and in times when automation was still well over the horizon. Despite the advent of not only the geared engine and the controllable pitch propeller on the one hand, and a good deal of automation on the other, they still remain basically valid as a comprehensive guide to how the machinery should be handled. Obviously with bridge control many of these checks cannot be done and reliance has to be placed on the reliability of the components of the engine and the various circuits coupled with the function of alarms, but it does no harm to be aware of what the protection and monitoring system is having to achieve.

Preparations for starting the engine

The recommended sequence of preliminary steps is summarised below. These steps should be taken methodically and unhurriedly.

1. The jacket cooling water should be heated slowly, by circulating fresh water from the auxiliary generator discharge through the main engine system. Where such a connection is not provided, steam may be arranged to percolate into the cooling water while it is being circulated by the fresh water pump. On some engines an electric heater may be installed for pre-warming especially on medium-speed engines. The fresh water cooler should be by-passed during this operation. The pipe system and cylinder jackets should be examined for leaks.
2. The amount of oil in the fuel service tank should be checked.
3. The low-pressure fuel oil filters should be examined and cleaned — if necessary; all connections to the fuel service pump should be opened.
4. The lubricating oil filters should be inspected and cleaned — if necessary.
5. The lubricating oil pump should be started, after ensuring that all the appropriate valves are open. If thermostatic valves are not fitted, the oil coolers should be by-passed; the oil system should be examined for leaks and the piston cooling oil flow returns checked.
6. If there is a crankcase vapour extraction fan, this should be started.
7. The turboblower lubricating oil pump should be started, if separate, checking the bearing oil temperature.
8. The cylinder lubricators should be filled and primed by hand pumping. A check should be made for leaks at pipe joints and plunger stuffing boxes.

426   OPERATION, MONITORING AND MAINTENANCE

9. All hand lubrication points, e.g. those on the manoeuvring gear, reversing gear links, etc., should be oiled.
10. The automatic valve should be moved by the handle provided; the air distributor operating piston should be tried, by moving the test lever, to ensure that it operates freely; the components should be lubricated.
11. After opening the indicator cocks, the engine should be moved through at least one complete cycle (two revolutions if a four-stroke) by means of the turning gear. This is important, especially if the engine has been stopped for an appreciable length of time, because if — for some reason — water has accumulated in the cylinders, damage is thereby obviated. (Cylinders have been known to be wholly or partially flooded, by way of the silencer and piping, when in a tropical port during the wet season.) wet season.)
12. The fuel oil system should be primed, as may be described in the builders' operating instructions.
13. The manoeuvring air compressor should be started and the air reservoirs charged to the required pressure.
14. The turning gear should be disengaged.
15. The auxiliary — or emergency — scavenge air blower should be started, if such a unit is part of the installation.
16. The air reservoir stop valve, the air distributor stop valve and the main engine stop valve should be opened. All air drains should be shut off.
17. If conditions external to the ship are safe, and if permission is given by the bridge, the engine should be tried, ahead and astern, on starting air.
18. The engine is now ready for manoeuvring. When 'Stand-by' is rung on the telegraph from the bridge, the indicator cocks should be closed.

Starting the engine

Immediately the first order has been given by the bridge telegraph, the engine reversing handle should be moved into the appropriate position, be it ahead or astern. (This only applies to direct reversing engines.) The manoeuvring handle is then pushed over to 'Start' or the air start button is pressed. This movement will cause the air pilot valve to be lifted, admitting air — by way of the automatic valve — to the air distributor and the cylinder starting valve.

Immediately the engine gathers speed, the manoeuvring handle should be pushed over to 'Fuel'. On medium-speed engines, or where a proprietary governor is fitted, the fuel racks are automatically opened to the fullest extent allowed as soon as an appropriate level of lubricating oil pressure is generated by the stand-by pump and is sensed by the governor. Some designs of engine provide in such cases a hand limiting control level and this should be set down to about half fuel during starting and lifted clear when the engine has started. When firing begins in the cylinders the handle is positionally adjusted, as may be appropriate to the required engine speed.

If the engine fails to pick up on fuel, a second or even a third application of starting air will be essential. After a third unsuccessful attempt it can be assumed that the fuel system is airlocked. Repriming will therefore be necessary before any further attempt to start the engine is made.

While the engine is running slowly, the pressure gauge readings should be observed; the exhaust pyrometers should especially be carefully watched for assurance that all cylinders are firing properly. The air inlet pipe on each starting valve should be felt; heat is an indication that a valve is sticking in the open position. If a leaking starting valve cannot be made gas tight by turning the spindle on its seat the engine should be stopped and the valve changed. If the engine is allowed to run with a leaky starting valve the oil on the spindle — and on the other starting valve spindles also — will become dry and ineffective. All the valves will therefore be liable to stick when the engine is started on the next occasion.

In no circumstances must the starting air valves be eased or freed by the application of petrol, paraffin or any kind of inflammable liquid when in position on the engine.

When the order 'Full Way' is received the auxiliary scavenge air blower should be stopped. The starting air stop valve should be closed and all air drains opened.

Keeping watch

On the way to the engine room the colour of the funnel exhaust gases should be observed.

The cylinder exhaust-gas temperatures should be noted and confirmation made that the power of the engine is well distributed among the cylinders. The cylinder jacket outlet-water temperatures should be observed and, by manipulation of the test cocks on the outlet pipes, it should be seen that there are no airlocks. The water level in the fresh water heater tank should be checked. If the air inlet pipe of a starting valve is hotter than the others the valve should be changed at the earliest opportunity.

Observation should be made of the turboblower lubricating oil temperatures and pressures, also the exhaust gas temperatures before and after the turbines. It should be confirmed that the amount of fuel oil in the service tank is ample for the duration of the watch.

The cylinder lubricators should be examined for satisfactory operation. The telescopic piston-cooling pipes and the scraper glands for the piston rods should be examined for leakage. All pressures and temperatures, as registered by the various gauges and thermometers, should be regularly observed for abnormalities. The amount of lubricating oil in the drain tank should be checked by sounding. If lubricating oil filters of the edgewise cleaning type are fitted the handles should be given a few turns. The sea temperature should be noted and the temperatures at lubricating oil and fresh water coolers observed.

A keen sense of hearing is desirable. It is by his ears that the engineer becomes aware of abnormal happenings in the running gear.

On vessels fitted with a controllable pitch propeller, the engineer should be careful to see that the engine is giving load sensitive readings consistent with the intended load. For instance, if speed appears to be on the low side, and at the same time readings suggest high loads, this may indicate a pitch setting error either constantly, or for the sea and hull condition, and this would point to a tendency to overload the engine unnecessarily. It is also inefficient, as is the converse condition where the pitch setting is too low.

Before a fuel service tank is brought into use it should be drained of water and sediment. If two tanks are installed the empty tank should be pumped up immediately.

Fuel oil filters should normally be cleaned once every week; but in bad weather, when the sediment may be disturbed, they may have to be cleaned several times a day.

During manoeuvres, the fresh water temperature should be regulated by throttling the salt water supply to the fresh water cooler.

In shallow water the high seawater suction — commonly termed 'the high injection valve' — should be used, for minimising the inflow of sand.

As far as is compatible with the vessel's duties, engines should not be run for prolonged periods at much less than 25% load, especially if the engine speed cannot also be reduced. If the need to do so is unavoidable, there is a risk of fouling the exhaust passages and turbine with carbon due to charred lubricating oil, etc. carried over from the cylinders. This risk will be much lessened if the engine can be fully loaded at least for a few minutes and at least once a day: also if the cylinders all fire as nearly equally as practicable.

Vital parts of the engine should be examined as frequently as is practicable, and special attention should be given to the rectification of leaks of all descriptions, immediately they become apparent. If the crankcase is opened for investigation, naked lights must not be used until the inside of the crankcase has been fully ventilated.

All tools and spanners should be maintained in good order and systematically stored.

Stopping the engine

At the end of a voyage, before 'Stand-by' is rung by the bridge on the telegraph, the air automatic valve and the air distributor should be lubricated and tested. On receiving the order to stand-by the main air stop valve should be opened, all air drains should be closed and the auxiliary scavenge air blower should be started.

It is desirable that there should be a *modus vivendi* between engine room and bridge whereby, before the beginning of engine manoeuvres, the engine revolutions will be gradually reduced. During this time care should be taken that the cylinder jacket outlet temperatures are maintained.

During the period of manoeuvring an engineer should be stationed on the top platform for observing the working of the fuel valves and for noting those which are faulty in operation or sluggish in firing. Careful attention must continue to be given to the jacket water temperatures, the salt water supply to the fresh water cooler being severely throttled or, if necessary, completely shut off, for ensuring a satisfactory temperature level.

On receiving the order 'Finished with engines' the main air stop valve should be closed; all the air drains should be opened; the auxiliary scavenge air blower should be stopped; the turboblower lubricating oil pumps should be stopped; the main fuel oil inlet valve should be closed. The starting air in the air reservoirs should be raised to full pressure; a full flow of lubricating oil should be maintained in circulation for at least 15 minutes before there is any thought of by-passing the cooler, reducing the pump speed or shutting down. By this means the deposition of carbon will be minimised, especially inside the pistons if these are oil cooled. Similarly, the flow of cooling water should be maintained for about 15 minutes, for minimising the rate of change of cylinder liner temperature. By observing the water jacket outlet temperatures a correct decision can be made regarding the moment to begin by-passing the cooler and reducing the pump speed.

If, when in harbour or at anchor, the engine room temperatures cannot be maintained above, say, 2°C or the coolant circuit cannot be separately heated, all cooling water should be drained from the engine jackets and piping system, and from the fresh water cooler and pump.

Reversing from ahead to astern (direct reversing engines)

If an order to reverse the engine is received from the bridge while the ship is manoeuvring in or out of harbour the ship will usually be moving slowly, and reversal can speedily be made without difficulty.

If, however, the vessel is proceeding at full speed the prolonged application of a powerful astern torque will be necessary. The starting air is only capable of turning the engine at reduced revolutions. It is obvious, therefore, that it will have to be applied over some specific period of time during which the force of the water acting on the propeller has to be overcome.

Only when the speed of the ship is broken sufficiently will the starting air power be able to overcome the above force completely and bring the engine to a stop. When the engine has actually been stopped its internal friction is generally sufficient to prevent it from beginning to run ahead again.

It should be realised that immediately the starting air is applied in the opposite direction to that in which the engine is running it has full effect on one cylinder at a time and for only one revolution or perhaps slightly more. After that the effect will diminish because the pressure of the trapped starting air, after the main pistons have passed their top dead centres, will counteract the original braking effect. It is therefore advisable in such conditions that the reversing procedure should be as follows:

1. The manoeuvring handle is brought to the stop position. The auxiliary scavenge air blower is started.
2. The reversing handle is moved, unhurriedly, from the ahead to the astern position.
3. The manoeuvring handle is pushed to the starting position, thereby applying the starting air.
4. After not more than about two revolutions the handle is pulled back to the stop position.
5. Movements 3 and 4 are repeated, the starting air being applied and then cut off.
6. On a third application of starting air the engine can usually be expected to stop and to begin moving slowly in the astern direction. (It is necessary to be absolutely sure that this is so before applying fuel.)
7. If, after the fourth air impulse, the engine is only creeping astern a fifth application of starting air, as described previously, is advisable. By then, sufficient engine speed astern will have been obtained for it to be safe to push the handle as quickly as possible from the starting to the full-fuel position without any risk that the engine will again move ahead.

8. Sufficient fuel must now be admitted for the engine to be able quickly to gather speed in the astern direction. If there seems to be a tendency to lag and slow down, the manoeuvring handle is pushed somewhat further on to fuel. One or two cylinder relief valves may lift slightly while the engine is moving slowly. If all the cylinder relief valves lift heavily it is a sign that the engine is running in the wrong direction, as will be observed from the direction indicator.

For a reversal from astern to ahead, with full way on the ship, the same procedure as described above will be followed. It is then likely that fewer starting air impulsive applications will be needed.

The time required for the complete operation of reversal need not be more than 30–35 seconds, at the most. Attempts to shorten the procedure can be dangerous and, indeed, will usually increase the time required.

Despite what has just been mentioned, it is not unusual for a longer time for reversal to be required, depending upon ship and propeller conditions.

Where the engine is non-reversing, it does not, of course, have to be stopped, but it has to be slowed to idle to enable the astern (or ahead) clutches to be safely engaged. With a controllable pitch propeller, of course, the procedure is much simpler.

Some working difficulties

When an engine will not start
Should an engine start on air but move in the wrong direction, one or more starting valves must be leaking or sticking.

If an engine will not start on air the reason may be one of the following: the starting pressure is too low; stop valves between air reservoirs and engine are shut; starting or distributor valves are sticking; a starting valve is leaking on one of the cylinders as the piston moves upwards under compression; or the engine is being braked, either by the main or bottom-end bearings being too tight, or because of entanglement at the propeller or because the turning gear is engaged.

When an engine turns slowly at starting
If an engine starts very slowly on air and there is no ignition, the cause may be one of the following: the starting air pressure is too low; the automatic starting valve does not lift sufficiently, or is leaking; the starting valves are leaking; the pistons on top of the starting valves are leaking; the pilot distributor valves are sticking; the air distributor chain is either slack or stretched; or the starting stop valve may not have been opened sufficiently.

When an engine will not fire

If an engine starts on air, but there is no ignition subsequent to the starting handle being thrown over to oil, one of the following causes may be responsible: valves between settling tanks and surcharging pump are closed; the fuel pump is air locked; fuel valves are leaking; unsatisfactory fuel oil, which may be too heavy or contain too much water; fuel valve by-pass is open; air is in fuel oil system; fuel oil filters are choked; no fuel oil in service tank; no fuel oil pressure because of service pump fault; air compression pressure too low because of faulty piston rings or loss of scavenge air pressure or leaking valves.

Should a cylinder refuse to fire when the engine is running, there may be air in the fuel pump, pipeline or fuel valves: the remedy is to open the priming valve on the fuel valve and the air valve on the fuel pump until there is a steady flow of oil. Alternatively, there may be a choked fuel valve, necessitating a change of valves.

Engine exhaust not smokeless

Should the exhaust gases become visible and contain either grey or black smoke, the cause may be: overloading of engine; unsuitable fuel oil; nozzle holes not suitable for fuel used; holes in nozzles partly choked; broken fuel valve spring; injection pressure too low; fuel valves leaking; air in fuel system; fuel valve lift incorrect; ignition too late; dirty inlet strainers; one or more of the fuel pumps may have ceased to function properly, or have become mistimed, because a fuel cam timing gear has slackened back, thus overloading the other cylinders; the air compression pressure is too low, because of leaking piston rings or scavenge air blower fault; the exhaust back pressure is too great; or the fuel oil temperature may be too high for its viscosity.

An admixture of blue smoke in the exhaust gases usually signifies that lubricating oil is being burnt in the combustion space. (Yellowish if it is excessive.) The cause may be a leak in an oil cooled piston, or it may be oil passing the piston rings. If it clears soon after reaching full service load, it may indicate a cylinder misfiring at low load. A white exhaust usually means unburnt fuel and indicates misfiring in one of the cylinders.

Engine knocking

If a knock is heard while the engine is running at slow speed, or on starting from cold, it may be only the characteristic knock which is liable to occur in most compression-ignition engines in these circumstances; it will probably disappear on the attainment of normal running speeds and temperatures. If, however, the knocking is persistent, or if it occurs while the engine is running at normal revolutions, an investigation should immediately be made and the cause discovered.

OPERATION, MONITORING AND MAINTENANCE

If the knocking is traceable to one cylinder the fuel supply can be by-passed, thus revealing whether the knock is due to faulty combustion or to a mechanical defect in a working part.

Apart from overloading, knocking may be caused by: too much 'play' in oscillating parts; ignition too early; injection pressure too high; leaking or sticking valves; unsatisfactory fuel atomisation; ignition too late; uneven distribution of fuel to different cylinders.

Engine slowing down
Should the revolutions fall, with the manoeuvring handle in its normal position, the cause may be: a seized or leaky fuel pump plunger; a component of the regulating gear having slackened back; a leaking joint, causing airlock; a leaking priming valve or sticking fuel valve spindle, causing air locks; a burst fuel delivery pipe; choked atomising holes; fuel valve lift too small; choked fuel filters; water in the fuel oil; leak in service pump by-pass valve; empty service tank; air leak on suction side of fuel system; choked air inlet strainers or filters; air compression pressure too low, because of faulty piston rings or deficiency in scavenging air pressure; seizure of an engine component; overloading of engine; fouling of propeller by extraneous object; increased propeller slip because of heavy weather, especially head winds; or increase of helm to turn ship. If the vessel has a controllable pitch propeller a fault or a mis-setting of the pitch control may be indicated.

Irregularity of running
If the engine runs with irregularity the cause may be: faulty governor or governor gear; air or gases in the fuel lines; water in the fuel; fuel too viscous; choking of fuel oil filters; faulty fuel valve or valves.

MONITORING

Maintenance encompasses all the work done to preserve the engine's ability to deliver its power reliably and economically. Monitoring covers both readings and interpretation of the behaviour of a running engine, and examinations of condition which can be made during brief halts well short of an overhaul, or in other ways which do not interfere with the engine's availability for service.

Monitoring data

As a guide to the condition of the engine, and as a warning against possible trouble it is invariable discipline to take regular readings of strategic pressure, temperature, etc., parameters from the machinery. (See again Figure 25.1.)

Obviously the values obtained, even for a steady uniform load, will be influenced by seawater temperature, intake air temperature (and, to be pedantic, by barometric pressure and humidity) but they will also be affected by many types of change in the condition of the engine or ancillary equipment.

These include not just the 'obvious' quantities like cylinder exhaust temperature and firing pressure (perhaps as indicator cards) but in essence the inlet and outlet temperatures and pressures of every fluid (air, exhaust, fuel, oil, water) entering and leaving each significant part of the engine, or its circuits. It also includes speeds of the engine itself and its turboblowers, pumps, etc., and flows. It may in certain cases include measuring metal temperatures.

The manufacturer's instruction book and test certificate will usually indicate the degree of instrumentation which the manufacturer regards as advisable, but this may be modified by the Superintendent or Chief Engineer to suit particular problems or conditions.

Whatever the degree of instrumentation, if two successive sets of readings that ought to be identical are not, *all* the differences should be identified and correlated and reasons should be sought. The search naturally becomes more urgent if subsequent readings continue the trend.

In modern data logging installations the facility is provided that when the vessel is proceeding under steady conditions, limits of variation are set close to the acceptable level of critical variables relevant to the engine output and operating condition. It can be arranged that, if the reading drifts away from its anticipated value, and infringes one of these limit settings, an alarm will be given.

It could well be that the reason diagnosed for a change in a reading is an instrument fault, in which case the remedy is simple. If it is not an instrument fault, and the warning is disregarded, it could be expensive. Only a very few major failures ever happen totally without warning. It follows that it is most unwise merely to 'correct' an awkward reading or change in reading without properly understanding why it is different, or has become different.

It should be noted that instruments of all kinds need regular calibration if they are to maintain their usefulness. This is not always easy to do, and engineers can, on the other hand, be carried away by a fetish for accuracy out of harmony with the real requirement — which is consistency.

If there is reason to doubt an instrument, it can be exchanged for another in the same or a similar position. With exhaust temperature, for example, if a high reading is unaffected by changing instruments, it should be taken seriously: if the high reading goes with the instrument, one, or both, of them is out of order. ('Unaffected', in this context, means that the change in reading is less than 20 °C. For other parameters different standards of repeatability may be expected.)

Instrument accuracy is often dependent on correct use and correct siting. This, particularly in the case of gas and air flows, is virtually impossible to achieve on a diesel engine, and once again it is consistency that one is trying to control. Correct, or at least uniform, immersion is vital, especially where the measured flow is likely to be highly stratified, i.e. space precludes the necessary length of straight approach and downstream piping to ensure uniformity across the whole section.

With the widespread — almost universal — use of soundproofed control rooms in modern tonnage, many readings are recorded remotely. In many cases readings are duplicated, and if this is so, particularly for parameters like cylinder exhaust temperature, and both indications are *not* derived from the same sensor, differences of reading method may be introduced. This is not important as long as it is understood, and comparisons are based on one reading location or the other only.

Monitoring conditions

Condition monitoring embraces checks on an engine's state of health without dismantling it, or at least only to the extent permissible in a brief stop. In fact, modern techniques can keep a wide range of engine data under continuous observation, including the facility to take corrective action automatically.

While a given engine type, and its components, can be shown statistically to achieve certain lives, there must be a spread, and it is the aim of condition monitoring to enjoy the benefits of the mean life without the hazard of the minimum.

Data monitoring is an essential part of condition monitoring, since changes in the condition or capability of crucial engine components will inevitably be reflected, however subtly, in one or more of the usual parameters monitored.

Cylinder pressures: indicators

Cylinder pressure is one of the major parameters monitored on all but the smallest engines. Before discussing pressures, it is necessary to consider the mechanics of measurement. Mechanically driven indicators, contrived to draw a diagram correctly phased to the crank rotation, are well proven on slow-speed engines, but are unable, because of inertia and wear in the linkage, to function other than in the draw card mode on medium or higher-speed engines.

Until recently, outside the laboratory, these engines have had to rely on maximum pressure indications only. Some of the indicators used have relied on the personal judgement of the user to determine when the maximum firing pressure was being recorded. This is because successive firing impulses in the same cylinder do not all have the same value and the user has to form a judgement as to what level, as he senses it, corresponds to the average firing pressure in that cylinder at that load. This is not too difficult for one operator, but can lead to difficulties if the readings of two or more people are being compared.

Recent developments have opened up the possibility of using, for all sizes of engine, an electronic transducer in the cylinder cock, usually based on a piezo-electric crystal to create an electrical signal which is fed to a cathode ray oscilloscope (CRO). A time base is accurately derived from the flywheel to provide indicator diagrams drawn either from TDC to BDC, or from mid-stroke to mid-stroke. A representative condition can be chosen by inspection, and then recorded as required using a Polaroid camera.

The CRO type of indicator invariably includes a ready adjustment for phasing, but the mechanically driven indicator depends on a cam which has to be adjusted mechanically. In either case the test is for both the expansion and the compression lines in a compression diagram to coincide exactly. To be pedantic, the expansion of the compressed air will be marginally lower than the compression because of heat losses. On a medium- or high-speed engine there may also be lags in the pressure wave in the indicator passages, which would tend to show the expansion curve very slightly above the compression.

On a direct drive engine using mechanical indicators, other, and different factors must be considered too. Slackness or tightness in the indicator or in the driving gear may cause the lines to cross and the compression card to show an enclosed area. This may wrongly indicate that positive or negative work is being done in the cylinder. The working card, obtained while firing is taking place, will therefore be either fuller or narrower by this amount.

A positive indicator compression card can usually be corrected by retarding the cam slightly; a negative card requires the cam to be advanced in the direction of rotation. On a positive card the expansion line will remain in front of, or above, the compression line, until nearly the toe of the card is reached; a negative card is recognisable by the expansion line joining the compression line near, or before, the centre of the card. That is, on a positive compression card the two lines tend to remain a longer time apart than when the card is negative. If the two lines coincide during the up and down stroke, until near the centre of the card, and then part slightly towards the toe, the card may be accepted as satisfactory. Any further retardation of the indicator cam will cause a definite loop on the down stroke, and the card will become negative. The tendency of the lines to remain apart, when running horizontally, is explained by the friction of the pencil against the paper, and by the slight clearance in the pencil lever rods.

The approximate amount the indicator cam should be moved on the circumference of the shaft is the measurement of the widest part of the loop, negative or positive, multiplied by two.

Cylinder pressures: taking readings

Whatever kind of indicator is used, if it is attached to the indicator cock, this must be blown through to remove carbon and other matter before the indicator is attached. The indicator itself must be kept clean, and on mechanical types, or maximum pressure indicators, the pistons must be taken out frequently for cleaning and oiling.

All cards from the same engine, except those from the top and bottom cylinders of double-acting engines, should be to the same hand, but there is no significance in this.

Mechanical indicators draw diagrams to a standard scale, or scales, depending on the stated and calibrated rate of the spring chosen. Electronic indicators can be set to a variety of scales, depending on the 'gain' setting, and the calibration of the pick-up. (The maker will furnish calibration information and procedure.)

References to diagram heights in the following text are based on the use of mechanical indicators, and should be interpreted accordingly, although the principles involved apply to either type of indicator, and in practice, the size of diagram on the CRO is likely to be about the same as in the mechanical case.

Cylinder pressures: compression cards

The ideal compression card is shown in Figure 25.2, diagram 1, where for the length of the card, the compression and expansion lines coincide. In diagram 2 the cam is probably correctly placed but the loops signify that there is a time lag in the drive, caused by tight gear or/and a loose indicator pencil lever. Diagram 3 — which is positive in area — shows the expansion line in front of and above the compression line for the length of the card; the cure is to retard the indicator cam 5-6 mm. In diagram 4 — which is a negative diagram — the expansion line lies behind and below the compression line for half the length of the card; for rectification the indicator cam should be advanced 3-4 mm.

It must be remembered that cutting out a cylinder affects the energy

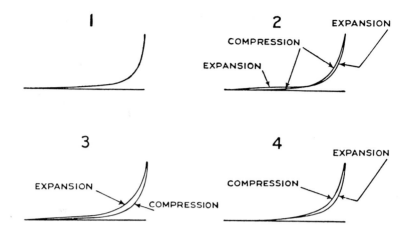

Figure 25.2 Compression cards

balance of the blower on a turbocharged engine and therefore the boost pressure and compression pressure. Unless the recording is done very precisely over a very brief interval, it will not give a true compression pressure. This is not too important if the diagram phase only is being set, but would be very important if the purpose of the diagram were diagnostic.

On a medium- or high-speed engine with jerk pumps, cutting out a cylinder is not simple and should not normally be done under way unless the engine is fitted with spragging gear. If the purpose is merely to check a time base signal the need to do this would probably only arise in dock or in port and the test could be carried out with a pump immobilised prior to running.

On slow-speed engines it is not as hazardous to cut out a cylinder and it is done quite simply by opening the by-pass valves on the fuel valve or lifting the plunger.

If an engine runs too fast the taking of a true compression card may be impracticable; the open by-pass valves may not prevent fuel from entering and firing in the cylinder.

If a fuel valve leaks, fuel may enter the cylinder during the stroke and fire at the end of compression. A small working card, instead of a compression card, will thus be obtained. When in doubt, the fuel pump stop valve should be closed and the fuel valve priming valve opened fully. Should a compression card show that firing is still occurring, there may be leakage from the piston-cooling oil system.

Examination should be made for vapour rising from the piston-cooling outlet pockets or smoke from the exhaust cocks.

Cylinder pressures: interpretation

Firing pressures will, of course, like other parameters, vary with the load on the cylinder, and with engine speed. Whatever the type of engine, the maker will have provided guidance about the levels expected at service and at maximum continuous rating (mcr) and perhaps at other loads and speeds. Traditionally these will be in the form of the figures recorded on the test certificate for the test bed run at the maker's works, and there will be a record of the settings and adjustments applicable to the engine at the time.

All of this information, plus the instruction manual, will be kept by the Chief Engineer, and in general the task of the ship's engineers is to maintain these values (allowing for any effects of seawater and air intake or ambient temperatures, etc).

Certainly firing pressures need to be checked regularly, and if the facility exists for an indicator card to be taken, mechanically or electronically, it materially assists the interpretation of the readings and the diagnosis of faults when or if they arise. The use of indicator cards is not only traditional, but more necessary with two-stroke direct drive engines which do not use proprietary jerk pumps, since it is more difficult to judge the load the engine is carrying without the indicator card.

Medium-speed engine loads can be assessed with a high degree of precision from the governor output or fuel pump control rod position *provided* that the rest of the engine readings indicate normal behaviour and that the calibration has not been disturbed.

If a card can be taken it is useful to take draw cards regularly, since the point where the firing pressure rise commences gives a good approximation to the compression pressure. If a card cannot be taken, the compression condition on any cylinder will have to be judged from other clues, including boost pressure, exhaust temperature, firing pressure, and water outlet temperature compared with other cylinders.

If firing pressure is not what it is expected to be, consider first whether conditions of operation might be responsible. For instance, if the vessel is operating in very cold ambients and the engine intake air is taken from the deck rather than from the engine room, the boost pressure will be higher, and so will firing pressure, regardless of the temperature of charge air after the intercooler. High seawater temperatures will tend to do the same.

If the fuel quality — or characteristics — is/are different, this could affect the pressures and temperatures quite noticeably. If no such factor suggests itself, or allowance has been made for them, the next thing to do is to make certain that the instrumentation itself is not at fault, by making cross checks with other instruments and/or by assessing whether other readings, which should support the warning interpretation, in fact do so.

The next step in the process of interpretation is to look for causes. If the pressure is low, the boost may be low, or there may be a mechanical reason for low compression in the cylinder such as a damaged valve or rings — or choked air parts in a two-stroke. The fuelling may be low because of mis-setting or of injector or fuel pump plunger leakage; misset, leaking, sticking or choked fuel valves (two-stroke) or the fuel pump may not be giving a full delivery or even a full stroke. Note whether other cylinders are affected, and look for the confirmatory evidence that should coexist.

If the pressure is high, there may be blockage of exhaust parts or reduced exhaust valve lift (these will tend also to cause burnt gas to flow back into the air manifold and enter the next nearby cylinder in the firing order, with serious risk of gross overheating). There may be mis-setting, or the fuel valve spring may be broken or set to too low a pressure.

If no fault is, in fact, found, and no explanatory circumstances exist, then the question of adjustment may be faced; but it cannot be too strongly emphasised that adjustment of the settings of individual cylinders, especially on medium-speed engines, is not a frivolous matter to be entertained casually.

The jerk pump is probably the most accurately set piece of equipment on the engine, and is normally more precise by an order of magnitude than the interpretation of any of the instrumentation likely to be used in judgement of it. If it falls under suspicion the first thing to do is to replace the injector, then the pump with a good spare.

Of course if the readings of all cylinders are high (or low) together, the load indication must be suspect and the confirmatory checks, and the remedy, are obvious. See also 'Keeping watch' above.

Light spring cards are taken for determining the pressures in the cylinders during the exhaust and the air intake periods. It is sometimes useful to keep records of the exhaust pressure heights from the atmospheric line, and the intake air pressure at the beginning of the compression stroke. From these records it can be seen, for example, when the exhaust ports in a two-cycle engine should be cleaned, or if the exhaust turboblower or air valves in a four-cycle pressure-induction engine need attention.

On two-stroke engines, particularly where the load has to be interpreted usually from cylinder readings, if there is any doubt about an indicator card showing the correct mean pressure — assuming that the indicator drive is in a state of proper adjustment — it is advisable, in the first instance, to rely upon the indicator card rather than upon deductions from the exhaust temperature as shown by pyrometer.

To assist removing or confirming the doubt, the exhaust temperature of each cylinder should be checked against that of the other cylinders. The same mercury thermometer should be used for all the cylinders. By checking all the cylinders in this way, the reliability of the pyrometers can be determined, and a useful comparison made with the indicator cards.

If, for any cylinder, the mean pressure appears to be low and the exhaust temperature high, the explanation is, usually, shortage of combustion air. This may arise from carbon-choked ports or broken piston rings.

Two-stroke engine indicator cards

In Figure 25.3, diagrams 5 and 6 are normal cards for single-acting two-stroke engines. In diagram 5 the mean indicated pressure is about 7.60 kgf/cm^2, the compression height about 38 mm, the firing height about 56 mm. The scale of all normal indicator diagrams is 1 mm = 1 kgf/cm^2. A light spring diagram is reproduced at 6; the scale of such a card is: 45 mm = 1kgf/cm^2. The atmospheric line is lettered A; B is the line of falling exhaust pressure; C the lowest pressure in the cylinder during scavenging; D the pressure at the beginning of compression; E the compression line.

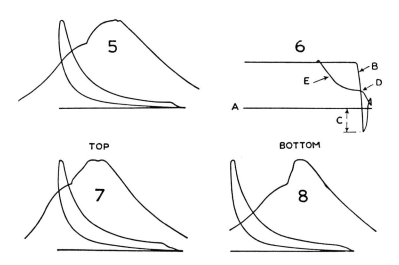

Figure 25.3 Two-stroke engine indicator cards

Diagrams 7 and 8 (Figure 25.4) are normal diagrams for two-stroke double-acting engines. In diagram 7 — for the top of the cylinder — the mean indicated pressure is about 7.1 kgf/cm^2, the compression height about 34 mm, the firing height about 47.5 mm; for the cylinder bottom the mean indicated pressure is about 6.7 kgf/cm^2, the compression height about 35 mm, the firing height about 48 mm.

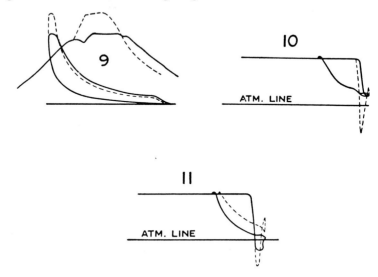

Figure 25.4 Two-stroke engine indicator cards

In the full load diagram 9 (Figure 25.4) the effect of a leaking fuel pump plunger is shown. The full lines are the actual, the dotted lines the normal shape of the indicator card. The compression height is normal; the firing height is too low. The fuel pump regulating arm has been adjusted to provide an increased fuel charge to make up leakage losses. If the fuel valve filter or nozzle holes are choked, the fuel injection pipe or the fuel valve leaking, the diagram will also lose height, and the appearance of the card will be similar to diagram 9.

The light spring diagram 10, which may be from the top or bottom of a double-acting two-stroke engine, shows the normal card in dotted lines, the actual card in full lines. Either the exhaust ports are partly choked by carbon or there is too much resistance in the exhaust system. Card 11 indicates loss of scavenge air; probably the blower change valves are leaking or are not properly closed.

Four-stroke engine indicator cards

Figure 25.5 shows normal cards at 12 and 13 for a four-stroke, single-acting engine. On the full-load working card 12, point Y shows where

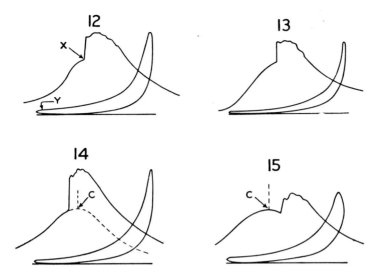

Figure 25.5 Four-stroke engine indicator cards

the exhaust valve begins to open. On the drawcard, taken by pulling the indicator cord by hand, X shows where fuel injection begins at the top of the compression stroke. Cards 13 are light-load diagrams.

Diagram 14 indicates the effect of too early ignition, probably caused by the fuel pump timing being too early, the fuel valve spring being set to lift at too low a pressure, or the fuel valve spring being broken. The maximum pressure is too high; the extended dotted compression curve shows firing taking place before top dead-centre C.

Card 15 shows the effect of too late ignition. The maximum pressure is too low; the draw card shows that firing has taken place after the crank has passed the top dead centre C. The fuel pump timing may be too late, or the fuel valve and its filter can be at fault.

Figure 25.6 illustrates light spring diagrams at 16, 17, 18 and 19. At 16, which is a power diagram taken with a weak indicator spring, 12 mm = 1 kgf/cm^2, A is the air inlet suction line, B the compression line, C the expansion line, D the exhaust line. The exhaust valve opens at E. The atmospheric line is dotted. In 17 the exhaust valve opens too late and closes too soon; the cam clearance is too great. In 18 the exhaust resistance is too great, perhaps caused by an obstructed exhaust outlet. In 19 the resistance in the air intake is too great, perhaps the result of an obstructed air inlet strainer or too small an opening at the air inlet valve. Alternatively, the cam clearance may be too great.

The cylinder air volume V is reduced by the amount S which the piston traverses before reaching atmospheric pressure. The dotted vertical line shows where the compression line C intersects the dotted atmospheric line.

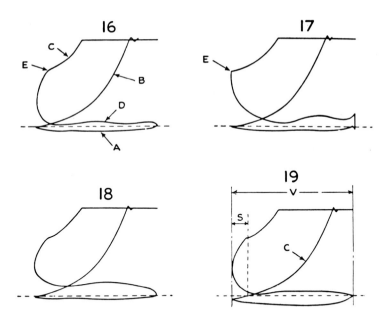

Figure 25.6 Four-stroke engine light spring cards

In Figure 25.7 diagrams 20 and 21 respectively show normal working diagrams and light load diagrams for single-acting four-stroke engines, pressure charged on the Büchi system. In 20 the mean indicated pressure is about 9.1 kgf/cm^2, the firing height is about 46 mm, the compression height about 35 mm. In 21 the mean indicated pressure is about 4.5 kgf/cm^2, the firing height about 39 mm, the compression height about 30 mm. Light spring diagrams, 45 mm 1 kgf/cm^2, are illustrated at 22 and 23, where A is the air inlet, B the compression and C the exhaust. Line C differs in the two diagrams — it depends upon the number of cylinders and the exhaust pipe length — but both are normal.

Cards 24 and 25 are typical for single-acting, four-stroke, under-piston, pressure charged engines. Diagram 24 is for overload working, with mean indicated pressure about 10.5 kgf/cm^2, compression height about 35.5 mm, firing height about 45 mm. Normal and light indicator cards for such engines are the same as for exhaust turboblown engines. Diagram 25 is a light spring card, 20 mm = 1 kgf/cm^2 for an under-piston charged engine. A is the air-induction, B the compression and C the exhaust line. The line C, in this type of pressure-charged engine, falls below line A (compare with diagram 22). The pressure at the beginning of compression is 0.25 kgf/cm^2 above atmosphere.

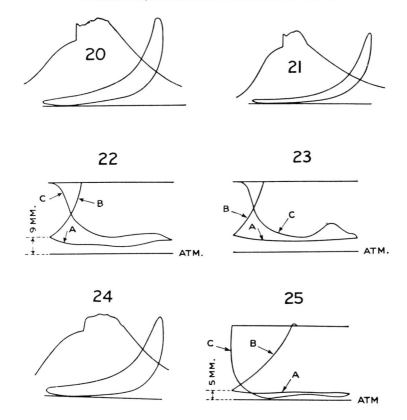

Figure 25.7 Four-stroke engine indicator cards

Exhaust temperatures

High exhaust temperatures may be an indication of overload (check the load indication, e.g. the fuel rack lever on a medium-speed engine), of cylinder damage affecting the compression pressure, or low boost (charging air) pressure or fouling or reduced flow in the coolant passages of the charge cooler. On a two-stroke engine it may indicate that the ports are getting fouled.

Exhaust temperatures will, of course, rise if the engine intake air temperature rises, and on charge cooled turbocharged engines, it may also indicate that ventilation to the engine room is inadequate so that the turbocharger intake has created a depression in the engine room; or equally it may indicate that there is an obstruction in the uptake giving rise to a back pressure on the turbine. It may indicate a timing fault, or on a turbocharged engine that the turboblower is dirty or damaged.

It has been found, too, that oil from certain sources of crude has markedly different ignition properties leading to appreciably slower burning in the cylinder, and this also causes substantially higher exhaust temperatures.

It should be noted that on turbocharged engines, even with accurately equalised fuelling on each cylinder, variations in exhaust temperature will inevitably occur (Figure 25.8). This is because of relatively unimportant variations in the way the scavenge air flow divides among the cylinders, as well as stratification in the flow past the pyrometer's sensor. (The pyrometer reading is highly sensitive to depth of immersion.)

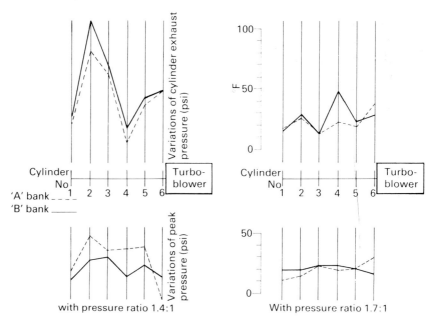

Figure 25.8 *Average cylinder readings of groups of similar twelve cylinder engines with different boost pressure ratios*

This can be checked by swapping equipment and instruments between high and low reading cylinders, and the makers will usually advise. If temperatures are needlessly 'levelled' and, as a result, fuelling rates are actually unbalanced, cylinders may well cut out on stand-by, and create oiling-up problems. Conversely, if the fuelling on one cylinder is increased beyond the level of the others, it will have to burn the extra fuel without the benefit of any extra air, and it will suffer higher metal temperatures in consequence. This practice has in extreme cases led directly to seizure on cylinders with 'low' exhaust temperatures.

At the other end of the scale it is worth noting that on stand-by, or at idle, the pyrometers will show a reasonably spread range of temperatures, depending to some extent on the engine, of about 60–80°C. If a cylinder ceases to fire for any reason its temperature will drop to about 10-20°C over the air manifold temperature. It is not an infallible sign of bad adjustment, though this is one possible cause. Rather it should be noted and interpreted with the behaviour at service rating, and with the checks described above.

Other readings

The turboblower readings are a valuable guide to engine behaviour. The speed N is plotted as $N/\sqrt{T}$ where T is the absolute air temperature. The base line, mass flow, is similarly corrected to $W\sqrt{T}/P$ where W is the mass flow, and P is the absolute intake air pressure).

The pressure ratio achieved for an engine load, with normal engine condition, is dependent on the cleanliness of the compresser and diffuser on the one hand, and of the turbine and nozzle on the other. The air side is susceptible to atmospheric dust and any smoke or fumes, particularly if there is oil vapour in the air. The turbine side can be fouled badly by excessive carry over of lubricating oil from the cylinders and particularly by burning heavy fuel, especially if this cannot be done completely. (Nowadays water washing facilities and routines are available to limit these effects on both sides of the blower.)

Turbines and nozzles can also sustain damage from debris coming from the cylinders. For a given pressure ratio, depending on compressor or efficiency, the air entering the charge cooler is expected to lose about 85% of the difference between its temperature and that of the seawater.

The measuring points for the air inlet and outlet temperatures are less than ideal because of installation constraints, but ought roughly to confirm this. Certainly any variation warrants checking. The water passing through the cooler must show an increase in temperature commensurate with the flow and its duty; this should be checked too: if too great the flow may be insufficient, or blocked.

Cooling water system

The same cross checks also apply across every component in the cooling system where inlet and outlet temperatures — and pressures — can be taken. It is very important that an engine is not over-cooled. This will encourage the formation of dew in the cylinder liner, and this dew, owing to the presence of sulphur combustion products in the cylinder gases, is likely to contain some sulphuric and sulphurous acids which (particularly the former) will cause unnecessary corrosion. On many four-stroke engines the problem is first betrayed by erosion on the fuel injector nozzle tip. (See Chapter 26, Figure 26.26.)

For modern engines, manufacturers tend to recommend that cooling water temperatures be maintained at nearer 80°C and development work on prototypes will be carried out at temperatures of this order. In such cases lower temperatures not only risk corrosion, but are likely to result in slightly reduced efficiency. A liberal flow of water through the engine is desirable. The temperature rise through the engine should normally be some 8° to 10 °C at most. The intended flow can be deduced, knowing the system pressure, if a pump characteristic is available. This can be cross checked with the heat transfer on any or all of the coolers.

The normal pressure at which the cooling water system operates will be of the order of 5 to 15 kN/cm^2. Once a temperature rise (for normal load) is established in association with the coolant pressure, and flow if this is separately recorded, any upset in the relationship between them should be viewed with concern and investigated.

If a cylinder jacket becomes so hot that steam issues from the test cock, then the fuel supply to the cylinder should be cut off and the engine revolutions reduced to allow the cylinder to cool. Only after the cause had been discovered and eradicated should an attempt be made to put the cylinder on load again. If neither water nor steam issues from the test cock, the cooling water pipe must be choked somewhere. The engine is then stopped and the obstruction dealt with. The unit must be allowed to cool slowly before any attempt is made again to circulate cooling water through the cylinder.

Lubricating oil circuit

Apart from similarly monitoring the temperatures at the coolers, etc., as with coolant, the pressures attained at normal load are of crucial interest. It is not sufficient merely to maintain the prescribed pressure at the engine busrail. In an extreme case this could be partially blocked and the 'correct' pressure set by opening up the relief valve so that 80-90% of the oil by-passed the engine.

It should not be difficult to ascertain the pump delivery, nor to assess the actual flow through the engine, referring to the temperature rise and the anticipated heat content removed by the oil. The pressure drop through each unit of the treatment circuit can also be checked and while it is difficult to lay down hard and fast rules except for specific installation, it should not be more than 0.1-0.3 bar for normal flows and viscosities.

Referring to makers' recommendations and test certificate values, it is again the changes that need suspicious investigation.

With all the fluids: air, coolant, oil, the diligent engineer will check thermometer indications by touch where possible. Even simple mercury thermometers have been known to go wrong — perhaps because they are sitting in a dry pocket which should have been filled with oil.

If the lubricating oil pressure is too low the cause may be: an insufficient amount of oil in the tank; leaky pipe unions or joints inside the crankcase; dirty filters; a defect in the lubricating oil pump; excessive clearance at bearings.

Other checks

Older engineers acquired as second nature a sensitivity to the 'feel' of the engines and machinery under their care, and could sense as much as recognise any change, including those discussed above, by unusual sounds, unusual appearances of oil or water (indicating possible tube leakage in coolers, or gasket failures) and so on. Hopefully this expertise will not wither, but it can now be supplemented by some valuable new diagnostic tools.

Continuous monitoring

With the advent of UMS operation on a fairly wide scale, many owners have felt uneasy about leaving the machinery protected only by the basic alarms, and perhaps alarms to indicate a drift in key readings.

Apart from the desirability of the kind of advance warning of impending trouble which more sophisticated monitoring might give, it ought to make the severity of the fault when eventually diagnosed much less. As ratings rise on all types of machinery, faults may develop to failure much more rapidly. Thanks to advances in electronics, especially microelectronics and microprocessors, the faculty now exists to monitor, directly, reliably and continuously, the most sensitive parameters rather than those like the pressure and temperature of the cylinder gases, or the lubricating oil, which react relatively late.

In a well-planned condition monitoring installation, diagnosis can be achieved without dismantling, so that overhauls can be safely deferred.

Proximity sensors

If a proximity sensor, usually an inductive coil with a high frequency supply, is mounted in the liner wall, the passage of the piston rings will give a recognisable signal pattern if they are functioning correctly, and differences in the pattern will indicate possible trouble.

Wear sensors

These operate by measuring the change in electrical resistance in the investigated component as it wears. They can be made very sensitive, and can give sensible indications of liner wear, for instance, or wear in other components, in very short periods. Their accuracy is unaffected by quite severe wear in the liner etc, where they are fitted.

Temperature sensors

Thermocouples can be adapted reliably to monitor liner and bearing temperatures, etc., and, indirectly, such parameters as ring temperatures.

Pressure sensors

Based on various principles these are mainly used to monitor cylinder pressure, and critical fluid pressures such as oil, charge air, fuel, etc.

Analysis

Various manufacturers offer self-contained display installations which can also include microprocessors to make deductions from the data to compare values with set limits for various conditions, and to take progressive, or simple, cautionary or preventative measures.

Spectroscopic analysis

This procedure entails regular samples of the lubricating oil from a representative part of the circuit, such as the bedplate scavenge pump to the sump tank, or the suction of the main pump if a wet sump system is used.

Engine makers usually recommend a suitable sampling point. Samples may be taken as often as desired, but weekly or fortnightly is about right. The samples may be collected at the next port of call for analysis by the oil companies who operate the service, or there are simplified kits for use on board ship (though these tend to be less comprehensively informative). Interpretation requires some skill, and is most sure as part of a continuous monitoring record preferably under the direction of the same laboratory.

Broadly, the spectroscopic examination will identify and quantify the various elements present in the lubricating oil charge. Allowance has to be made for metals such as calcium or tin which may originate in certain additives, but, for instance, copper/tin/lead may well originate in bearing linings, aluminium from the piston (depending on the material used in the construction of the engine) chrome from rings or liners — or from coolant leaks. Iron may originate from wear of liners, rings, crank, etc.

Over a period when all is well, the quantities of each element will remain fairly stable, but any rise may be a warning of abnormal wear, or even a developing failure, and such a warning warrants investigation as soon as possible. How soon it becomes imperative is impossible to say, but it is not unusual for an examination carried out four to six weeks after drawing the first suspicious sample still to be in time to prevent major damage. The report on each sample will usually indicate the degree of urgency with which its findings should be observed.

Vibration survey

Older engineers will remember the practice of using a stethoscope, or even one improvised from a screwdriver or a length of steel rod, to try to pinpoint the source of a suspicious noise heard while the engine is running.

Vibration monitoring has taken this much further and involves first using a signal sensor, usually a piezo-electric transducer, at any one of several nominated positions likely to be of interest. Secondly, the signals may be fed to the analyser, where either displacement, velocity or acceleration may be chosen as the criterion, and then the whole of the recorded spectrum of vibration can be analysed at various discrete frequencies.

It is found that the spectrum, or vibration signature, thus recorded is distinctly different for each particular component or point of measurement. If a library of vibration signatures is built up when the machinery is new, subsequent readings can be compared with them and a change at one point between successive readings is a *prima-facie* indication of possible trouble. As a rough guide and to avoid unnecessary disturbance for what may in fact be acceptable changes, a change at a particular frequency should be considered serious if its amplitude doubles the original value.

Other refinements of the technique include, for instance, shock pulse monitoring, which is particularly useful in detecting problems with ball or roller bearings. It is based on the analysis of the vibrations triggered by an external pulse in the bearing being investigated, while the engine is running. Side band analysis (of the frequencies adjacent to the fundamental) is particularly helpful in checking the condition of gear teeth.

Static tests

These are techniques which can be applied when the engine is standing and involve a minimum of dismantling, perhaps taking advantage of simple servicing work such as changing injectors. For instance, on a two-stroke the ring condition can be examined through the ports. The condition of the injectors as they are withdrawn should be assessed for the quantity and type of deposits, erosion, excess heat, etc.

Optical aids are particularly useful and the Borescope, sometimes called the Endoscope, is well known for examining the condition of liner and valve seats via the injector hole. It has been joined now by the more versatile Fibroscope based on fibre optics.

Techniques are also available for monitoring the thickness of plating and for detecting bonding failures between layers of material. Whether an engineer reposes his faith in specific checks, on consideration of engine readings, or on the general 'feel' of the engine to tell him whether all is still well with his charge or not, let it be repeated that he must always be suspicious of the least change. Anything odd should be investigated, however trivial, no matter how many times this exercise proves to be a false alarm. However easy to cancel the symptom by adjustment, the prudent engineer will always seek to understand the cause first.

OVERHAULING AND MAINTENANCE

Maintenance requirements differ somewhat between slow, medium- and high-speed engines, and inevitably as regards detail, the frequency and complexity, of maintenance will be in accordance with the age of the engine and/or the time of its conception.

Each manufacturer provides full maintenance and overhauling information for each engine in his range. Maintenance starts with daily attentions like checking oil level, hand lubrication, and taking readings. It goes on through weekly, monthly and then longer checks, although these are more usually quoted in terms of hours run. The attention required or recommended naturally mounts in complexity, from filter changes, cleaning the cooler, cleaning and resetting the injectors (currently at something between 3000 and 5000 hours), to exhaust port cleaning or exhaust valve withdrawal at 5000 to 10 000 hours, through to major overhauls at perhaps 30 000 hours. Maker's recommendations are inevitably on the pessimistic side compared with average achievement, so that even the operator who has the worst conditions to contend with can be pretty certain of undertaking his first overhaul well in advance of any indication of trouble.

Among the maker's recommendations will inevitably be the ship owner's preference to undertake a cycle of maintenance which breaks up the schedule into a number of tasks which can be distributed among times when the vessel is in port anyway, so that it is never, if possible, off hire for engine reasons alone. This 'staggering' effect is achieved automatically by starting the overhaul cycle on a new vessel with one or two cylinders somewhat earlier than the engine builder's recommendation, and then timing the overhauls on the remainder in similar groups over a period which eventually enables a reasonable estimate to be made of the safe life between overhauls which the components on this particular engine will achieve in normal service.

There is also the requirement of the Classification Societies, which stipulate minimum frequencies for particular inspections and overhauls. The societies naturally place a premium on safety, but they do take a reasonable attitude to demonstrable capability to extend overhaul periods. In general, they tend to require surveys of cylinders and other major parts at four-or sometimes five-yearly intervals.

The reason for any overhaul (unless a problem has been found) is to correct for wear, to eradicate deposits where they interfere with efficiency of components, and to restore surface treatments. It is also an opportunity to look for signs of trouble, such as incipient cracking, evidence of any malalignment, or undue wear, corrosion or leakage.

Manufacturers of engines and their components, and manufacturers of fuels, lubricants and coatings, have achieved among them impressive gains in the time between overhauls. Even on inferior fuel, many engines now achieve 10 000 hours or more before cylinder heads need removal. Some approach this figure even for the withdrawal of exhaust valve cages. Piston ring and groove wear does not demand correction in some designs up to 15 000 hours or more. Crankshaft bearings are capable of being safely left for 30 000 hours if oil treatment is correctly carried out.

Modern economic conditions have obliged ship owners to seek the longest possible overhaul and maintenance periods, and so the need for frequent checks has been designed out of modern machinery as far as possible. Since every time an engine is dismantled it disturbs the bedding and invokes the need for modified running-in procedures, it is possible to examine unnecessarily frequently and to prejudice life overall by so doing. Conversely, if a problem or a weakness is shown to exist, then intervals must be shortened to suit in order to preserve reliability what ever the cost. Modern additives and treatment for coolant and lubricants have cut down substantially the deposition of solids and contaminants and therefore the need for frequent cleaning of their respective circuits.

Tools and fixtures

Most manufacturers have given a great deal of thought to simplifying maintenance. Many have achieved the maximum freedom of access to make it unnecessary to call for special spanners. Many also go to some trouble to minimise the variety of size of spanner required. For large highly stressed bolts manufacturers have, to a large degree, replaced the extended spanner or slogging spanner (or even the torque wrench) with hydraulically loaded tensioning gear. Special tools are included where necessary, as well as lifting tackle and extractors.

Many examples can be seen in those chapters devoted to particular makes of engine showing clever redesign, so that, for instance, air, coolant and lubricant connections are made automatically when a cylinder head is refitted.

The need to maintain and store properly all the tools should be obvious. Owners will have different ideas on the detail of the way this is achieved, but most provide for the purpose a centrally sited store, to which, of course, they should immediately be returned after use. The provision of spares varies with the duty of the vessel, and to some extent, with the preference of individual owners. As a minimum the Classification Society for the vessel will stipulate a holding dependent on whether the vessel is engaged on short or long voyages. In the former case it is smaller, because it is likely that adequate provision can be made from a central spares depot.

General

Maintenance of two-stroke engines has tended to be rather generous in the past, partly because their unchallenged reign developed while engine rooms were fully and generously manned, partly because a two-stroke can be (dare one say) more temperamental than a four-stroke in some regards, and partly because the sheer size of these machines imposed their own special conditions. The advent of the four-stroke marine engine, usually derived from designs, or according to the practice, of engines intended for land-based service where manning levels available have for a long time been much less generous, must have done much to encourage the basic concept of the unmanned engine room.

The manufacturers of two-stroke slow-speed engines, however, have not been slow to evolve simpler maintenance techniques, reduce the degree to which they relied on manual attention, and increase the intervals between service attention. The differences between the two types of engine are pointed out where necessary in the following sections.

Pistons

In two-stroke engines pistons, and their rings, must pass over ports, which inevitably interrupt the oil film and create difficulties for the ring ends and edges. On the other hand, rings can easily be examined through the ports without significant dismantling.

It does no harm to take advantage of the facility on a new or overhauled engine, and examine pistons and rings monthly through the ports until they can be seen to have achieved satisfactorily stable condition. Thereafter six-monthly examinations would suffice.

Gas leakage past the pistons should never be allowed to become pronounced. At best it prejudices lubricating conditions and promotes wear; at worst it could initiate a seizure. It obviously does not improve efficiency though its effect on performance is most noticeable at low speed, and it may make starting very difficult. (See also 'Piston rings,' below.)

Wear on liners will create problems, and the maximum permissible liner wear before replacement is about 1% of the diameter (i.e. 1 mm per 100 mm). In general the first ring groove is located as remote as is necessary from the piston crown to ensure that the temperature in the groove can be withstood by the oil, though with cooled pistons this is now much less of a constraint. The condition of the piston ring grooves is, of course, important. If there is any indentation of the groove surfaces in way of the ring ends, the ring groove will require to be machined out to take an oversize ring. If this is done, the finish and profile of the ring groove must be correct, since, if it is not good, gas can blow past the ring via the back of the groove.

Where there is a common system for bearing lubrication and piston oil cooling, it is important that all bearings and bearing liners are maintained in good condition, since if oil escapes from the bearings too freely, it may be difficult to maintain the correct oil pressure, and this may in turn affect the amount of oil available for piston cooling.

On larger engines, at any rate those using a separate oil circuit and telescopic tubes to feed the oil to the piston, a regulating valve is usually fitted to the bearing oil system to control the temperature. This control also varies the division of oil flow between the bearings and pistons, but it is always arranged so that the bearing oil supply can never be cut off completely.

On the great majority of medium-speed engines the piston cooling oil is fed, with the piston lubricating oil, through the connecting rod from the bearing oil supply. Many, but not all, engines incorporate the facility for measuring the temperature of the piston cooling oil at outlet, and this must, of course, be maintained at the recommended figure, and uniformly, for all pistons.

On engines incorporating holes in the cylinder liners for piston lubrication, these should be cleaned periodically. In crosshead engines a scraper ring device for the piston rod is fitted on the crankcase top. Occasionally it should be confirmed that the segments of the ring are not butting, thus preventing the ring from functioning efficiently.

Piston rings

If there are problems with piston rings, particularly on the larger two-stroke engines, it could be that the liner finish is not correct. The kind — as distinct from the quality — of finish adopted is perhaps not as vital as was once thought. For instance, at one time many manufacturers of all types of engine preferred a turned finish to produce a definite grooving, but nowadays a controlled honing operation is found advantageous, particularly in medium- and high-speed engines, the object being to produce a surface after running in akin to a smooth plateau criss-crossed by the deeper honing marks which perform the oil retention function. (See also the section below on 'Running in'.)

Ring packs, their type, location and complexity, may be seen to differ appreciably between ostensibly similar engines, but each solution will be what the manufacturer concerned has considered appropriate for his design and should not be altered without reference to him.

Note that in many cases compression rings are no longer of plain rectangular section. In some designs they are slightly tapered, in some there may be a circumferential cut-out on the inner diameter (usually on the top face) to bias the pressure exerted by the ring on the liner wall to the bottom edge of the ring to promote downward scraping. Always check, therefore, whether the rings are directional (usually marked 'top' on the upper surface) and make sure that they are fitted the correct way up (Figure 25.9).

Scraper rings vary considerably, too, both in their number, position and design. As ratings have climbed, more and more makers have tended to drop the second scraper ring and rely on one only, and this is often placed above the gudgeon pin. The choice is influenced by the design chosen and, while some makers still remain faithful to the simple slotted type ring with two lands, sometimes directionally chamfered, many makers have taken advantage of the availability of two-part rings with a peripheral spring mounted in the groove on the inner periphery to promote conformability. The width of the land of each of the scraping faces of the scraper ring and the pressure of the sping itself are both deployed to achieve the oil control necessary, and it is obviously very important that the correct type of ring is replaced in the scraper ring groove.

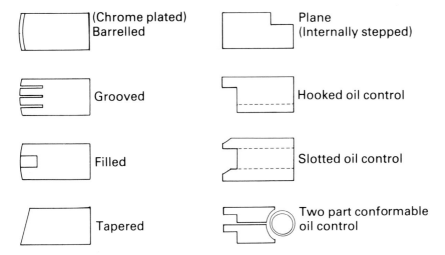

Figure 25.9 Piston ring types

If combustion gases are blowing past the pistons, and there is no problem with groove damage or with the piston rings sticking in their grooves, the cause may be wear in the rings so that the gaps have opened up enough to allow significant passage through the ring pack. This tendency can be very considerably enhanced if the ring gaps down the pack are not staggered to alternately opposite sides of the piston.

On larger two-stroke engines it is general practice to radius or bevel the upper and lower edges of the piston ring working faces, ostensibly to assist in retaining the lubricating oil film; but this is definitely *not* recommended for four-stroke engines.

On two-stroke engines the rings that pass the port have to be positionally constrained so that the ring gap will coincide with one of the part bars. This is done to prevent the ring ends being caught on the port edges and possibly breaking. It is achieved either with a pin to locate the ends of the gap or with an anchor piece which locates the body of the ring. Otherwise the rings sometimes have a tendency to move circumferentially round the groove.

Piston ring manufacture is a specialised business, and to achieve a truly cylindrical ring working surface when fitted in the cylinder calls for fairly complex pre-forming of the ring in the unstressed state. If in emergency a ring has to be obtained from an unauthorised supplier, it would be prudent at least to check for circularity by placing it in the liner with a light behind it to see that it fitted all the way round the circumference.

When new rings are fitted care must be taken to ensure that the gap is correct (four-stroke engines) or that there is sufficient clearance in the laps (two-stroke engines) to allow for the expansion of the rings. If this is checked by placing the ring in the liner, the check should obviously be made at an unworn part of the bore. If enough wear has occurred at the top dead centre position to leave a distinct ridge, either it must be removed or a stepped ring should be fitted. This particularly applies if, in addition, the top ring groove has had to be machined out and an oversized ring fitted.

If the engine is using chromium plated liners, then on no account must any of the rings have a chromium plated surface. If the liner surface is cast iron, then it is prudent to break the glaze on the liner to assist subsequent bedding-in (naturally taking great care that any debris generated is prevented from entering crankshaft oilways or other sensitive areas).

Running in

At one time the experienced engineer considered it prudent to give any new or disturbed combination of sliding surfaces an opportunity to bed-in before applying full load to them. Modern technology has rendered this less vital as far as crankshaft and other bearings are concerned, but, while it has modified the requirement for the piston and liner combination, a running-in procedure remains highly advisable, if not vital, whenever a new liner or new rings are fitted, and even to some extent when the pistons are withdrawn and replaced without renewing any component. In large engines, running in parts of one or two cylinders is achieved by cutting back the fuelling on these cylinders for a few hours.

The need for running in arises:

1. Since the distribution of temperature around, across and down both the piston and liner is not uniform, neither is the expansion, and neither, therefore, is the shape.
2. For obvious reasons only a minimum of lubricating oil may be used.
3. It would be uneconomic, if not virtually impossible, to try to produce an ideal bedding surface on the components before they are assembled. The bedded-in profiles of rings, for example, are quite complex.

When new components are run together, or are introduced in combination with older components, the mating surfaces are likely to bed unevenly. This will produce harder contact, and hence heat, but also wear which will tend towards an even distribution of the contact pressure round the periphery. In addition, even when high spots have been worn down, it is likely that asperities all round the initial surface of the two components will break through the lubricating oil film and also cause heat and wear.

If the engine is run at full load immediately, the heat generated is likely to build up until serious damage and wear debris results. The point of running in is to control this heat and the generation of abrasive debris, while the wearing-in process continues and therefore the generation of heat tends to diminish. The progressive raising of the speed and load is, in effect, maintaining the heat of the wearing-in process as high as is safe until bedding-in is complete and fully lubricated contact is achieved.

It follows that, within reason, rough surfaces will bed-in more easily than smooth ones. This is why it is good practice to break the glaze on a used liner when new rings are fitted. If two polished surfaces are to be bedded the high spots are much more likely to lead to scuffing, because without the asperities of a roughened surface the surface is less likely to retain enough oil to control the temperature and to flush away wear debris.

Timing of cylinder lubricators

The difference between the genuine timing of cylinder lubricators — where this is achievable — and the indefinite injection of lubricating oil is reflected in rates of cylinder liner wear, in the carbonising of ports and in the incidence of scavenge fires. The correct instant for injection is when the piston is moving very slowly relative to the angular movement of the crank. This occurs when the piston is approaching and when it is leaving the top and bottom dead centres. Near the top dead point the cylinder gas pressure is too great for oil injection. Lower down, also, the gas pressure behind the piston rings can be sufficiently high to prevent the discharge of lubricant by the lubricator plunger. There remains the lower end of the stroke, and this is the most practicable region for the purpose.

Because of the uncertainties, some engine builders make no attempt to time their cylinder lubricators.

In four-stroke, single-acting, crosshead engines a timing effect can be obtained. The point to which the crank is set when the lubricator plunger has just reached the end of its pumping stroke is 30° before bottom dead centre. Experiment shows that, with such an engine running normally at, say 100–110 rev/min, there is a time lag between plunger movement and oil injection of approximately 20° of crank angle; i.e. injection has been completed by the time the piston is about 10° of crank angle from the bottom dead point. But in experiments on lubricating oil injection settings at 30° after bottom dead centre, under conditions closely simulating those in a four-stroke trunk engine running at 135 rev/min, there was at best a continuous dribble, with a minor periodic pulse superimposed upon it.

As ordinarily made, the cylinder lubricator has a very short effective stroke and, in ships which travel from cold to hot and from hot to cold climates, it is in continuous need of adjustment, because of the oil viscosity variations which are inseparable from the changing ambient temperatures. The incorporation of a small, thermostatically controlled, heating element into the lubricator overcomes the need for viscosity adjustment. A speed-reducing gear of say 1/3, by which the effective plunger stroke is increased and the number of strokes per minute reduced, can be helpful. The injection lag is, however, thereby increased.

The non-return valve, which is always provided on the lubricator discharge pipe, should be placed as near the cylinder as practicable, to assist the timed impulse of oil. If the non-return valve is mounted on the cylinder there may be carbonisation difficulties.

Cylinder liners and jackets

The piston side of the liner has been discussed under 'Piston rings' above. With adequate control of coolant and lubricant quality, treatment, pressure, temperature and flow, it is not necessary to withdraw pistons in a modern design engine unless the condition monitoring indicates a developing fault, or until the interval recommended (or modified in the light of experience), has elapsed.

When they are withdrawn the condition and wear of the liners can be noted and, in two-stroke engines, carbon deposits removed from the ports. The sign of port clogging is a steady rise in exhaust temperature over a period, accompanied by an increase in air pressure.

Cylinder jackets in large engines are provided with inspection doors through which the water spaces can be examined. However, with correct water treatment this should be unnecessary, short of major overhauls. In addition, frequent inspection can be expensive if the treated water has to be discarded, or tedious if it has to be stored for re-use. If water treatment is not in use, then the examination should certainly be carried out fairly frequently, for instance at intervals of about six months.

It is important that rubber rings used to seal liners to jackets or water space in cylinder blocks should be of the correct size and material. If the ring is too large for the groove there is a serious risk that it may be volumetrically compressed in the groove when the liner is in place and this will produce a constriction in the bore which will upset the behaviour of the piston. The ring should not project more than one millimetre above the metal surface before fitting the liner, for a 10 cm groove.

Originally these rings were made from natural rubber, but it has been found that with the advent of higher jacket coolant temperatures such rings tended to deteriorate. So many manufacturers now specify that the rings should be of nitrile rubber, and usually the development will have been done with a single proprietary type. It is most unwise, therefore, to use an unauthorised spare: if, in emergency, one has to be used, it should if at all possible be of the correct material. If it is oversize in either circumference or cross sectional diameter, or both, it is possible to cut a piece out of the circumference and close the ring again with a scarfed joint. If the diameter is 10% oversize, then 20% of the length should be cut out and pro rata. It should be verified that the adhesive material used to remake the joint is suitable for the temperature and environment of the ring.

If a natural rubber ring has to be pressed into service where a nitrile one should have been used, its life may only be of the order of a few hundred hours, (more if coolant temperature can be reduced to about 65°C), and it should be replaced at the earliest opportunity. Some water treatments affect rubber too.

Bearings

It is traditional on the large direct drive engines that the clearances of connecting rod top and bottom ends should be checked every six months. At the end of each long voyage the nuts should be tested for tightness. Once a year the main bearing clearances should be checked. Wear down readings should be compared with the crankshaft deflection readings.

The thrust block, with its bearings and thrust pads, should be examined annually and the clearances checked. All oil passages should be thoroughly cleaned and tested for a full and free flow of oil.

Although makers of medium-speed engines do not call for bearing examinations or alignment checks before at least two years' service, it is always prudent to make a check. However, provided the previous assembly was conscientiously done, and the lubricating oil treatment correctly carried out, experience confirms that it is quite safe to allow such a period of time to elapse before examination. If spectroscopic analysis is used, it will give ample warning of the need for an earlier check. Any difficulty in maintaining oil pressure, particularly at low speeds if an engine driven lubricating oil pump is used, is also cause for concern.

Camshaft drive

In a medium-speed engine with gear drive the maker's instructions will outline a simple routine whereby all the gears are meshed at distinctively marked teeth.

Where a chain is used, the points to be observed when connecting it up are:

1. The camshaft is turned to the reference position indicated by the maker's instruction manual as appropriate to the type of engine; and from the fuel cam position it will be noted which crank is required to be on top dead centre;
2. The crank is placed on top dead centre, according to the flywheel mark or gauge supplied;
3. The chain is put on in such a way that the camshaft setting mark remains in the reference position when the chain is tightened up;
4. The fuel pump lifts are checked with the respective cranks on dead centre and should correspond to the adjustment sheet records.

Periodically it is necessary to tighten the chains. The procedure is:

(a) the locknuts underneath the tightening bolt boss are slackened back; the limit movement studs are screwed back; the nut on the top of the bolt is tightened until the first pressure mentioned on the engine builder's adjustment sheet is reached;
(b) the locknuts, with their ball washers, are screwed up against their faces and locked;
(c) the spring is tightened further, until the second and higher pressure stated on the sheet is reached;
(d) the top nut is secured by its locknut;
(e) the limit movement studs are adjusted to the clearance stated on the engine builder's drawing.

After the chain has been retightened it is necessary to check the position of the camshaft relative to the crankshaft. With the crank on top dead centre the corresponding fuel pump cam should have raised the plunger the amount stated on the engine adjustment sheet.

The driving chains should be inspected at the end of each long voyage. Particular attention should be paid to wear of rollers and also chain tightness. The chain lubrication sprayers should be examined carefully and the flow of oil from each sprayer checked.

The chain wheels should be examined for wear of teeth, as frequently as possible. Any variation in alignment will become evident by wear on one side of the teeth. The flow of oil to all jockey wheel bearings should be checked.

The amount of chain wear which takes place is directly proportional to the percentage elongation. In general, an elongation of 2% is regarded as marking the end of the useful life of a chain.

Governors

The proprietary governors, now standard on all medium- and high-speed engines and many others, should require little attention, but any work done on them must be done strictly to the maker's instructions. They require scrupulous cleanliness but repay it in trouble-free service. The governor drive should be checked for wear as this could upset the response of the governor.

Cams

Cams nowadays give little or no trouble for very long periods though the loaded faces should be examined at routine overhauls for any signs of cracking. Cracks, if present, would indicate a failure of the case, perhaps because it was insufficiently deep. Wear or scuffing could indicate a lubrication fault and possibly roller damage, or perhaps a tight roller.

In direct reversing engines the roller edges should be examined too for possible chipping sustained during manoeuvring. Many makers still provide a facility for the fuel cam to be adjusted on the shaft for fine setting the injection timing. This may be provided by clamping it to the shaft, or to the valve cam group through slotted holes fitted with clocks or wedge pieces when finally set. Some modern designs use the oil injection method with a shallow tapered bush. Other makers recommend a single timing setting and only allow fine adjustment by follower height adjustment, and that mainly to compensate for tolerances. (See also Chapter 5.)

On direct drive engines particularly, the fuel cam usually consists of two main components, one being used for ahead running and the other for astern running. The components are mounted on a taper on the camshaft, and are held in place by fitted bolts through slots in the cam. The position of the cam for optimum combustion conditions is determined during the shop test of the engine. When the test is ended, wedge pieces are fitted to prevent the cam from shifting.

To provide a continuous surface for the cam roller to work upon, the joint between the ahead and astern cams is tongued and grooved. When the wedge pieces are fitted and the retaining nuts tightened up, locking washers prevent the nuts from slackening back. A tapped hole for a starting screw is provided in each half-cam, for easing it off the camshaft taper, should any alteration in timing be required.

To adjust a fuel cam: the nuts are removed from the fitted bolts; the locking plates are removed; the wedge pieces are removed from the ahead or the astern side of the cam — or both sides — depending upon which cam requires adjusting; the cam is moved until the fuel pump lift at the corresponding main piston top dead centre is equal to the required figure; then new wedge pieces are fitted and the cam reassembled.

Care of scavenge belts (two-stroke engines)

Cleanliness in scavenge belts is of paramount importance. To prevent fires occurring in these spaces the precautions to be taken are:

1. Piston rings to be maintained in good order to prevent blow-past.
2. When cylinders and piston surfaces are dry — although the cylinder lubrication appears to be normal — the piston rings should be attended to.
3. Scavenge belts and spaces should be cleared of carbonised and oxidised oil as frequently as practicable.
4. Care to be taken to keep the exhaust ports clear of carbon.
5. One cylinder unit must not become appreciably overloaded in comparison with the others.
6. If, despite all precautions, combustion occurs in a scavenge space, the fuel valve by-passes for the cylinder should be opened, the drain cocks closed and the lubricating oil to the main piston increased until combustion has ceased.

As long as lubricating oil in the liquid state can be drained from scavenge spaces there is little likelihood of fires starting.

Connecting rod bolts

With hydraulic tightening equipment there will be a few problems, unless an emergency such as a seizure occurs. (See Chapter 26.) It is, however, always useful to examine the connecting rod bolts critically when dismantling, especially in engines where they are torqued other than by hydraulic stretch.

If an engine has been running with connecting rod bolts insufficiently tightened, examination of the shanks of the bolts will show bright belts of rubbing; the undersides of bolt heads and nuts will also show bright hammering marks. Bolts should be rejected if they show: thread stretch; excessively bright marking under head or nut, indicating prolonged and severe hammering; or excessively bright marks on the shank, caused by long and heavy 'working'.

Piston rod nuts (crosshead engines)

The same care applies to these as to connecting rod bolts, especially in double-acting engines.

During crankcase examinations the locking devices should be inspected, and all nuts tested with feelers for slackness. During the overhaul of cylinder units, or when time permits, the locking plates on the piston rod nuts should be removed and the nuts hammer tested. As with all important fastenings, double tightening up of nuts is necessary, to guard against possible defects on the pressure surfaces. The first tightening of the nuts will smooth out such faults, and thereby provide a reliable datum line for the final securing of the piston rod during the second tightening-up operation.

Fuel valves (two-stroke direct drive engines)

Fuel valve designs necessarily differ according to the proprietary engine types. A characteristic design is described below.

Two fuel valves per combustion space are usual in two-stroke engines. Normal practice is for the spindles and nozzles to have flat seats except in small engines, when they are conical. In some engines the valves are oil cooled by a separate circulating system. Fuel valves are always fitted with air vents and air release ball valves, which are used for priming. Normally each ball valve is screwed down tight, but when priming or forcing the air out of the fuel system, the plug on top is released; the oil and air then pass out through a drain pipe.

There is no special delivery valve, or other non-return valve, between fuel pump and fuel valves. It is therefore essential that fuel valves do not leak, otherwise the fuel injection will fail. The upper face of the valve spindle sleeve forms a seat against the lower end of the valve box, and is carefully ground in. The top face of the nozzle, or pulveriser, is also ground and forms a seat against the bottom face of the valve spindle sleeve. Both are firmly secured to the valve body by the bottom coupling nut and are prevented, by steady-pins, from turning.

The nozzle, except in small engines, is made in two parts which are pressed together. The outer part has a diagonal hole for the fuel oil inlet, and the inner part is provided with grooves which form passages for the oil to circulate around the nozzle piece, flowing out through the outlet groove into the fuel valve spindle chamber. This arrangement helps to keep the nozzle cool when no separate cooling system is provided. When the fuel valve is being assembled it is important to guard against the steady-pins being damaged or sheared, thus resulting in the oil flow through the head of the spindle case being blanked.

The fuel valve spring is arranged in a space above the spindle. A guide washer at the bottom of the spring transmits the spring pressure to the fuel valve spindle top. The spring is adjusted as mentioned when dealing with pressure testing. An indicator rod is arranged in the hollow adjusting screw and, passing through the spring, rests on the fuel valve spindle rod. By pressing a finger on the end of the rod it can be ascertained if the valve is working properly.

Fuel valve examination (two-stroke direct drive engines)

When examining fuel valves, all parts must be washed clean with paraffin or with a proprietary flushing oil such as Ensis. They should be dried by warmth or air circulation only. It is essential not to use any kind of cleaning rag because particles of fluff may adhere to the cleaned parts, eventually to find their way into the injection holes, impede the spray and cause the injection pressure to rise, thus reducing the quantity of fuel oil injected. Increasing the fuel delivery, by adjustment of the length of the pump regulating rod, will only assist slightly. If rags and cotton waste have been used for cleaning fuel valve parts during the overhauling the consequences may be burst pipes, leaky joints and the necessity for the frequent changing of fuel valves.

After being washed in clean paraffin, all parts should be placed on warm dust-free plates, to dry before reassembling.

During the first voyage of a new ship all fuel filters should be frequently removed for cleaning, even if the valves seem to be working perfectly. Later, with the continued use of properly purified fuel oil, the filters can remain undisturbed until the fuel valves themselves require overhauling: this may be from, say, six to eight weeks. When a filter is taken out to be cleaned the housing must be washed out also. Clean paraffin should be used, with a small brush, to remove the dirt. The use of cotton waste or rags is to be avoided, as previously stressed.

When changing a nozzle piece great care should be taken not to damage the valve spindle or its sleeve. While the nozzle piece is out of the fuel valve, the centre hole should be cleaned with a spiral drill and the pulveriser holes with a special cleaning needle. Should the nozzle face be pitted or scratched, the defects can be ground away against a surface plate or against the face of another nozzle. A nozzle with a damaged face should never be reused.

Each fuel valve spindle and sleeve are matched. Should the spindle be damaged, the sleeve should be returned to the makers for matching with the new spindle. Hard spots on valve spindles can be polished away by the use of a fine hone, after which the spindle may be lapped into its sleeve with jeweller's rouge. Traces of scoring on new valves can be treated in the same way.

If, during normal running, a cylinder should stop firing because of a defective fuel valve, priming valves of both fuel valves should be opened and the outflow of fuel oil into the drain funnel observed. The fuel valve which discharges oil vapour, or aerated fuel, will be the faulty one. A steady stream of darker-coloured oil will be discharged from the tight fuel valve. If the faulty fuel valve is leaking but slightly it may be possible at full load — by shutting the priming valve of the satisfactory valve and leaving slightly open the priming valve of the faulty fuel valve — for the satisfactory fuel valve to continue to operate; the gases entering during cylinder compression and firing can then find an easy outlet between the strokes of the fuel pump.

It should be superfluous to state that each fuel valve should be perfect, in all respects, before being used. The spindle should be a tight fit in its housing, but at the same time — when perfectly clean and held at an angle of about 45° to the horizontal — it should be able to slide out of the housing by its own weight. Before trying a fuel valve in this manner all oil must be washed off with clean paraffin. If the spindle surface is not absolutely smooth and free from all tool marks and blemishes the fuel valve will operate satisfactorily for only a very short time; fine, solid matter in the fuel will lodge in the surface undulations and blemishes, and cause the spindle to operate sluggishly. If the valve does not close snappily after injection is completed the combustion products will flash through the fuel valve. This may cause overheating and distortion of the spindle, so that leakage begins and the fuel injection pump becomes inefficient in its working. The pump must then be replaced at the earliest opportunity.

Fuel valve lift (two-stroke direct drive engines)

The springs of fuel valves are adjustable; the opening pressure is stated on the adjustment sheet supplied with the engine. The valve lift should never exceed the lift stated on the adjustment sheet, otherwise the springs may be overstressed and break. If the lift is below 75% the fuel may be wire-drawn at full load and the combustion be impaired, because of reduced injection pressure at the nozzle. In these circumstances the indicator cards will show a reduction in height.

Before replacing a fuel valve in an engine it should always be tried on the test pump. The lift height should be checked, by ascertaining how much the lift stop-screw can be tightened up. After the priming valve has been opened it is possible — while running — to check and readjust a fuel valve having incorrect lift. Maintenance of correct lift will reduce spring breakages and also ensure minimum load being carried on pumps and pressure pipes.

As previously mentioned, when assembling a fuel valve great care must be taken not to shear the dowel pin in the nozzle end, because the result may be either to blind the oil supply to the nozzle entirely or wrongly to direct the fuel spray, whereby an important part — such as a piston rod sleeve or a cylinder cover wall — may become burned and cracked.

Fuel injectors (medium-speed engines)

The makers of these engines almost invariably use proprietary brands of fuel equipment, and even if they do provide a design of their own, it follows very closely the practice of the proprietary designs. (Reference may be made to Chapter 5.)

When the clamp or strongback is removed, the injector normally remains tight in the cylinder head. It is bad practice to lever the inlet connector (feed pipe) or injector body from side to side in order to break the carbon seal and then to lever the injector from the bore in the cylinder head. A length of 10 mm or 12 mm threaded rod welded to a spare injector cap nut can be used, in conjunction with a simple bridge piece, to withdraw the injector from the bore. Modern designs usually make specific provision for injector withdrawal tools. When an injector is removed from the cylinder head, some loose carbon will fall to the bottom of the cylinder head injector bore. A profiled spade tool will remove any hard carbon and a similarly shaped felt cleaner may be used to wipe the bore before the injector is refitted. A check should be made to ensure that extra sealing washers are not left in the bottom of the injector bore in the cylinder head. Only one sealing washer should be fitted.

The condition of the injectors as removed from the engine should be recorded to indicate the external appearance, the performances when tested, internal condition, etc. These can be pointers to engine faults as well as injector problems. Testing is described below.

Next slacken the adjusting screw and release the spring tension before removing the nozzle nut. This is particularly important when dealing with dowelled nozzles and holders as the angular load on the dowel can raise a burr on the edge of the dowel hole when the nozzle face separates from the holder face while the nut is unscrewed.

Damage to the thrust face (nozzle joint face) of the holder, if only slight, can be rectified by lapping with greenstick and a small lapping plate. Large and relatively heavy holder bodies can present difficulties on holding square and steady. It is therefore preferable to clamp the holder rigid and move the relatively light lapping plate with a firm pressure.

Severely indented or chipped thrust faces require rectification by the makers and will include rehardening of the face followed by relapping.

Injector springs should be checked occasionally for squareness of the ends by standing them on a flat plate beside a try-square. Any bowing will be indicated by rubbing marks on one side of the coils. Evidence of pitting on the wire surface requires the spring to be replaced.

Early injectors incorporate a ball or roller pressed into the lower end of the spring spindle. This should be checked for chipping or wear, which in turn could damage the thrust pin or the nozzle needle.

Any abnormal amount of metal dust or swarf in the injector feed pipe or edge filter indicates possible damage or a fault in the fuel feed system and this should be checked, particularly the filtration equipment.

Hard carbon, inside the nozzle nut, must be removed as this can prevent the nozzle centralising itself on the holder or even cause nozzle distortion when tightening the nut. The nozzle nut tightening torque must conform to manufacturer's recommendations.

If the injectors are water cooled, the following measures are necessary. If the coolant is oil rather than water, the checks still apply except for 5, which would have to be made as appropriate to the fluids involved.

1. The coolant passages of each injector should be blown out — immediately on removal from the engine — with a compressed air line.
2. The coolant circuit should be clean and unobstructed. Any dirt, pipe scale or congealed grease from the gland of the coolant circulating pump must be flushed out to ensure that coolant flow is unrestricted.
3. It is advisable, after reassembly, to blank off one end of the coolant circuit and pressurise with compressed air. Subsequent immersion in the test tank will reveal any leakage across the sealing faces indicated by bubbles.
4. The complete coolant circuit can be further checked, after installing the injectors, by running the circulating pump until the circuit is at the recommended pressure. The outlet cocks should then be closed simultaneously; any leakage would then result in a rapid pressure drop indicated by the circuit pressure gauge.
5. Leakage of fuel into the coolant circuit results in an oily scum on the top of the coolant tank and remedial action should be implemented to prevent further contamination.
6. If an engine normally equipped with cooled injectors is required to operate on distillate fuel for prolonged periods, corrosion of the nozzle can occur due to overcooling. The nozzle coolant circuit should be disconnected if the operating conditions are of a tem-

porary nature, but nozzles without a coolant passage and associated parts should be fitted to the injectors if these conditions are likely to last for several weeks or months.

7. The coolant temperature should reach the normal value (approximately equal to the jacket water outlet temperature) quickly after starting. This value should be maintained regardless of engine load or fuel pre-heat temperature.

Next the nozzles require to be considered. Nozzle assemblies should be rinsed in 'injector test oil' or fuel oil after removing all traces of external carbon. Paraffin must not be used. The nozzle can then be dismantled and examined (Figure 25.10).

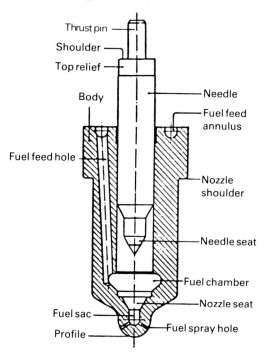

Figure 25.10 Basic nozzle details

On the nozzle body:

1. The sealing face should be clean and of uniform colour. It is permissible to restore the flat surface by lapping with a suitable plate or glass.
2. If the nozzle body had been previously serviced to correct the needle lift, it is essential that:

OPERATION, MONITORING AND MAINTENANCE

(a) the depth of the fuel feed annulus has not been reduced excessively;

(b) the depth of head has not been reduced to a point where the nozzle nut bottoms before clamping the sealing faces.

3. All feed holes must be clear and not obstructed by dirt or pipe scale.
4. All carbon should be removed from fuel chamber and sac with special brass cleaning tools. Compressed air can be used to dislodge and clear the remaining carbon particles.
5. Spray holes can be checked with cleaning wire (using very short wire in a 'Bray' chuck if dealing with blocked holes).
6. The exterior of the nozzle profile should be from flats or damage, and should not indicate any cold corrosion. (See Figure 26.26.) If it does, the nozzle should not be reused, and, depending on the nature of the damage, corrective action is needed to forestall a repetition.
7. Ensure that all carbon under the head shoulder is removed as this can cause nozzle distortion when the nozzle nut is tightened — leading to sluggish needle motion.
8. If possible, after washing the nozzle, the condition of the seat should be checked with a probe light or microscope. The seating line must be narrow and not pitted. The earlier performance tests will have given a lead if the seat is excessively damaged.
9. All observations should be recorded to help plan the rectification work needed.

On the nozzle needle:

1. Any chipping or damage to the thrust pin usually renders the nozzle assembly unfit for further service.
2. Feather edges on the thrust shoulder — caused by excessive impact on nozzle holder face — can, in extreme cases, cause jamming of the needle and should therefore be trimmed off with a suitable stone.
3. Entrained dirt can be identified from small clearly defined rectangular patches of polish marks on the needle surface. The length of the dirt patches usually equals the needle lift.
4. Blackening or seizure of the needle is a clear indication of gas blow-back due to a broken spring, low nozzle opening pressure, incorrect timing (causing overheating) or excessive needle lift.
5. All nozzles on a given engine should conform to specification. An incorrect nozzle with a different differential ratio can result in discrepancies in needle opening and closing pressures compared with adjacent injectors.

6. A satisfactory nozzle will have a clean, bright seating line at the top of the seating cone. The extreme needle tip may be black but the parallel portion above the seat should be clean and bright.

Nozzles suffering severe mechanical damage or excessive back leakage should be replaced. Back leakage passes between the needle and the nozzle body: it is used in some smaller engines to provide the cooling effect needed when operating on heavy fuel. It is measured by timing the pressure drop, usually from 150–100 bar, on the injector test stand. The manufacturer will provide guidance, but leakage times should be at least 10 seconds.

The needle should be free, but this has to be interpreted according to the size. The needle tip should be ground to the correct angle if necessary *only* on a precision grinder or on the Merlin or other specialist nozzle reconditioning machine. The body seat is restored by lapping the seat with a cast iron lap and lapping paste. The lap itself wears and must be checked frequently and reground to the correct angle as necessary.

Needle and seat must *not* be lapped together (refer to Chapter 5 and Figure 5.4) but may be lightly touched together to obtain a witness mark if there is any doubt about the seating line.

After re-establishing the seat, the needle lift must be checked, as this can have a crucial effect on injector behaviour and life. Too high a lift can result in spring failure, impact damage between the needle shoulder and the trust face, and at the seat. It can also promote blowback of combustion gases into the nozzle body by prolonging the opening period while the fuel pressure drops below cylinder pressure. Too low a lift restricts the flow and may overload the pump.

A nozzle body and needle should always be kept together, as a matched pair.

After servicing, the assembled injector must be tested.

Fuel valve testing

To pressure-test a fuel valve: the valve is connected to a test pump and then, after fuel oil has been pumped through the line and all air forced through the valve air vent, the pressure at which the valve actually opens is observed on the pressure gauge. The correct lifting pressure is stated on the adjustment sheet for the engine. The adjusting screw for the spring is now set so that the spindle lifts at this pressure: the screw is then locked in position and the lift pressure rechecked. Next, the nozzle is wiped thoroughly clean and pressure reapplied, this time to 10 kgf/cm^2 below the injection working pressure. If the pressure remains steady for a few minutes the valve is tight. A trace of oil at the nozzle holes is of no importance, as the valve will normally be worked-in completely after a few minutes of running. However, should the nozzle become wet or should drops appear, then replacement or regrinding of the valve is necessary.

Where the injector has a conical seat (including all four-stroke engines) it is essential to test the action of the injector by working the handle of the tester fairly briskly and observing that a good spray issues from each hole of the injector. The relative condition of the spray holes can be checked by placing a piece of cardboard just below the tip (*not* held by hand) and depressing the tester handle briskly once. The pattern (for a symmetrical nozzle) should be symmetrical.

While tightness and quality are the main points it is worthwhile to operate the injector test slowly and observe that the nozzle 'chatters'. That is to say, it operates in a rapid series of short bursts of atomised fuel and does not imitate the action of a watering can. The latter case would imply that the needle was seating on the inner rather than the outer edge of the seat.

Injectors should be tested when removed, as a diagnostic check on their condition, and after reconditioning.

In two-stroke engines having two valves per cylinder further tests may apply when next the engine is run.

If the injection pipe to a fuel valve becomes hot, and the valve is not leaking, there may be choking of nozzle holes, a dirty filter or too small a valve-lift. While running slowly, it is possible to determine if the two fuel valves in a cylinder are set to open at the same pressure. A pressure difference will cause uneven fuel distribution and inferior combustion at full load. The valve which is set to the higher pressure will cease to operate, or will react feebly, when the engine is running light. If a pair of fuel valves on the same cylinder stop functioning it is usual to find only one valve leaking. As already mentioned, if the leak is small it is possible to ascertain which valve is at fault from the gases which blow out from the open priming valve or by-pass drain pipes. If the leak is heavy both drain pipes may be blowing out air; it will then be necessary to feel the valve bodies for temperature, to determine which one is at fault. When a leaky fuel valve has been withdrawn for overhauling the lower part may show heat discoloration. This discoloration should be removed by emery cloth.

A record should be made showing the behaviour of each fuel valve; working and spare valves must be maintained in perfect condition — there is no alternative. A valve which repeatedly fails after only short periods of service is useless.

Fuel valve priming

This is not normally necessary on four-stroke engines except when refitting injectors, or fuel pumps. In the case of two-stroke engines this may not be so.

When preparing an engine for starting it is most important to have the air cleared from the fuel lines and every fuel valve properly primed with fuel oil. For this purpose, the engine-manoeuvring gear handle must remain in its stop position, so that the cylinder fuel pump plungers do not block the free passage for the fuel oil. All intermediate shut-off valves between the fuel service tank and the fuel valves are then opened. The priming, or air release, valves on the fuel valves are opened up, and fuel oil passed through by hand pumping until the oil is completely air-free. As soon as a steady stream of oil is issuing from a fuel valve its priming valve may be shut.

When changing a number of fuel valves it is desirable to prime only one cylinder at a time, because, if many priming valves are open together, the force of the oil through airlocked fuel pumps and fuel valves may not be strong enough to dislodge the air in all the pockets. While priming with the hand pump, a pressure of at least one bar should be maintained and a careful search should be made, preferably with a torch-light, to discover any leaking high-pressure fuel oil joints. Defects should be rectified immediately. In no circumstances can even a very small leak be tolerated anywhere in the fuel injection system.

If fuel oil should refuse to flow through any particular pair of priming valves the cause may be that either the stop valve of the corresponding fuel pump has not been opened or that the regulating lever of the pump has not been brought back to the no-fuel position by its spring, with the result that the pump plunger has been lifted by the fuel cam and the free passage of oil through the pump has been blocked.

If, during manoeuvring, an engine is running dead slow for some time it is strongly recommended that the priming valves be opened slightly on any pair of fuel valves which, for some reason, may not have been operating satisfactorily at light load, and thus release all air and aerated fuel which could accumulate in valves and pipelines. On closing the priming valves, when the next order from the bridge is received, all cylinders will then fire. For this and similar reasons, at least one engineer should be stationed on the upper platform until all manoeuvres in and out of port have been completed. Before entering a port, when the engine is at 'Stand-by', the engineer should open the priming valves on both fuel valves of all cylinders and carefully inspect the fuel oil outflow. He will then be able to pick out any valve which it may be advisable to change when reaching port. He will also be forewarned of the valves which may be likely to be unsatisfactory during manoeuvres.

Fuel injection pumps

Whereas (see Chapter 5) all four-stroke fuel injection systems incorporate delivery valves at the pump, the fuel valve is the only delivery valve in the fuel injection line of a normal two-stroke engine. There is, therefore, a relatively large volume between the top of the fuel pump plunger and the fuel valve. Moreover, the suction efficiency is entirely dependent upon the vacuum created during the downward stroke of the plunger. When the plunger top edge begins to open to the fuel pump suction holes the fuel must rush in and fill the space thus created. If any air — either from a leaky fuel valve or from a leaky joint — enters the system during this period, the vaccum will be proportionately vitiated and a reduced charge of oil will enter the pump delivery chamber. During the delivery stoke the entrained air must necessarily be compressed before the pump plunger can force any oil through the fuel valves into the engine cylinder. If the air leakage becomes great enough the pump will not be able to cope with it, and fuel injection will stop.

Only one fuel valve seat need leak for both valves to cease operating. If a pump is unable to deliver fuel the air or gas from a leaky fuel valve or valves will accumulate, and as the system cannot then rid itself of the air, it may be necessary to shut the stop valve on the pump concerned — to prevent its neighbours from becoming airlocked.

If a fuel pump stop valve is shut off the pump should be hung up, to prevent the plunger from running dry and seizure. If, however, it seems undesirable to hang up the pump the stop valve should be left slightly open to ensure lubrication of the plunger. The priming valve of the leaky fuel valve should also remain open for the escape of the accumulated air and gas. If a fuel pump plunger suddenly shows signs of sticking in its sleeve it will frequently be found that a leaky fuel valve is the cause; i.e. the pump has ceased to deliver fuel normally, overheating by hot gases has occurred and the plunger has seized. As soon as the fuel valve has been located and changed, the pump will immediately begin to function properly again, unless the abnormal conditions have persisted so long that the plunger, when seizing, has scored the sleeve.

After a fuel pump has been overhauled its plunger height should be rechecked by means of the fuel pump mandrel, as directed on the engine adjustment sheet. If the plunger height has not been disturbed when taking the pump adrift it is not always necessary to use the mandrel. Then, all that is required when linking-up the pump lever, is to adjust the length of the regulating rod so that, with the starting handle in the stop position, the pointer indicates the normal number of degrees on the scale. When running the engine it will be observed, by comparing indicator cards and exhaust temperatures, if the fuel delivery of the pump is correct, or if adjustments are needed.

In the case of four-stroke engines, the maker's instructions will advise on dismantling. The control rod/sleeve meshing should be noted and recorded. Be especially careful when compressing or releasing the tappet spring. Whatever the reason for the examination, whether dictated by the life since the last overhaul or by a problem such as leakage, it will be necessary to examine all the components carefully.

Check that the delivery valve holder sealing face is not scored and the delivery valve stop (if fitted) is tight in its bore. Examine the high-pressure seal for signs of tracking (formerly a copper washer, occasionally fibre or nylon, latterly a tin-plated steel washer). If the high-pressure lapped joint between the top of the barrel and the underside of the delivery valve seat has been leaking a darker discoloration of the normally clean surfaces will be seen.

Black markings on the lower parts of the plunger diameter and in the barrel — especially if the surfaces are becoming matt — and/or an even grey-green discoloration of the steel surfaces exposed to the fuel, such as the outside diameter of the barrel, is an indication that the water content of the fuel has been high. Note that this may not necessarily apply if the fuel, with water and perhaps other impurities, has been homogenised or emulsified and can be kept permanently so.

A cracked barrel can indicate excessive pumping pressure. Possible causes are inadequately heated heavy fuel or a blockage at the nozzle, injector edge filter, pipe ends, pump delivery valve holder, etc.

A grey shading on the plunger surface above the helix can sometimes be seen during servicing. This is caused by spill erosion but can be ignored as long as the area does not extend down to the control edge of the helix (Figure 25.11).

The dog on the fuel pump plunger should be checked in the control sleeve slot to confirm that there has been no excessive wear which would affect the accuracy of control. For the same reason, the amount of visible wear on control rod and control sleeve teeth should be checked to see if there is any appreciable difference between those in the normal running position and those at the end of the travel.

The spring coils should be examined for flats caused by a bowed spring rubbing on the bore of the fuel pump tappet or the control sleeve. This spring length should be checked against a new spare. If the old spring is more than 10% shorter than the new one it should be renewed.

Bright marks on the bottom of the pump tappet are made by the engine operating mechanism on the outside and by the foot of the pump

OPERATION, MONITORING AND MAINTENANCE 477

Figure 25.11 Grey shading on plunger surface above helix is caused by spill erosion. (Lucas Bryce Ltd)

plunger on the inside. These areas should be checked for heavy indentation or cracking of the surface.

By standing the pump plunger and the lower spring plate on a dead flat surface, with the spring plate pressed firmly down, a check can be made that the plunger head is quite free to turn on its axis. The plate should be renewed if it tends to limit free movement of the plunger foot.

The pump body should be checked for the following:

1. Damaged threads on delivery valve holder or inlet connections.
2. Ovality or excessive wear in the tappet bore.
3. Damage or indentation on flat or conical element seating surfaces.
4. Signs of erosion in the fuel gallery or on the erosion plugs. The plugs should be renewed if the erosion has penetrated the face.

If the pumping element is being replaced, and it comes from an authorised source, no anxiety need be entertained about its high standard of interchangeability. However, careful comparison of the original component and the replacement is essential. (Engines have been known to 'run away' when a set of left-hand helix elements were fitted to pumps from which a worn set of right-hand helix elements had been removed).

When rebuilding a fuel pump, after the barrel has been removed and replaced, it is advisable to carry out a pressure test to check the seating of the barrel in the suction chamber. After rebuilding the top portion of the pump, the plunger should be replaced in the barrel and retained by a pin or bridge piece across the bottom of the pump which is otherwise left open. The fuel gallery should be connected with air at a suitable pressure, e.g. 4–5.5 bar (60/80 lbf/in^2). The pump should then be immersed upside down in a tank of fuel oil or test oil, when any leakage under the shoulder of the element (or past the 'O' ring on some pumps of later design) will be indicated by bubbles of escaping air.

An 'O' ring seal is a convenient and reliable means of providing a static seal at various points in the fuel pump. It is important that the correct type and hardness of elastomer are selected for the various applications. A conventional low-nitrile compound is suitable for use with distillate fuel, but special materials are needed for heavy fuels requiring a pre-heat temperature of up to 120°C (250°F) (e.g. 3500 secs Redwood No. 1 viscosity Class G). Frequent renewal may be necessary for some arduous applications. Close attention is required and replacement of the seal should follow immediately any increase in leakage rate is observed.

The pump manufacturer's instructions should be followed implicitly when remeshing control rod sleeve (rack and pinion). An additional check to verify that tooth meshing is correct can be made by ensuring the gash (vertical slot) in the plunger aligns with the port in the barrel at the 'stop' control rod position (no load). This can be observed by viewing the plunger-barrel relationship from the open top of the pump. To neglect this precaution invokes the risk of not being able to shut the engine down.

The timing should be done according to the engine manufacturer's recommendations, though the most accurate method is that based on measurement. Refering to Figure 5.10, if dimension T is known from measurement during assembly or by design, Z can usually be measured by jacking the pumping line off the cam and using feelers, and dimensions Y can be set by measuring the fall of the follower on the leading face of the cam.

The spill timing method, also described in Chapter 5, is a quicker but rather less accurate way of achieving this result.

Some manufacturers pre-set the cam and tappet; others allow adjustment of either. If pumps can be cycled ashore to a service centre which has the equipment, dynamic tests can be run, and this is desirable though not vital. If a brand new or fully reconditioned pump is fitted on an engine where there are longer service pumps some differences in cylinder behaviour may be expected. If these are significant something is wrong and checks should be made.

When the pump is refitted to the engine, whatever timing guidance or procedure is allowed, it is essential to see that the tappet height is checked, with the follower resting on the base circle of the cam, to be sure that it will not cause the plunger to contact the underside of the delivery valve holder at maximum lift. This would cause very serious damage.

Some makers offer a setting gauge which, in effect, reproduces the profile of the underside of the pump at minimum lift. It also ensures that the plunger line remains in contact with the cam and is not 'hung up' at the circlip which retains the tappet follower.

High-pressure pipes

Special care must be taken when remaking or replacing a new pipe with proprietary compression type fittings. With these fittings, high spanner torques are not required to make a satisfactory joint. It is important that the end face of the pipe is square and that there is the correct pipe protrusion through the ring or olive before tightening the pump nut.

The use of unnecessary force can malform the ring or damage the threaded connection on the pipe or injector. The high-pressure pipe should always be bent or formed to the correct shape before fitting. In particular the ends should align with the pump and injector fittings, without strain.

When high-pressure pipes are stored, their ends should always be protected by tape or a blank plug in the pipe nuts. Before fitting to an engine, the pipe should be blown through with compressed air to remove any dirt that may have entered during handling or transit. Check too that the sheath is not damaged.

Shock adsorbers (damping vessels)

The shock adsorbers are intended to prevent the fuel oil — which during the delivery stroke of the plunger is spilled back into the suction line — from causing excessive surging in the line. Depending on the construction, the shock adsorber piston or diaphragm is displaced by the fuel oil pressure and returned by the spring load. By reducing surge, they assist in maintaining a steady charging pressure for the fuel pumps. If more than a few drops of oil should leak from these, it indicates a probable malfunction.

If difficulties arise through broken shock adsorber springs, the lift of the pistons may be reduced, within reasonable limits; but their free movement must be ensured, to obviate hammering in the suction line.

The spill line carries away the air which is entrained with the fuel or which is blown back through a pump. If a valve leaks badly, so that it cannot operate, the non-return valve in the spill pipe must be unable to release the air or gas through the line. The fuel should then be cut off as described earlier.

Fuel surcharging pump

The fuel pumps are usually charged by a service or surcharging pump, located between the fuel oil service tanks and the fuel pumps, and driven direct from the engine. It is fitted with a by-pass valve which lifts at about 2.5 kgf/cm^2 service pressure. A hand gear is incorporated, whereby the surcharging pump can be operated and the pipelines primed and freed from air before starting the engine. An independent, motor-driven, rotary pump is sometimes fitted instead of the hand pump. A spring-loaded hand regulated by-pass valve is then usually arranged on the manoeuvring platform. During stand-by periods and prolonged stops the fuel should be by-passed to the service tank. In trunk engines a cog-wheel type of surcharging pump is often arranged on the end of the camshaft or on one of the blower rotor shafts.

A fuel oil filter is arranged between the settling tank and the fuel service or surcharging pump. This filter consists of two or more compartments, each fitted with a cartridge. By means of a change-cock these cartridges can be isolated, one by one, and taken out for cleaning. Indicator plates show which filter is in service. The service pump discharge pressure should be maintained at about 2 kgf/cm^2; it can be regulated by the by-pass valve. If the filter should become choked with dirt it may cause the engine fuel pipes to become airlocked from minor leaks in the surcharging pump suction line.

Turbocharger

Maintenance is specialised and should be carried out as detailed in the manufacturer's instructions. Insofar as cleaning is a prime reason for overhaul, the effect of deposits can be controlled if the facility for 'water washing' is provided on both air and exhaust sides, and is regularly used as directed.

Starting gear

The starting gear normally comprises: an automatic valve; a pilot valve; a distributor; and starting valves. The automatic valve, operated by the pilot valve, is worked by the starting handle. The starting valves are also automatic and are operated by air pressure from the distributors. These, in turn, are operated by circular cams in phase with the engine cranks.

Before starting the engine, the main stop valve for the automatic valve is opened; air passes through a small pipe across the pilot valve top and then through another pipe to the top of the automatic valve. The air, pressing upon the large area of a differential piston, keeps the automatic valve shut. When the starting handle is moved from the stop to the starting position, the pilot valve is lifted so that its air inlet is blocked; the pipe connection to the top of the valve piston is vented to atmosphere; the air pressure on the valve differential piston is relieved; the air pressure below the piston opens the automatic valve and air flows simultaneously to all starting valves and to the distributor.

The starting valves and their pilot valves in the distributor are held off the cams by springs under the spindle heads. When starting air reaches the distributor it tends to open its pilot valves; but the cams prevent this for all except that pilot valve whose spindle end can enter into the cam depression. The pilot valve opens and air passes through the connecting pipe into the appropriate starting valve, entering above the piston on the valve spindle. The starting valve lifts and air enters the engine cylinder. As soon as the engine has turned through the appropriate angle, the distributor cam shuts the pilot valve, and the air trapped on top of the starting valve piston is vented to atmosphere through an outlet pipe on the distributor pilot valve. The starting valve is immediately closed by the spring at the top. The next cylinder unit then receives its quantum of starting air in similar manner.

As soon as the engine has attained the necessary speed, the starting handle is pushed further over, from the starting to the fuel position. This causes the pilot valve for the automatic valve to be tripped and the air pressure to be transmitted to the top of the differential piston again, immediately to close the valve. The air trapped in the starting air lines, between the automatic valve and the starting valves, is vented through a special outlet in the valve, leading through a pipe into an air silencer. The air between the valve and the distributor is released to atmosphere through the same outlet. All air used for starting the engine is thus cut off and released, rendering the starting valves inoperative.

The engine maker will indicate whether any part of the starting gear should be lubricated by hand or otherwise. Normally any lubrication necessary is now provided automatically, but if lubrication is provided it is essential that it is allowed to drain from the distributor housing quite freely, and this freedom should be regularly checked. If oil or any other combustible liquid is allowed to enter the starting line there is a very serious risk of explosion.

On medium-speed engines the attention required will be set out in the maintenance schedule, but may be expected to involve examining the automatic valve and the pilot valve at about 2000–3000 hours, grinding the seats if required.

On many engines, particularly in the medium-speed category, the starting system is somewhat simpler; for instance, the functions of the distributor and the pilot valve may be combined. Where it is not, it is possible to test all starting valves simultaneously in the following manner:

1. The shut-off valve to the distributors is closed.
2. All indicator cocks are opened.
3. The main stop valve to the starting service is opened.
4. Compressed air is admitted to all starting valves, by placing the manoeuvring handle in the starting position.

Care must be taken to ensure that the turning gear is not in mesh during this test.

A leaky starting valve will become apparent from the air which will blow through the indicator cock of the corresponding cylinder. Such a valve should be exchanged for a spare valve, and immediately reconditioned, to be ready again for use.

When working on starting valves, the main stop valve on the manoeuvring platform — as well as the pipe lead to the distributors — should be kept closed; all drain cocks on the pipelines should be opened.

Manoeuvring gear

Usually, a pilot starting air distributor is operated from gear driven from the camshaft, a lost-motion coupling being provided in the drive. If the engine turns freely over the dead centre in either direction on starting air, but does not pick up on fuel, then the fault will be in the lost-motion coupling movement. If the engine turns sluggishly on starting air and all starting valves are in order, then either the timing or the condition of the pilot valves in the air distributor is at fault. The distributor valves are operated by negative cams. The ahead and astern cams are arranged alongside each other so that, in the astern position of the reversing gear, the astern cams take up their positions below the distributor valves.

If any of the gear has been disconnected, a check on the distributor timing should be made as follows:

1. Turn any crank, preferably the crank of No. 1 cylinder, if its distributor valve is easily accessible, to its top dead centre.

2. Measure the distance, on the corresponding air distributor valve, from the spring collar to the distributor block face.
3. Push the valve as far down as it will go, by pressing a finger against the spindle end and measure the compressed length between the spring collar and the distributor block face.
4. Without turning the engine, put the reversing gear into the astern position and measure the compressed length between the spring collar and the distributor block face. If the difference in the ahead direction between the compressed and uncompressed lengths is only a few millimetres, and if the corresponding difference in the astern direction is, say, 12 mm or more the deduction is that the intermediate wheel of the distributor driving gear is not in correct mesh.

Lubricating oil filters

If the lubricating oil filters contain wire gauze elements, these should be changed and cleaned regularly. After a crankcase overhaul, or the cleaning of lubricating oil drain tanks, the filter elements may have to be washed several times at short intervals. Where other types of filter are used, the precautions specified by the makers as necessary for continuous service must be carefully followed. The pressure differences in the gauges before and after the filters will normally be sufficient indication of when the filters are dirty and require changing.

Where cloth is used as the strainer material, it is important that the oil pressure before the filter should be watched. Excessive pressures will cause by-passing of the oil, or bursting of the cloth and its cages. Dirty lubricating oil will then enter the bearings, to the detriment of the engine. Filters of the fabric type accordingly need frequent cleaning: normally their use should be avoided.

In no circumstances must gasoline, petrol or low-flash spirit of any kind be used for cleaning oil filters.

In general, two-stroke large engines are able to continue to use wire gauze type elements, since the general use of white-metal for bearing material associated with relatively large clearances permits the engine to be operated without detriment even though particle sizes of the order of 25 μm might be present in the lubricating oils. In the case of four-stroke medium-speed engines, on the other hand, it is necessary to filter out particles greater than 10 μm to avoid unreasonable damage to the hard bearings used in these engines, in association with relatively fine bearing clearances. To this end these engines are obliged to use throw-away type filters. The life of these filters can, however, be extended if by-pass centrifuging is used to control the impurities present in the lubricating oil charge.

Lubricating oil coolers

For propelling engines the oil cooker is usually arranged with a by-pass on both the water and oil sides; these should be judiciously used during manoeuvring, especially in cold climates.

For small engines, in which the lubricating oil pump is engine-driven and there is no independent stand-by pump, and in which the cooler is arranged with a by-pass pipe for the oil; the regulating valve should be fully opened during starting operations. It is important that the lubricant should reach the crankshaft bearings with the least possible delay, after the engine has started to turn. When the oil is cold and viscous the resistance in an effective cooler can be very great, and the resulting increase in pressure can therefore cause the lubricating oil to be discharged back into the sump, through the spring-loaded by-pass regulating valve, instead of being delivered to the bearings.

On tonnage built since the 1970s, it is highly likely that thermostatic valves are fitted. It can be verified by the feel of the pipes leading to the cooler that the oil is being allowed to reach its correct temperature as quickly as possible, and that it is afterwards maintained.

Formerly, the recommended maximum was of the order of 70°C, but nowadays many makers tend to prefer somewhat higher figures, nearer 80°C.

Lubricating oil coolers should be cleaned, on the oil side, at reasonable intervals; the decreasing temperature difference between oil inlet and oil outlet will be a sufficient guide.

There are many proprietary cleansing fluids on the market. An effective cleansing medium is coal-tar naphtha, but trichlorethylene may be preferred. The cooler to be cleaned should first be drained of oil, through the drain cocks or bottom plug. The drains are then closed and the oil side of the cooler is filled with the cleaning medium, which is admitted through the air cock or thermometer boss at the top, by means of a funnel.

In large coolers the filling is more conveniently done by a hand pump, which draws from a drum and discharges through a flexible steel hose to the funnel. The fluid should remain in the cooler for at least three or four hours; then the dirty fluid is drained away to an empty drum.

The fluid may be used again, after being cleaned in the oil purifier, but care must be taken that it is not mixed with the lubricating oil in the system. All pipes between purifier and lubricating oil tank should, therefore, be disconnected. Expensive cleansing fluids should be sent ashore and purified by partial distillation.

Should a cooler be very dirty, it may be necessary to repeat the cleaning process.

Cooling water systems

If plain fresh water is used for cooling the cylinders and turboblowers, it is advisable that all accessible spaces should be periodically inspected for deposition of scale. There are, however, many proprietary treatments for water to prevent any deleterious effect on the sufaces of the cooling space, and engine makers will have tested some of these and will make recommendations. They will usually advise at some suitable frequency that a sample of the cooling water is taken so that it may be checked to see that the treatment is still effective.

Air reservoir

Manoeuvring air reservoirs are constructed of welded steel plates and strong dished ends. They are fitted with stop, safety and drain valves; a manhole and door are arranged at one end. Reservoirs are inspected regularly; precautions must be taken against internal corrosion and pitting, especially at the top and the bottom. It is of great importance that reservoirs should always be well drained, and that a protective coating should be applied as necessary.

A spare bottle is often supplied, so that if all the air is lost from the manoeuvring reservoirs an auxiliary engine may be started by the bottle and the reservoirs filled by the manoeuvring air compressor. The spare bottle should always be fully charged to the maximum pressure. Should the air be lost from this bottle also, the emergency compressor can be started up for recharging. Double drain valves are fitted on the bottom of the bottles, for draining away condensed water and oil. To manipulate a double drain valve, the inner valve should first be opened fully, and then the outer one should be opened slightly, for blowing-off. Bottles should be emptied at least once in two years, the bottle heads taken off, the inside cleaned with boiling caustic soda solution, then well washed with fresh water.

Lubricating oil

This humble but vital fluid has been the subject of a great deal of development work during the 1960s and 1970s. Apart from blending from a wide variety of base stock to produce the correct viscosity and viscosity index, additives are now available to resist oxidation, to disperse carbon particles, to improve pressure carrying capacity, to resist foaming, to control acidity, etc. Additives themselves are now into their second generation. Many of the earliest additives tended to be based on compounds of calcium, but it was sometimes found that these themselves gave trouble because of the ash build-up as the additive was burnt in the cylinder, sometimes depending also on the degree to which it had been spent. This has resulted in a search for alternative formulations of additive to achieve the same results but without the adverse effects.

In addition, engines are seldom naturally aspirated, and the cylinder conditions obtaining in highly supercharged engines demand something more sophisticated than the pure mineral oils once favoured. It is common practice for engine builders to test their prototypes to determine an oil specification which will cope with the rating, the duty, and the fuel which is intended to be used. In the great majority of cases it is open to any reputable oil company to furnish an oil to a user which corresponds with the specification recommended by the manufacturer.

It is extremely unwise to consider the use of oils which are not known to conform to this recommendation or to any previously agreed variation from it.

It is not only the main engine which has had the benefit of the oil technology. Auxiliary engines, of course, benefit too, but it must be remembered that if these are not of a type which is capable of burning the same fuel as is used in the main engine, it may be inadvisable to use the same lubricating oil as is used in the main engine. It is frequently possible to bunker a single oil type for all engines, and for the gearbox and propeller; but this must be agreed with the respective manufacturers beforehand.

Compressors will sometimes also accept a formulation of oil designed for the internal combustion engine, but the conditions demanded for efficient long life operation of compressors frequently require special oils. If the manufacturer's recommendations in this regard are not followed it may well be that the compressor will fail to achieve its recommended overhaul life because of valve fouling and malfunction, leading ultimately to serious damage.

The circulating oil should be purified as often as possible and regular samples taken both for chemical analysis to determine whether additive depletion is greater than its replenishment in make up oil; and also, as mentioned earlier, for the purpose of spectrographic analysis to monitor the condition of the engine.

Acknowledgement

The editors and publishers would like to thank Lucas-Bryce Ltd. for permission to base the sections on medium speed fuel injection equipment on their Service Manuals.

D.A.W.

26 Significant operating problems

It is perhaps a sign of the times that the range of problems which now merit the term 'significant' is so different in relative emphasis compared with those which did so when earlier editions of this work were prepared. Torsional vibration, once a baffling and unpredictable disaster, is now an almost exact and certainly well-understood science. That is not to say that failures do not occasionally occur which involve torsional vibration, but these are much more likely to have their origins in an error of build or of operation (or in a defect such as a crack or flaw) which undermines the basic values on which the TV calculation was done.

Torsional vibration calculations can now be made routinely and reliably, which estimate vibration frequencies within better than 1%, and amplitudes and stresses within 10% at the outside. Also, it is mandatory for the maker's calculations to be submitted to the vessels Classification Society who make their own independent assessment of the torsional vibration conditions.

Fatigue, although still a lurking potent problem, is a lot better understood at least empirically. The incidence of fatigue failure is not now considered to warrant a whole chapter of examples. Nor are crankcase explosions such a spectre as they were in the 1950s, when the loss of 78 lives from this cause on the liner 'Reina del Pacifico' was still a very recent memory. Such explosions can, and do, still occur, despite all the measures and routines taken to avoid them, but the techniques developed to contain their effects nowadays make serious injury or damage from this cause, let alone loss of life, very rare indeed.

However, despite three generations of design improvements and refinements in the light of experience, ships' engines cannot function without engineers to sail with them, to monitor their behaviour, to be watchful for signs of trouble, and to take whatever steps become necessary to keep the vessel on hire without undue cost or risk. However clever monitoring systems are, and however reliable machinery becomes, there are things which only a trained human being can do. The aspiring marine engineer would do well to remember that 'nothing happens for nothing'. In other words, every slight departure from the ordinary, whether it is a change of temperature, pressure, flow, noise or appearance, should be investigated. It may only be an instrument aberration, but it may be the first sign of potential trouble.

While it is true that many of the problems we now find significant would have been considered less so by earlier generations, a broken cam follower, for instance, can be just as effective in stopping a ship as can a broken crank.

It may be noted that changes in other disciplines have had a considerable effect on the way the problems arising in ships' machinery are handled. No longer need the Chief Engineer be entirely alone in judging the best solution of a problem on his ship. The widespread use of the radio telephone means that, wherever the ship is, he can draw immediately on the resources of his Superintendent's Department, and, through them on those of the manufacturer concerned.

The computer and its associated data storage enable copious information to be called up, whether to interpret design criteria, advise repair limitations, or to advise the availability of replacements.

Air travel means that a specialist from the manufacturer or the Superintendent's staff can be on board anywhere in the world, even at sea, in no more than 48 hours.

Whether these developments are wholly beneficial can be, and often is, argued. There are many who regret what they see as a diminution of the responsibility, and therefore even of the status, of the sea-going engineer. On the other hand, the economics of shipping are seldom such that an owner can easily afford his vessels to be off hire for even a day, and he cannot neglect the advantages of modern communication. It does not mean, however, any change in the value of competent sea-going engineers, who can use their initiative shrewdly to avoid problems altogether as far as possible, to solve on the spot those problems which are so soluble, and to discern quickly the nature of those where he must call on wider assistance without delay.

The following pages set out and describe as comprehensive a list as space allows of the problems which experience tells us can tax the ship's engineers. Like all phenomena, they are all avoidable in a perfect world.

The first duty of the Chief Engineer is to see that the disciplines of prudence and preventive maintenance are clearly established and faithfully observed; and that every one of the engine room staff should be watchful for, and inquisitive about, any unusual occurrence; however trivial. Let us repeat: nothing ever happens for no reason.

Many problems, of course, have their roots in the basic design of the machine or component; some would say all problems are avoidable at the design stage. Examples are found even in basically excellent machinery of brackets, pipe clips or other details whose design is downright sloppy. There are designs which pay insufficient heed to either the basic needs of service attention, or to the effects of the working environment. Examples include difficult access, finicky adjustment, hotspots, dirt traps (particularly near components which have to be removed for servicing). These sins are still sometimes found in the products of even the most reputable manufacturers, and the marine engineer who suffers from them is not renouncing his manhood if he complains loudly and succinctly via his Superintendent to the manufacturer's service department. This is often the only way in which the manufacturer realises that his design is wanting. Sometimes nothing can be done immediately without creating other problems; but any reputable design engineer will usually be very grateful for such feedback.

Another range of problems is that which stems from initial installation, or from a major (or sometimes any) overhaul: the risks of dirt, loose fastenings, leaks, maladjustment. But there are many problems which could occur at any time, and these can only be contained or avoided by vigilance and disciplined routines. The order in which they are set out below is rather arbitrary, but approximately in descending order of difficulty of treatment.

TORSIONAL VIBRATION (TV)

Chapter 1 includes a section detailing the nature of TV and how it can be controlled. Today it is well understood and it is extremely rare for TV to become a problem during service. If it does, it is because something has changed: one of the basic assumptions of the calculation has been undermined.

If TV does arise as a problem, it will often betray itself first as noise or vibration at the shaft section, which is most susceptible to do so. This depends on the shape of the elastic curve for the mode of vibration which is proving troublesome: but how audible its effects are depends on the design. For instance, if one-node vibration becomes excessive, it will be most noticeable at the forward end of the engine, especially if there are any gear drives taken from that point. Otherwise, it betrays itself, though less sensitively, as a greater roughness which disappears if the speed is altered up or down.

Two-node vibration usually betrays itself as a problem at the main reduction trains in the gearbox in the form of tooth separation noise.

Three-node vibration, too, is usually more noticeable at the free end of the engine, but at a much higher frequency than the one-node mode.

It should be stressed that TV problems nowadays are extremely rare and as most engine designers strive to incorporate detuning or damping in order to avoid having barred speeds anywhere in the operating range of speed, there may be little sign that anything is wrong until enough damage has been accumulated to be noticeable.

If a TV problem is suspected, all that the ship's staff can do is to examine the shaft system for loose fastenings, the gearing for tooth damage, the shafts for incipient cracks, or the vibration damper for damage. (Sometimes for its absence after overhaul!)

Viscous dampers have occasionally been found with casing damage which has gripped the inertia ring, thereby destroying the damping effect. They also suffer from deterioration of the fluid, but all manufacturers of these components now include instructions and means to sample the fluid at 16–20 000 hours to check if it is still serviceable.

Pendulum or spring type dampers very occasionally suffer broken elements. Beyond these checks, if there is still anxiety about possible TV damage it is best to discuss it with the Superintendents who will be able to organise a frequency check from a Classification Society, a University, or from the maker.

A major change in the power balance between cylinders — either the loss of one cylinder, or a serious difference in timing of one or more cylinders — can upset the basic assumptions on which the TV conditions have been calculated. The TV implications of any such change must always be checked with the maker if it cannot be corrected immediately.

Balancing

The reciprocating and rotating balance is a design matter affecting not only the vibration levels present in the ship, but the lubricating conditions of the main bearings.

All that the ship's engineers need normally do is to see that balance weights remain secure and that at major overhauls *no unauthorised change* affecting weight is made to any part of the crankshaft/piston assembly without its full implication for balance having been fully considered.

This may involve changes from, say, an aluminium to a two-part piston, a new connecting rod design, or even a replacement crankshaft, or balance weights of a different design. If such a change is done with the maker's knowledge, it is inconceivable that the balance implications have not been considered and advised to users, though it is no crime to check that this is so. But if the change arises because of an emergency, particularly if it affects less than all cylinders, it would be prudent to ensure that the maker has vetted the balance change (and of course the TV implications.)

Virtually all modern engines incorporate line-by-line balance, or at the very least internal couple balancing to minimise internal couples which could otherwise in, say, an unblanced six throw four-stroke engine with symmetrical crank sequence, seriously overstress the centre main bearing as speed was increased — and create thereby noticeable vibration.

Apart from the purely mechanical forms of vibration described above, cylinder output balance does have some influence. The piston, as it moves down the bore on the power stroke, exerts a lateral force on the liner in order to push the connecting rod as it inclines to follow the crank. This creates a reaction couple tending to turn the engine frame in the opposite direction to the crankshaft, and roughly proportional to the force on the piston. Clearly, if all pistons exert an equal force, the reaction is fairly constant. But if one piston exerts much less, or even no, force, its absence will set up variation in the reaction torque which may set up noticeable vibration, perhaps at some speeds more than others. (The effect can be appreciated by shorting one cylinder of a motor car engine.)

For any or all of these factors, it may be found that a particular engine installation is 'rough'. If malfunctions are eliminated and investigation reveals any effect of resonance, the remedy may be to search out and then avoid certain speeds while a longer term solution is prepared in conjunction with the Superintendent and the engine manufacturer.

Other vibrations

Apart from vibrations from the above causes, or occasionally from axial movement of the crankshaft because of linear flexing of the crankwebs, vibrations manifest themselves in relatively small components in the simple form of plates 'drumming' or brackets vibrating resonantly in a simple cantilever mode excited by some component of engine vibration (or propeller vibration).

It is sadly true that small detail components in an engine room can seldom afford to receive the quality and quantity of design attention that is bestowed on large critical ones. As a result they have a distressing tendency to 'happen'. Any faults in the design are only put right when they show up, hopefully on the development or prototype runs or during commissioning, but perhaps, in a relatively new design, or an uprated one, during service. It is often, in fact, more efficient to resolve these problems by experiment than by analysis.

Resonant vibration may well entail using up most of what is misleadingly called the 'factor of safety' — used to sanctify over-simple design, but what is more accurately described as the 'factor of ignorance'.

The remedy is not just to replace what fails, or just to stiffen it, but to identify the resonance, by observing the effect of speed changes, and then to break it. Weight can be added or subtracted; stiffening or flexibility may do the trick, or damping can be introduced.

Note that vibration can only be transmitted through a member up to the natural frequency of that member. This is the principle of the anti-vibration mountings. Consider again the four cylinder four-stroke car engine whose heavy unbalanced secondary force can only be felt at idling speed, which may be just below the natural frequency of the mounts.

As a rough rule of thumb, brackets usually need stiffening (perhaps by additional mounting feet, struts or bolts). Instruments should be on anti-vibration mounts (or off the engine altogether) and flat plates can usually be silenced by welding a rib on them, or sticking a damping membrane on them. In any instance of undue vibration it is useful to the diagnosis to gauge where amplitudes are greatest, and how they vary along and around the engine. Portable reed type vibrometers are useful but a qualitative assessment can be made very quickly by holding the thumb or finger nail close to the vibrating component and in the anticipated plane of vibration.

Vibration is not always a bad thing. The author recalls a problem with a proprietary brand of oil filter, which tended to build up a flow vortex whose pressure difference was opposed to the engine pump. Vibration destroyed the vortex so the cure was to mount the filter back on the engine.

CRANKCASE EXPLOSIONS

Put very simply, an explosion based on chemical reaction requires a correct mixture of the reacting ingredients (at least locally) and an ignition source near the region of correct mixture. The correct mixture of air to fuel is termed 'stoichiometric'. Outside limits of approximately 100:1 (lower explosive limit) to 7:1 (upper explosive limit) explosive ignition are unlikely. The normal proportion of oxygen present in ordinary air is 21% by volume.

What happens to cause a crankcase explosion is that:

1. The proportions of the mixture of air and oil vapour or mist in the crankcase gets close enough to stoichiometric to be able to ignite.
2. There is a source of ignition present.

Normally crankcase vapour is too thin to burn, but it can reach and may pass through the ignitable proportion as a hot spot within the engine is warming up; at first locally, then generally, and when it is cooling down; or if a crankcase door is removed before it has cooled; or if starting air is used while the engine is still hot and while there is excessive blow-by past the piston rings.

Normally nothing in a correctly functioning engine presents an ignition source. But if a seizure takes place, or a bearing fails, or if a moving component becomes partially detached and can strike against the entablature, or vice versa, or if there is a severe blow-by, a part of the engine may attain the red heat not only sufficient to vaporise oil, but to act as a glowplug; or sparks or even flame may be introduced.

Given such conditions, what then follows is an explosion which is almost too rich, and is therefore slower burning and less severe than a stoichiometric one. However, it can attain sufficient force to break cast iron crankcase doors, or to buckle aluminium or steel ones. If no explosion relief doors are fitted, the vacuum which follows the first explosion (Figure 26.1) draws atmospheric air into the crankcase through the shattered doors, and if the ignition source is still present, and the crankcase vapour still rich, a very much more severe explosion occurs.

The problem tends to be worse on large engines purely because of the inherently greater volume of gas, and hence greater possible increase of pressure, but the Classification Societies have quite rightly laid down that every engine whose bore is above 300 mm shall be fitted with explosion relief doors. They are not necessary on smaller engines which are too small to suffer rupture if an explosion occurs. This is partly because of the reduced scope for developing destructive pressure, and partly because the smaller structure, with smaller spans, is relatively stronger.

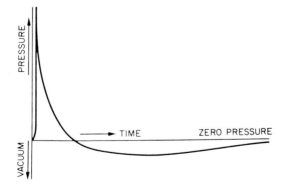

Figure 26.1 Diagrammatic representation of explosion curve

The Classification Societies have laid down formulae which stipulate minimum areas of explosion relief door for given volumes of crankcase. At present the rules are that there shall be one relief door in the way of each crankthrow, that its free area shall be not less than 45 cm^2 and that there shall be a minimum of 115 cm^2 per cubic metre of crankcase volume.

It is not enough to protect the engine structure, and to protect the engine room staff from flying debris. Even an effective explosion relief door does not prevent large volumes of flame which can still cause serious injury, so a flametrap is necessary. The essence of a flametrap is to interpose a large amount of cold (or cool) metal gauze in the path of the flame, so that in heating the gauze the flame dies. Oiling the gauze improves its efficacy by improving heat transfer from the flaming gas to the gauze.

Nowadays the BICERI door, developed by the British Internal Combustion Engine Research Institute, and licensed by them for manufacture in several countries, is almost universal among builders of all sizes of diesel engines. It is made in several sizes and is illustrated in Figure 26.2. As can be seen, it incorporates a spring-loaded valve, a flametrap and a deflector which can be turned to channel the issuing material where it is safest. As the flametrap is inside the crankcase and inherently oil wetted, it is several times more effective than an external flametrap.

Normally, the crankcase of a diesel engine is provided with an open vent pipe, or other form of breather, to allow for the expansion of the warm air inside the crankcase.

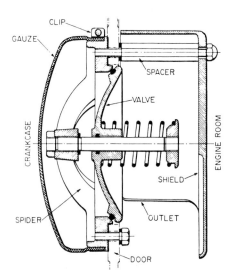

Figure 26.2 BICERA explosion relief valve

To reduce the explosion risk on shipboard, systems of exhauster fans have been applied fairly widely. With this arrangement, pipe connections from the top of the crankcase are led to a motor-driven exhauster fan. This fan may, or may not, be fitted with an oil trap. The discharge from the fan is led to atmosphere. There is no free access of air to the crankcase. The action of the suction fan, therefore, is to induce a slightly sub-atmospheric pressure in the crankcase, experiments having shown that, with such an arrangement, the risk of explosion is reduced.

If the atmosphere in a crankcase is below the inflammability range, and normally it is well below the lower limit, the admission of fresh air is admitted to a crankcase which is already filled with a heated over-rich mixture.

For engines with lightly attached crankcase doors, the exhauster system reduces seepage from the joints of doors, frames bedplates and so on. In a normal engine the leakage of air into the crankcase — as at, say, the sump tank air pipes, or the small annular space which may be found in some types of trunk engine where the cylinder liner passes through the crankcase top — can be beneficial; it can help to keep fresh the crankcase atmosphere. If the atmosphere grows dangerous by the creation of oil vapour the fan should assist in staving off trouble.

Even with BICERI or other explosion relief doors, and whether the crankcase is ventilated or not, it is necessary to be aware of the risk of admitting air if an ignition source is present. It is an essential safety procedure, therefore, *not* to remove any doors from an engine until at least 20 minutes have elapsed after shut down. This will give any hot spot present time to cool down.

It goes without saying that crankcase explosions do not occur unless something has gone wrong. An explosion may be an indictment of maintenance procedures, or it may be an indication of an unforeseeable malfunction. Either way it is prudent to intercept the problem before an explosion is caused. Even though these have today been rendered relatively harmless, they still disrupt. Many engines are, therefore, now fitted with oil mist (or smoke) detectors.

The apparatus used is a modification of the photo-cell equipment used in steam boiler plant for the detection of smoke in flue gases or that used in the cargo spaces of ships.

The detector can be made completely automatic by providing a trigger in the bell circuit, which releases carbon dioxide from bottles and floods the crankcase as soon as the oil mist reaches a prearranged density. Alternatively, the trigger can be hand-operated by the watch engineer as soon as he hears the bell and satisfies himself that it is not a false alarm. The time interval between the ringing of the alarm bell and the incidence of an explosion is ample to enable suitable action to be taken.

BSRA System. The apparatus developed by the Graviner Co. and the British Ship Research Association is shown in Figure 26.3. It has received wide and successful application.

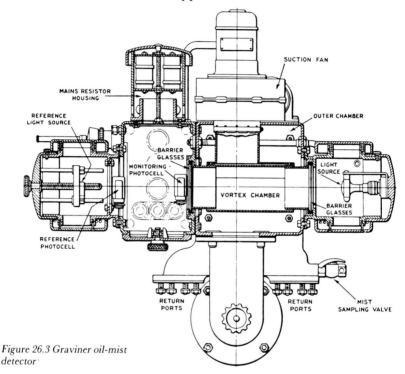

Figure 26.3 Graviner oil-mist detector

The mist detector operates on the principle of measuring the degree of mist density by photo-electric cells which are normally in electrical balance. The value of the out-of-balance current is indicated either on an instrument or on a chart recorder. The instruments provide a continuous indication of the condition of the crankcase atmosphere and warning will be given audibly in the control room or wherever necessary, so that the engine can be shut down. If preferred, it can be arranged to shut the engine down directly.

Explosions in diesel engines are normally associated with the crankcase and its adjuncts, including the chain-drive casings of engine-driven pumps and the castings of thrust blocks. But explosions are not unknown in scavenge belts and in high-pressure starting-air lines. lines.

In these places, as in the crankcase, there are two prerequisites for the generation of an explosion, namely, an inflammable mixture and a hot spot. In the starting-air line the accumulation of lubricating oil from the cylinders over years of running provides the inflammable mixture; the flash-past of hot cylinder gases through a leaking or sticking valve in the air line provides the hot spot. Explosions in starting-air lines can be particularly violent. Periodical stripping, examination and cleaning of the pipes and fittings is the only certain method of obviating them.

Scavenging air belts, sometimes called entablatures, are more prone to fires than they are to explosions. These fires are most frequently of a mild nature, but sometimes they are serious and prove most difficult to overcome. There are on record instances of fires of such severity that the heat from the bottom plates of the scavenge belts — which were also the top plates of the crankcases — was responsible for raising the temperature of the crankcase contents to explosion point, with fatal results. There will be no scavenge-belt fires if the main pistons are maintained gas-tight, if there are no points in the cycle at which the exhaust-gas back pressure is greater than the scavenging air pressure, and if the scavenging belt is maintained in a clean condition. It is necessary that each scavenge belt should be fitted with a relief door.

FIRE

For all the obvious catastrophic implications of any fire at sea, and the considerable volume of experience codified in stringent regulations, fires are still responsible for a significant proportion of shipping losses, and much damage. Many of these start as engine room fires and over two-thirds of these are due to oil leaks or fuel leaks.

Unlike an explosion, a fire needs only an accumulation of combustible material, enough oxygen (which is always present in the normal atmosphere), and either a spark, a flame, or some metal above the ignition temperature. It should be remembered, too, that pre-heated heavy fuel in the lines is above the flash point.

It has been mandatory for many years that all new main engine installations (and auxiliary engines) have the high-pressure fuel lines jacketed so that in the event of a rupture, the escaping fuel is channelled safely to a receiver equipped with a level switch and an alarm, and cannot spray hot exhaust pipes. It has also been laid down that exhaust pipes themselves must be jacketed, both to cut down unwanted radiation, and to reduce the risk of fuel or oil, leaking, dripping or gushing from a ruptured or overflowing line or tank, actually touching hot exhaust surfaces.

It is not easy to cover every square millimetre of exhaust heated metal: flanges, for instance, often have to be left free. Also, if any considerable flow of oil or fuel can reach the cover over exhaust pipes, it can never be guaranteed that some may not accumulate inside the cover, merely because of the inevitable compromise with maintenance access. At the time of writing, too, there are in service still many installations which pre-date these enclosure regulations.

Apart from ensuring that fuel lines are not strained, or subject to vibration, (which may cause rupture) and that joints and connections are secure and fluid tight (and that protective coverings are maintained), it is prudent to ensure that level switches in daily service header and overflow tanks are in good working order. Disastrous fires have been attributed to overflowing fuel (not shut off by the level switch) cascading over running machinery.

Note that while fuel is readily associated with fire, the potential flow of lubricating oil from a supply line rupture is many times greater than the flow of fuel from a ruptured HP pipe — and not much less inflammable.

There is no real substitute for continual vigilance against potential sources of spillage and for disciplined observance of safe handling procedures of all combustible liquids, even as waste.

FATIGUE

At first sight the term 'fatigue' is sometimes taken to imply a deterioration in the material itself. It should be understood that no change takes place in the material except that the crack propagates through it as the material is taken through enough cycles of enough stress to exceed the S–N curve. Test pieces taken immediately adjacent to crack surfaces betray no difference in behaviour from those taken in unstressed material.

While knowledge of fatigue in metal increases slowly but steadily, and has done since it was recognised last century, it is probably still most accurate to describe our knowledge of it as empirical. That is to say, we understand how to deal with it more as a result of calibrated observation than by analysis. Fatigue data on any new alloy is obtained by running a series of tests, not by calculation.

Basically, while a metal component may have for a single test cycle a yield point, at which it suffers permanent strain, and an ultimate tensile strength (UTS) where it will rupture, these values cannot be safely achieved when the component is subjected to repeated stress.

Most metals have an appreciably lower stress limit, called usually the 'fatigue stress' or 'fatigue limit', which it will endure for an unlimited number of cycles without rupture.

It is usual to regard 100 million cycles (10^8) as a safe (or infinite) life. Test results are usually plotted on what is called an 'S–N curve', though more correctly — since more normal — it should be an 'S–log N curve' since the number of cycles is usually plotted on a logarithmic basis to make the safe limit clearer. Once 10^8 cycles are achieved at a given stress level, further cycles do not produce failure for the majority of normally used metals. The larger the applied stress, the fewer cycles the component will withstand before it breaks. That is to say, the point lies on the sloping part of the S–N curve.

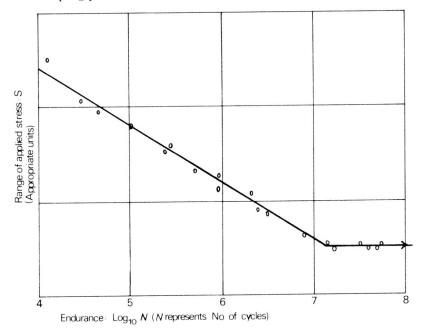

Figure 26.4 S — log N curve

The actual values attained with a given component vary with the surface finish, and the presence of notches, as well as with the nature of the material, and also with its previous history.

Some materials can accommodate the presence of notches better than others. Cast iron, for instance, particularly the lower grades, are virtually insensitive to the presence of notches and to the surface finish. In general, steels tend to have fatigue limits of about half the UTS, but, depending on the grade, it can vary from 30–60%. Non-ferrous materials vary widely.

The actual level of stress which will bring about failure does depend also on the combination of average load and the range of the alternating load. Each application has to be considered separately, but the discussion on the stress effects of tightening up bolted joints below indicates the factors involved.

In certain circumstances the effect of overloads may enhance the fatigue life of a component. Exceeding the design stress in a component briefly in an emergency need not therefore necessarily mean inexorably that it will fail later. Unfortunately one can never be sure in practice that the qualifying condition has been met. If the stress level sustained has exceeded the S–N curve for the number of cycles involved, fatigue damage may have been sustained, and the component will then ultimately fail — perhaps many hours later. Since one can never be sure, it is a matter of prudent judgement what to do if reciprocating or other components of machinery subject to cyclic stress suffer excessive stress.

If an engine suffers a major seizure, it is prudent to renew the large end bolts of the cylinders affected, because the cost of new bolts is a small premium against the catastrophic consequences of the big end coming to pieces. If the crankshaft or shafting is suddenly obstructed it would be similarly prudent to renew all coupling bolts; but most engineers would continue to use the shaft components subject to a regime of regular inspections until sufficient further cycles had been accumulated either to show up incipient damage, or to permit it to be declared safe. Inspection would be with dye penetrant or magnaflux, or ultrasound, as appropriate in those areas identified as the most susceptible to fatigue. The engineer should be wary of components which have been abused or dropped, or which have sustained surface damage. If a blemish is sustained on the journals of a crankshaft, its most potent danger is of creating a stress raiser. This is its literal effect. The stress (f) in the layers of a uniform beam in bending varies with the distance (y) of the layer from the neutral axis according to the well-known formula:

$$f/y = M/I = E/R$$

where M = the applied bending moment
I = the section moment of inertia
E = Young's modules
R = radius of curvature

One may therefore draw lines of equal stress at each increment of stress across the section, and the frequency of these lines across a unit of the cross section indicates the stress gradient.

The lines of equal stress crowd together in any case at a section change like a crankshaft fillet, producing a higher stress gradient. A stress raiser forces these equi-stress lines further into the component,

SIGNIFICANT OPERATING PROBLEMS

and thus to an even higher stress gradient. In addition, the maximum stress may even be higher if the blemish is particularly sharp.

It is therefore not enough merely to smooth off material thrown up proud of the surface, but to raise the ruling radius of the blemish as far as is possible (without losing any more of the bearing surface if it is a journal, or it will prejudice the oil film). This is particularly important if it is in the most heavily stressed area, such as near the inner fillet of a crankweb. Unfortunately, design or manufacture may leave sharp edged discontinuities in highly stressed components.

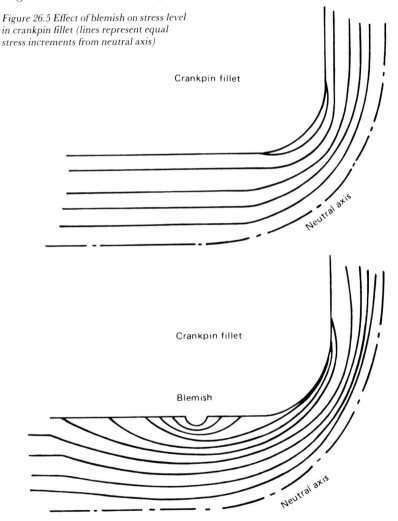

Figure 26.5 Effect of blemish on stress level in crankpin fillet (lines represent equal stress increments from neutral axis)

When cracks do occur, and they are opened up, it is usual to find on microscopic examination (or sometimes with the naked eye) a blemish near the origin. This may be damage caused by a blow with a sharp edged implement (perhaps a tool dropped, or clumsy refitting of a bearing keep) or it may be a metallurgical fault like a slag inclusion. Occasionally no identifiable blemish is found and one is left to speculate whether some aspect of normal operation involves greater stress than anticipated, or there has been a period of overstress either due to maloperation, or to maladjustment (bearing alignments for instance) or perhaps to locked in stress caused in manufacture. The author has known gudgeon pins to fail from origins in the centre of the wall due to the interaction of working load, with the tension caused in the core by the compressive stress set up by hardening the skin.

The appearance of fatigue cracks is often instructive, especially in a ductile metal where stress alternates between tensile and compressive, or where it is a torsional stress that causes failure. In such cases, once the S–N curve has been overstepped at the point of initiation, rupture commences, and progressively extends a little every so many cycles, spreading out radially from the origin in the plane of maximum stress. If the stress alternates as above, the surfaces chafe together in the compressive phase and produce approximately concentric polished rings centred on the origin. Naturally as the crack progresses it first tries to relieve the original overstress, but eventually as the remaining unbroken material becomes insufficient to carry the load, failure becomes more rapid until the component breaks.

In thicker sections the break shows no evidence of yield, but a rough crystalline appearance. If the damaging stress is purely tensile, the appearance differs in that polishing of successive increments of the crack cannot take place because there is no compressive phase to make it happen. This aspect of appearance is not, therefore, very relevant.

Cracks in cast iron seldom polish to any noticeable extent because of the nature of the material, and origins are then sometimes more obscure. The direction of cracks is always indicative. Bending stress failures proceed radially, or normally, inwards from the surface; torsional stress failures in a shaft proceed in a helical direction at 45° to the axis of the shaft. Towards the end of the period of propagation, the applied stress pattern may, of course, be different, and a crack will then change direction to reflect this change.

To live with machinery without fatigue it is not necessary to go beyond the empirical rule that, with some easement in the case of cast iron, stressed components should have a good surface finish, and be free of scratches, metallurgical defects, locked-in stress, sharp radii or corners (internal or external) or blemishes. The higher the UTS of the material involved the greater may be the notch sensitivity, and the greater the importance of these dicta.

It should be noted that a corrosive environment will lower the safe fatigue strength of most metals.

The foregoing has been written mainly with iron and steel in mind, which are of course the principal materials found in marine machinery. Non-ferrous metals have similar fatigue characteristics except that some do not have a definite fatigue life (certain aluminium alloys for instance) and a designer using them has to specify a safe life as well as a safe stress. However material of this kind is practically never used in a stressed role on board ship.

Some examples are given of fatigue failures: Figure 26.6 shows a torsional failure originating from an ill-formed but certainly badly sited tapped hole. The 45° progress in the anti-clockwise direction, becoming increasingly rapid as it approached the course of the secondary propagation at bottom left, is quite distinct. The final break is seen at the upper left half of the section.

Figure 26.6 Torsional creeping crack

If a hole, for example, for an oilway has to break out in a component where a high cyclic stress is carried, it is vital that its mouth is very carefully and smoothly blended. Figure 26.7 is a good example of bending failure in a fulcrum pin due to the effect of inadequate smoothing of the mouths of oil holes. The markings clearly show these to be the origins. The enlargements (Figure 26.8) show the appearance of, first, a less extensive crack at another oil hole (marked '*m*' on Figure 26.7); and, second, the extent of that crack when the material has been split open to reveal it.

Figure 26.7 Fractured slave-rod pin

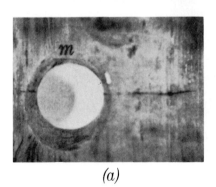

Figure 26.8 Macrograph of oil holes

Metallurgical flaws have been mentioned, but welding is often a source of trouble if imperfectly done. This is only controllable by the skill of the operator, but good appearance, absence of voids, spatter, discontinuity, cold rolling or undercutting are reassuring. In the last resort X-rays, (where practicable) are the only non-destructive check possible.

Welding repairs should never be attempted on highly stressed components without specialist advice. Welding affects the metallurgical properties of the parent metal (the 'heat affected' zone) and can sometimes cause considerable stress concentration.

Components subject to heat such as cylinder heads are prone to crack if design conditions are exceeded, particularly where a high temperature gradient is involved. Such regions are subject to a tensile stress on the cold side and a compressive one on the hot side. If deflection is not allowed the material has to withstand these, but at high enough temperatures (over 300°C in cast iron) the hot side material will tend to creep. That is to say, it will flow plastically to relieve the compressive stress. Unfortunately, when it cools, it is likely to develop an equally high tensile stress due to shrinkage, and in a rigid component like a cylinder head this may lead to cracks. Obviously insufficient water treatment and hence scale formation will aggravate this tendency. Most designers ensure that temperatures in such locations are below the creep limit, and newer designs of engine aim at having a relatively thin flameplate (or a very well cooled one) to withstand temperature, well supported by an upper or intermediate deck which carries little thermal stress. (Cracked cylinders or heads entail the risk of filling the cylinder with water leading to hydraulic lock when restarting unless the much neglected practice is followed of barring round with cylinder cocks open for one complete cycle.)

Finally, a word about tests: that is to say, tests before the failure of the component, to avoid possible associated damage or hazard to the ship due to loss of power.

Where the suspect surface can be reached easily, probably the best check is the dye penetrant method, also known as the Ardrox test. The suspect area must be thoroughly cleaned, and any adjacent areas or recesses — which must not be contaminated — must be suitably sealed. The area is then painted with the red Ardrox dye, left a few minutes, then wiped clean.

A white chalk suspension is then sprayed on to the surface. If a crack is present, it will retain dye, and capillary action will cause it to emerge and suffuse the chalk clearly indicating its presence. The deeper the crack the more definite the line of the dye.

Care is needed when applying the test to non-machined surfaces. Paint must be removed first, and if fine 'cracks' appear consider whether these may only be casting or forging marks, or just scale. Ardrox can be used on any non-porous material.

If the end of the section containing the suspected crack is accessible, and facilities are available, an ultrasonic test may be useful. This sends an ultrasonic signal into the material, this is portrayed on a cathode ray oscilloscope (CRO). If no crack or discontinuity is present, only one reflection will be seen from the opposite wall (assuming the shape permits this). If a crack is present, it will show as an intermediate echo. This test is rather a specialised one and needs a specialist to interpret it correctly, but most major dockyards should have, or be able to borrow, such a facility.

FRETTING

Fretting takes place when two surfaces under high intermittent load are permitted any movement. The movement upsets the uniform transfer of load from one surface to the other and local microscopic welding tends to take place due to movement. When further movement breaks these tiny weldments, they tend to break off and gradually form a powder of micronic size, which adds to the destruction of the surfaces. When the joint is opened, there will be, depending on the degree of movement which has developed, a patchy black iron oxide staining progressively leading to distinct shallow black pits. If oil or fuel has reached the area of contact, and especially if humidity is high, or water for any other reason has reached the contact surfaces, a brown paste or fretting product (sometimes termed cocoa) may be present in quite noticeable quantity.

If this takes place in bearing housings behind the steel backs of the shells, there is insufficient nip either because the bolts are not tight or because the shells are too short, or the housing too big. The consequence is insufficient heat transfer to the housing from the shells and linings which may suffer premature failure.

If it takes place in bolted joints it means that the bolts are being unloaded completely, or the structure is not dimensionally stable, and this enhances the risk of failure. In any case, fretting is progressive and if it occurs in a stressed region it may well bring on the onset of fatigue damage because of the effect it has on the surface finish. If fretting is found in a joint, machining is usually essential to restore the surface finish if the parts are to be reused. If this is not practicable they must be replaced.

The tightness of the joint should be checked carefully when it is remade to ensure that at the minimum stress in the cycle a positive pressure still exists. Since this will have to be discerned by measurements, it will usually be necessary to interpret these with the design data.

There are some locations where fretting sometimes takes place and improvement can only be found in metallurgy: one example is the contact between the fuel pump return spring and the inside of the lower plunger guide. Fretting is more a visual offence than a danger at such places, but harder surfaces will effect an improvement.

SEIZURES

It is fortunately fairly rare for piston seizures to occur nowadays. Scavenge fires, if they occur, in two-stroke engines can lead to a considerable risk of seizure unless the lubricating oil supply is immediately increased. If a Graviner mist detector is fitted, a warning will be given, but perhaps only minutes or seconds before it reaches its full retarding effect. It is imperative that such a warning is heeded, and the engine power reduced at the governor output to slow ahead if not shut down altogether with the least possible delay.

The forces involved are considerable, especially if, as is quite normal, the hand throttle (if any) is set clear of the governor output lever. The governor, sensing the fall in speed as the seizure develops, opens up to maximum fuel to correct the speed fall. As a result all the surplus power of the remaining cylinders on the affected engine — maybe up to nineteen cylinders — will be diverted to drive the seized piston up and down.

If the seizure occurs when running at or near mcr it is likely that insufficient reserve of power remains to drive both the propeller and the seized piston and the engine will slow.

An earlier tell-tale for the watchful engineer to notice (or for a properly reset data logging installation to sense) is a rise in the cooling water outlet temperature on the affected cylinder, or a general rise in exhaust temperatures. Such signs may increase by a few minutes the warning of impending seizure.

If the control room is manned at the time, the officer of the watch may notice the change in engine — and particularly turbocharger — note. But the engine must be shut down instantly. Otherwise enormous strains may be put on the cylinder running gear particularly near TDC, and something may yield. Either the piston will pull apart, the liner may break, the connecting rod bend, or the large end tear apart with heavy consequential damage, and the vessel will be disabled.

Cast iron pistons are worse than those with aluminium skirts in this regard. In fact in dire emergency a seized aluminium piston, if stopped in time, may be run again at reduced speed for a limited period after being allowed to cool down.

However, even if the engine has stopped without further damage, the affected line parts — including the large end bolts — must be renewed or refurbished. It is obviously essential to try to determine what was the cause: blocked cooling oil supply, overfuelling, valve damage, ring damage, timing at fault or whatever it may be: otherwise the phenomenon may well be repeated.

If there has been consequential damage, it may still be possible to continue the voyage at reduced power by removing all working parts from the crankthrow and cylinders affected, blanking the oil supply and valve ports and (assuming the frame is still, or can be made, oil tight) choosing a speed and output to suit balance, vibration and turbocharger characteristics.

It is essential to check the connecting rod for visible bending immediately an engine has suffered a seizure (or should one say when it has cooled enough to be safe to open the crankcase without risk of explosion!). A bent connecting rod may have also damaged the liner.

This check is also essential if the engine has suffered a hydraulic lock in spite of all precautions.

In either case even if no visible bend is present, it would be prudent to have the rod checked between mandrels at the earliest opportunity.

BOLTS AND BOLTED JOINTS

When a bolted joint is tightened a tensile stress is induced in the bolt or stud according to Hooke's Law.

$$E = \frac{f}{e}$$

$$= \frac{P}{a} \times \frac{l}{\delta l} \qquad (26.1)$$

Where E = Young's modulus N/mm^2
  f = stress N/mm^2
  e = strain
  P = force applied N
  a = area on which P acts mm^2
  l = unstressed length mm
  δl = extension mm

Therefore
  $f = E \times e$

$$= E \times \frac{\delta l}{l} \qquad (26.2)$$

Conversely

$$\frac{\delta l}{l} = \frac{f}{E} \qquad (26.3)$$

$$\delta l = \frac{fl}{E} \qquad (26.4)$$

A converse compressive stress is, of course, induced in the materials joined by the bolt, or at least in an arbitrarily defined column of material trapped between the washers under the nut and bolt head. (This is a slightly simplified account, hopefully in the interest of clarity.)

The bolt can be stressed up to just below its elastic limit and the strain (or stretch) can be calculated and measured. This is, in fact, far and away the safest way to tighten it, but not always practical. A skilled fitter tightening a suitably sized bolt with a spanner and the power of his arm can 'feel' when he has stretched the bolt to a safe extent in terms of yield (this does, of course, mean that he may well tend to under-tighten high tensile bolts). He is helped by the design of spanners whose length is so chosen to match the size that it is unlikely that a reasonably strong man using it *as it is intended* can over-strain an ordinary steel bolt of that size.

Where stretch is impractical as a criterion, torque spanners are, of course, widely used. It should be borne in mind that a damaged or tight thread will give a false reading. Always check that the nut can be run down finger tight first. Torque wrenches also depend on regular calibration and, in some cases, on correct handling.

This is quite reasonable for simple fastenings, but where vibration, high (and highly fluctuating) stress, and high-grade steel is involved, the inadequacy of the simple approach is revealed in all kinds of failure.

For large diameters the slogging spanner has held sway for many years, and where space and size permit it, the extension tube (possibly slogged up as well) with enough success to prolong the practice, but generating habits that boded ill for the much smaller threads on more finicky equipment.

The realisation of the possibility and the advantage of using hydraulic pre-tensioning (sometimes electric pre-heating) of important bolts to set the desired elongation and stress precisely and easily has led to their adoption in virtually every new design. (Pre-tensioning has the further advantage of eliminating any extraneous torsional stress in the bolt itself, due to using a spanner, where it uses up some of the safe stress limit.)

On vibrating equipment a short bolt needs a shake-proof spring washer or a tab washer to lock it, because a very slight relaxation — or any differential heat expansion — will cause it to lose tension and become loose.

Consider a bolt 10 cm long in steel stressed to 200 MN/m². Steel has a modulus of approximately 200 000 MN/m², therefore, from Equation

$$\frac{\delta l}{l} = \frac{2000}{200\ 000} = \frac{1}{1000}$$

Therefore, $\delta l = \dfrac{l}{1000} = \dfrac{10}{1000}$ cm $= 0.1$ mm \hfill (26.5)

0.1 mm or 100 μm might be little greater than the surface finish. The slightest fretting or settlement of high spots will eliminate the initial load. Tightening this joint may entail only one-tenth of a turn, even with a fine thread. A lock washer will extend this tightening phase to a full turn or so.

Here the attraction of longer, more resilient bolts in newer designs, becomes immediately obvious. If the active bolt length had been 50 cm the required pre-load now entails a stretch of 0.5 mm, which is obviously much less likely to be upset by slight settlement. In fact when this much resilience is present, a locking washer becomes superfluous, or, to be fair, a second line of defence and a means of visually checking that the nut *has* been tightened.

Resilience does not eliminate the need to see that bolt and nut seats are truly flat. If they are not bending stress will be added to the bolt load, and any localised seating may settle and shed load.

As an example of bad design, Figure 26.9 indicates a collar on a tie-rod. The collar c is provided with flats f for holding the collar when tightening the nut at the lower end of the rod. The arrangement is shown in Figure 26.9(a) after initial tightening of the nut. That is, the load is sustained by the area of collar, diameter d, minus the two segments x.

Under working conditions the collar will tend to rub a slight recess into the surface at the area of contact. If the positioning pin, shown in Figure 26.9(b), is sheared, the flats after subsequent tightening can assume a different position, as indicated where the segments have moved to y. This means that the load is then carried by the two small segmental areas x, in Figure 26.9(a). As these areas become fretted away, the attachment will hammer and become slack. The difficulty can be avoided if flats f do not extend the full depth of the collar.

Note that screw threads do not automatically run co-linearly. A nut can tilt on the bolt threads by as much as 2° purely because of the working clearance between the threads in each component. Of course if the joint in question is tightened hydraulically first, the nut is much more likely to centre on the threads than if it is torqued up under pressure.

SIGNIFICANT OPERATING PROBLEMS

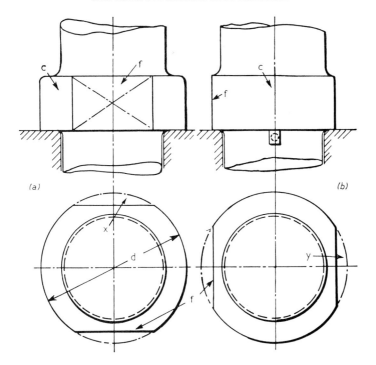

Figure 26.9 High spots from fretting

Stress effects of tightening up and working loads

It is sometimes not clearly understood how a bolted joint is stressed. Figure 26.10 shows diagrammatically, albeit in an exaggerated form, the effect upon a bolted flange joint of the tightening-up stresses and the working stresses. The initial conditions are indicated at 1, respectively

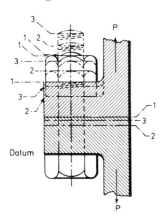

Figure 26.10 Effect of tightening and working loads

for joint face, flange top face, top of nut and end of bolt. The effect of tightening-up the bolt is shown at 2. That is, the bolt is lengthened and the flange is compressed. Upon the application of the working load p there is a partial relaxation of the flange and a further elongation of the bolt, as indicated at 3. It is under the last named condition that the flange joint must remain pressure tight.

Figure 26.11 shows, in outline, a simple bearing block.

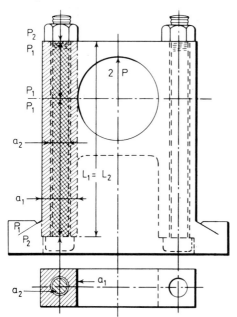

Figure 26.11 Effect of tightening and working loads

In the above $a_1 a_2$ = cross sectional areas of block and bolt
$L_1 L_2$ = initial length of block and bolt. $L_2 = L_1$
$E_1 E_2$ = moduli of elasticity of block and bolt material
$\delta l_1, \delta l_2$ = alterations in length of block and bolt
$f_1 f_2$ = stresses in block and bolt
P = working load on block
$P_1 P_2$ = tightening-up loads on block and bolts; $P_1 = P_2$
$p_1 p_2$ = alteration of loads on block and bolt; $P = p_1 + p_2$.

If a tightening-up force $P_1 = P_2$ is applied by the spanner, the bolt elongation $\delta l_2 = P_2 L_2 / a_2 E_2$ and the block compression $\delta l_1 = P_1 L_1 / a_1 E_1$. The distance moved by the bolt end beyond the nut = $\delta l_2 + \delta l_1 = x$.

If $\delta l_2 / \delta l_1 = K$, where K = constant; $\delta l_2 = \delta l_1 K$
$\delta l_2 (1/K + 1) = x$.

SIGNIFICANT OPERATING PROBLEMS   513

Given x, derived as above, and knowing the a and E values, and therefore K, the magnitude of δl_2 can be obtained. Also $\delta l_1 (1 + K) = x$

Tensile tightening up stress in bolt
$$= f_2 = \delta l_2 E_2 / L_2 \tag{26.6}$$

Compressive stress in block due to tightening-up
$$= f_1 = \delta l_1 E_1 / L_1 \tag{26.7}$$

The area of block under simple compression is empirically chosen.

Application of the external load P causes the block to relax a certain amount and the bolt to stretch a like amount the block and the bolt being bound together. It can be shown that the additional tensile stress in the bolts caused by the working load $= p_2/a_2 = P/(1 + K)a_2$. The reduction in the compressive stress in the block $= p_1/a_1 = P/(1 + 1/K)a_1$.

$$P = p_1 + p_2$$

The total tensile stress in the bolts, due to tightening-up plus working load
$$= \delta_2 E_2 / L_2 + P/(1 + K)a_2 \tag{26.8}$$

The total compressive stress in the block, due to tightening-up plus working load
$$= \delta_1 E_1 / L_1 - P(1 + 1/K)a_1 \tag{26.9}$$

Example. A cast iron bearing block is held down by steel bolts as in Figure 26.11. Determine the stresses in the bolts, given the following particulars:

Cross sectional area of one bolt $= a_2 = 13 \times 10^3$ mm^2.
Cross sectional area of stressed region of block $= a_1 = 39 \times 10^3$ mm^2
Length $L = 1780$ mm. $E_2 = 20 \times 10^5$; $E_1 = 12 \times 10^5$ bar.
Upward thrust per bolt due to external load $= P = 45\ 500$ kg.
Bolt is initially tightened by turning the nut through 80°, the pitch of screw thread being 6.00 mm.

$\delta l_1/\delta l_2 = a_2 E_2/a_1 E_1 = 13\ 000 \times 2\ 000\ 000/39\ 000 \times 1200\ 000 = 0.556$
  $\delta l_1 = 0.556\ \delta l_2$ and $\delta l_1 + \delta l_2 = x$,
i.e. $1.556\ \delta l_2 = x = 80 \times 6.00/360$
  $= 1.333$ mm
$\delta l_2 = 1.333/1.556 = 0.857$ mm

Stress in bolt tightening
$$= l_2 E_2/L_2 = 0.86 \times 2\,000\,000/1780 = 966 \text{ bar}$$

Stress in bolt due to tightening plus working load
$$= 966 \text{ bar} + P(1+K)a_2 = 966 + 122 = 1088 \text{ bar}$$

The pre-load in a bolt subject to fluctuating load is set so that there is no separation of the surfaces it is intended to hold together.

Apart from the risk of fluid loss if they are sealing a cylinder, the loss of closing pressure may interfere with necessary heat flow, the separation or relaxation may induce fretting and, because a load applied as an impact has the effect of intensifying the value of the load, it may hasten the failure of the bolts.

This may be appreciated from a Goodman diagram.

In Figure 26.12 $6u$ = ultimate stress
                   $6y$ = yield stress
                   $6n$ = fatigue stress limit established by rotating bar test
                   $OA = AB$ = mean load
                   BC = positive alternating load
                   BD = negative alternating load

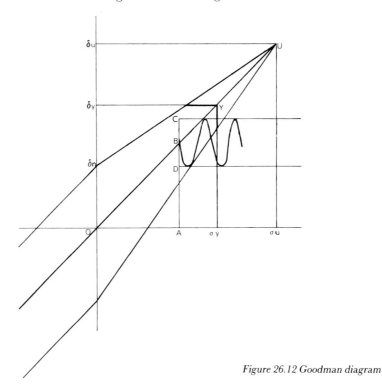

Figure 26.12 Goodman diagram

DA indicates that a positive load of sufficient magnitude will be maintained. (Similarly for the joint material, which, being compressed, is shown in the opposite quarter.)

Note also that the line 6n–U allows the total applied stress to rise as the steady load increases and the alternating load decreases up to the intercept of 6y Y 6y where the material would yield rather than fatigue.

Goodman diagrams are sometimes drawn slightly differently, as in Figure 26.13. In this case B coincides with A.

Also shown for interest in Figure 26.13 is a more detailed curve (x) for a typical high tensile steel. This is more generous than the simple straight line, which therefore errs on the side of safety.

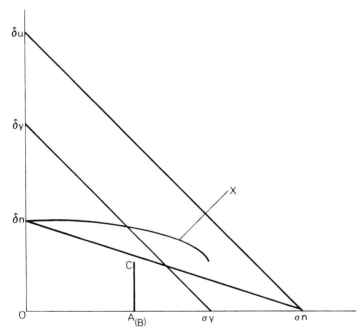

Figure 26.13 Goodman diagram (alternative form)

CRANKSHAFT ALIGNMENT

Initial alignment is discussed in Chapter 24, but the medium-speed engine, being much shorter and stiffer, requires some extra understanding. It should be understood that however carefully journal bearings are aligned, journals of a crankshaft cannot remain in line with the crankpins between them carry torque. The journals would move parallel to the original alignment — by amounts increasing towards the propeller end — but varying as the firing order unfolds. As the bearings constrain their movement, the crankwebs have to twist to allow the journals to take up their positions in their bearings. As a result the journals are not parallel with the bearings, and they move about within the clearance during the cycle.

Manufacturers go to some trouble to minimise the effect of imbalance, but they can do nothing to reduce the effect of twist. They are handicapped, up to the time of writing, by the absence of a rigorous mathematical procedure for evaluating the detail strains in a crank element and the movement of the journal in its bearing. Even with a computer, the process of evaluating the bearing loads and clearances has to make some basic simplifying assumptions.

However, all engine types are subject to many thousands of hours of development running to verify, among other things, that the bearing system as a whole is functioning correctly. This is further correlated with experience in service. In a sense the calculation is calibrated by subsequent experience, but the results in most cases have been such that unexplained crankshaft and bearing failures are very rare indeed.

It is, however, of the utmost importance to preserve the alignment of the bearings, since this is a basic assumption made in all the design and development work on the engine. When alignments are checked, it is important to remember that merely turning the unloaded shaft may not detect the absence of support at any one bearing. On a large direct drive engine, the shaft elements are usually heavy and flexible enough to find support. On a medium- or high-speed engine the shaft will be quite strong enough to span it when at rest, but may be dangerously overstressed when, under running conditions, it seeks the support of the low bearing.

Most medium-speed engine manufacturers will recommend that a check is made of actual bearing support or alignment during a major overhaul. This may take the form of using soft lead discs to record the actual clearance, or the use of 'nipped' deflections. This, however, is a matter for Chapter 24.

Apart from bearing damage or excessive wear, alignment is not a major problem within an engine, which presents a very stiff structure and sits on very stiff seatings (except for the effect of temperature gradient from cylinder head to tank top, which is to cause the engine frame as whole to take up a slightly 'hogging' attitude).

Where alignment problems usually arise, unless a flexible alignment coupling is incorporated, is in the last two bearings of the crankshaft because of deflections in the after section of the hull, and hence of the shafting. The inevitable consequence of neglect of actual alignment when running is a bending failure at a crankweb.

MAIN ENGINE CRANKSHAFT DEFECTS

While manufacturers endeavour to produce to standard tolerances and quality, components like crankshafts are very expensive, and it is economic to try to salvage departures from standards up to a point. That point is further for large built-up shafts than for, say, an auxiliary engine crankshaft; but no reputable manufacturer will risk the integrity of his product.

Nonetheless, if minor forging blemishes like inclusions are revealed in manufacture they are usually (if not too deep) completely ground out and the edges of the depression blended smoothly with the surface. Occasionally a pin or journal will be turned to a standard undersize to remove blemishes, or to true up damage or wear.

It is vital that any such variation from standard is recorded, and preferably that a warning plate is affixed to the engine in way of the affected shaft section(s) to guard against incorrect bearings being fitted subsequently — and to ensure that the required non-standard shells are available.

Alignment has been covered already, and its correction, unless the bearing shells are incorrect, or have been incorrectly fitted, or it concerns the end bearing — may well entail removing the shaft. In engines incorporating a bedplate the real solution is to rebed the housings to a mandrel, but in very large engines special bearing shells may be more economic. If the shaft rests in underslung saddles, or keeps, the problem is simplified in that correction may be possible by adjusting the effective centre height of the keep or saddle in question.

The vital question, however, would be not the correction method but the reason for the loss of alignment, which is a very rare event. When it does occur there is every likelihood that there is a frame crack present, and this must first be identified and repaired.

Built-up shafts have their own problems. It may be instructive to quote an account of an actual repair to a direct drive main engine whose crankweb had slipped 9.5 mm relative to the crankpin following a hydraulic lock caused by water leakage into the cylinder. (Note that this expensive problem would have been discovered and avoided if the engine had been routinely turned through one cycle before starting.)

The account is derived from an article by R.J.F. Hudson in *Marine Engineers Review* for September 1976, based on a paper presented at the University of Hong Kong.

The crankshaft repair was effected as shown in Figure 26.14 Brackets made from mild steel plate were welded to the top flange of the bedplate as buttresses for the hydraulic jacks. Restoration of the crankshaft back to its correct alignment mark was achieved by the 100 tonne jack. The

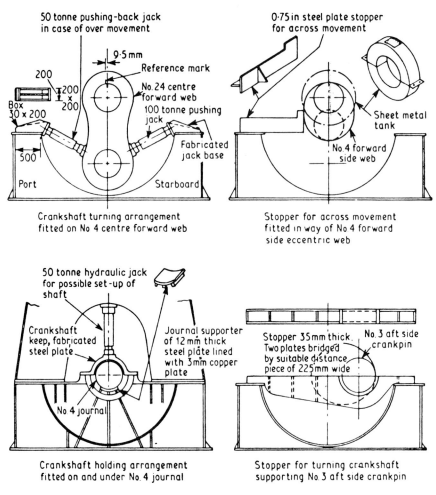

Figure 26.14 Repair procedure to direct drive engine

50 tonne jack opposite served two purposes: (a) it was to provide solid opposition to the movement, to prevent the web being uncontrolled; (b) since accurate observation of the fine alignment marks originally chiselled to mark the position of web to journal was difficult under the working conditions, if the web unfortunately moved too far it could immediately be repositioned by this jack. A further 50 tonne jack was employed to press the crankshaft securely down within its bearing pocket to counteract any upward reaction produced by the restoring couple. Also, as shown in the sketch, a mild steel girder was made and installed underneath No. 3 unit aft side-rod crankpin. This was to prevent the forward section of the crankshaft rotating when hydraulic

pressure was being applied. The girder was held in position by a few tack welds.

The sheet metal tank was fabricated to surround the slipped side-rod crankpin. This tank was to hold dry-ice used to pre-cool the crankpin in readiness for safe introduction of liquid nitrogen internally through the crankpin lubricating oil hole. About 30 hours of pre-cooling took place while other necessary work went on. Figure 26.15 shows the connections and pipes which were made up for conveyance of the liquid nitrogen from the tanker truck down into the engine room and to the engine. These and other parts fabricated for the tank were made ashore.

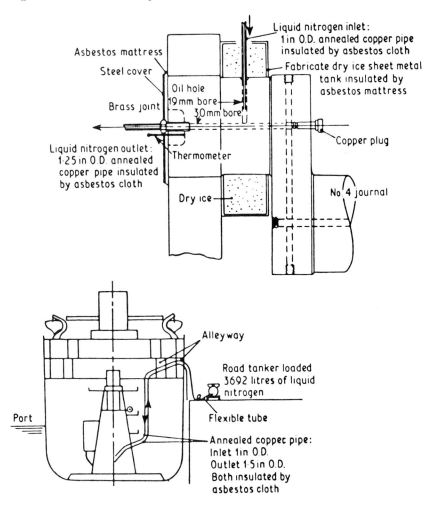

Figure 26.15 Piping liquid nitrogen to main engine to 'freeze' the crankpin

To reset any crankshaft is no easy task. Even in the clinical conditions of first manufacture the work is very critical. Furthermore, during manufacture it is possible to use closely controlled heating methods to build up the shaft piece by piece. The great shrinkage forces produced when the crankweb locks upon the pin secures the parts together.

In the situation being described, it was not considered that any heating method was safe or practicable. Apart from the dangers involved in using extreme heat in the confines of the crankpit, prolonged heating produces changes in the grain structure of steel. In general, therefore, Classification Societies avoid uncontrolled heating as an approved method of repair. It was for these reasons, that liquid nitrogen was chosen. Liquid nitrogen has a critical point of $-146°C$. As a gas it is slightly lighter than air, is colourless, tasteless and odourless. Since it was not known exactly how much nitrogen would be required to produce the required contraction in crankpin diameter, a truck load was requested. Also, since it would be cooler for the work at night, it was arranged for the supply to be made at 6 pm.

The tanker had a volume of 3700 litres, about 3.6 water tonnes, all of which capacity was utilised. A calculation suggested that a temperature difference of approximately $150°C$ would produce relaxation of the shrink, and this could be achieved by applying gentle heat to the web in the final cooling stages of the crankpin. Since the alcohol thermometer being used was only graduated to $-90°C$, the graph was projected to indicate when $-120°C$ would be reached (Figure 26.16). Very little hydraulic pressure was required to move the crankweb slowly back to its position. Only four minutes were taken to restore the two alignment marks as accurately as the eye could do. Happily this proved quite good enough. The shrink reapplied itself through the remainder of the night, assisted by some gentle heating to the pin. Later on, the shrink was tested using the jacks, but everything was satisfactory and was approved by the surveyor attending.

CHAINS

Difficulties with chain drives in service are remarkably few. The safety factor of chains should never be less than 25, and the loading upon which this factor is based is the maximum which is realisable under the most irregular conditions of working.

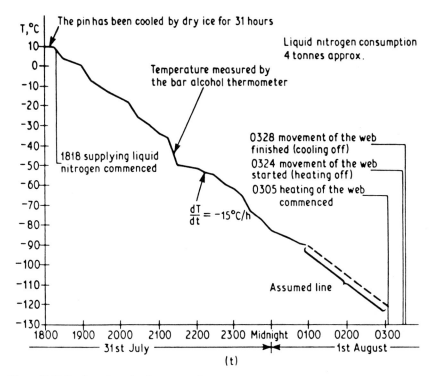

Figure 26.16 Cooling of crankweb over two days

Figures 26.17 and 26.18 show a typical arrangement of a camshaft chain drive. There is a 0.45 carbon cast steel sprocket wheel constructed in halves and keyed to the crankshaft, a forged steel intermediate wheel — where the chain lead requires such a wheel — and an adjustable spring loaded wheel for chain tightening. The chain is of duplex design. In large engines, where there may be a double duplex drive, the pairs of chains are matched by the chain maker to ensure that the load is equally shared.

It should be stressed that not only is it necessary for pairs of chains such as these to be matched, but also for each strand of a duplex or triplex chain. If one strand of the links differs enough to upset the load sharing on any link, the chain might just as well be a single strand. Reputable manufacturers try to work to tolerances of microns, sometimes even with a differential fit between the links of inner and out

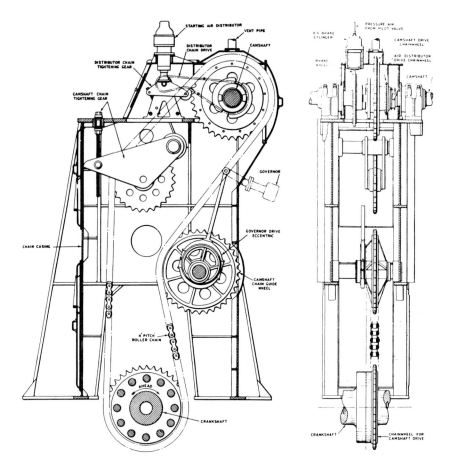

Figure 26.17 A typical chain drive   *Figure 26.18 Chain drive*

strands. It is hazardous to venture into the unproven.

It will be observed that the chain tightening sprocket wheel is located on the slack side of the chain. This is the preferred arrangement for the ahead direction of operation.

Note that although it is not always significant — depending on whether it is valves only, or fuel injection pump, etc. that are being driven by the chain — wear and thus taking up slack, does inevitably alter the timing. It is important not to allow slack to develop as this damages both chain and sprockets, and even nearby parts of the frame. It is also vital to maintain close alignment of sprockets and especially of the jockey and shafts.

The tensile strength of a chain varies as the square of the link pitch, or of the angle between adjacent sprocket teeth. Increase of tensile strength thus involves greater increase of accelerating forces.

Polygon action of chains

In addition to the normal load which a chain connecting two sprocket wheels has to carry, there is a momentary load due to the alteration in chain length as a link passes over the tangential point. This is the so-called polygon action. It can be shown from first principles that this alteration in length = $0.62\ I/n^2$,

where I = length of link in mm;
       n = number of teeth in the wheel.

The polygon action of chains in a chain drive is not always negligible. The absolute strain on the chain will depend upon the rate of change of momentum arising from sprocket diameter, chain pitch and chain velocity. The strain ratio, and consequently the stress, will vary inversely as the chain length. In the extreme example of a horizontal chain between two sprockets each having six teeth, the relative angular velocity of the wheels will vary throughout each pitch from R_2/R_1 to R_1/R_2, where R_1 = vertical distance of chain pin, with pins on vertical centre lines, R_2 = vertical distance of chain pin, with pins on horizontal centre lines (Figure 26.19). Acceleration at each engagement is a function of the angle between adjacent teeth of the sprocket and of the chain velocity. The resulting forces are the product of this acceleration and the specific weight.

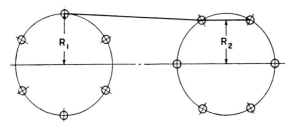

Figure 26.19 Polygon action of chain drive

Polygon action makes it desirable — really necessary sometimes — for chain wheel centres to be made a multiple of the chain pitch.

The matter is sufficiently important to merit attention in greater detail below.

Chain drives are an attractive alternative to the trains of gear wheels that were commonly used for the camshaft drives of tall four-stroke engines. The form of the gearwheel tooth provided constant relative velocity of the driving and the driven members of the gearing train. That is, there was no relative acceleration at engagement and disengagement of the individual tooth.

The chain does not comply with the above-stated characteristic. Acceleration at the engagement of each chain roller is a function of the angle between adjacent teeth of the sprocket wheel and of the chain velocity. The resulting forces are the product of this acceleration and the specific weights involved. A short chain drive will aggravate the effect of the accelerating forces.

The acceleration effect of the angle between adjacent sprocket-wheel teeth is illustrated diagrammatically in Figure 26.20. For simplicity, the smaller wheel shows the extreme example of a sprocket with only six teeth. In the upper diagram (a) the link joints are on the vertical centre-line for the large chain-wheel, and in the lower diagram (b) the link joints are on the vertical centre line for the smaller sprocket wheel. Inertia of the driving and the driven shafts, and their gear, prevents them from reacting to the varying ratio of angular velocities required by the chain as the wheels rotate. The ratio varies from S/R to Y/X during each period of engagement and disengagement. The accelerating effort of the chain results in abnormal stress in its parts.

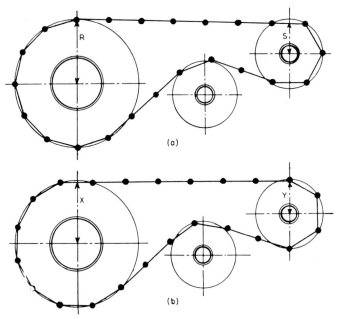

Figure 26.20 Chain drive. Acceleration effects

GEARS

Gearing is much more the concomitant of the medium-speed engine rather than the direct drive engine. Apart from the reduction gear, the great majority of medium-speed engines, and a lot of the higher speed engines, prefer to use gears rather than timing chains for the camshaft phasing gears.

It is not the purpose of this volume to provide a treatise on gear design, but rather to set out for the practising marine engineer the salient points he needs to know.

The great majority of gears met in marine propulsion installations take the form of spur gears or single helical, though double helical tends to be found in the largest reduction gears; and bevel drives and occasionally worm drives, are sometimes used for auxiliary purposes.

Except, of course, in worm drives the object of gear designers is to secure as close an approximation as possible to pure rolling contact between the teeth of mating gears. Manufacturing tolerances and deflections under load mean that this can only ever be an ideal objective, but it means that the teeth of virtually every power transmitting gear are generated in involute form for externally toothed wheels, and are straight cut for racks or internally toothed wheels. (Involute tooth form is not the only gear tooth profile which will achieve pure rolling contact, but, because it meshes with straight cut teeth on a rack, it offers considerable manufacturing advantage.)

Any departure from the ideal of pure rolling contact is bound to lead to a decrease in efficiency and therefore an increase in the heat which has to be carried away by the cooling oil. It is obvious that a decrease in efficiency from 98–96% doubles the duty of the gearbox oil cooler. It also obviously increases the likelihood of wear. For all these reasons it is important to maintain as closely as possible the design condition, and therefore the limiting tolerances particularly on centre distance, backlash and the physical condition, of the teeth themselves.

Noise is often an indication of wear or damage, though it will, of course, be borne in mind that spur gearing is always noisier than helical. Variations in the tooth pitch of an individual wheel, perhaps of only a few hundredth of a millimetre, often give rise to audible noise at the gear's rotational frequency, and in extreme cases the shock loading involved could give rise to tooth damage or even failure.

It is most unlikely (though not totally unknown) that a manufacturer's own spare would betray such a fault, but it is a point to bear in mind if, for any reason, a spare gear has to be bought from an unauthorised source.

Manufacturers normally try to ensure that the mating gears are of dissimilar materials in the interests of wear, for instance cast iron against alloy steel, or, if both gears have to be of alloy steel, that they are not of identical specification. If the transmitted pressure is high it is more often necessary that both gears are case hardened. Case hardening may be done either by induction or by flame, but it is important in either case to ensure that only the flanks of the teeth are hardened, that is in the zone most subject to contact. The hardened zone should not normally reach the fillet at the root (dedendum) of the tooth, nor should the cases on the opposite sides of the teeth meet at the top of the tooth. This is to preserve an adequately tough core.

Normally it should only be necessary to check the appearance of gear teeth from time to time. Correctly meshed, loaded and lubricated gear teeth present a smooth polished appearance. Any heat marking in marine gears would indicate trouble (other than the markings left on the edges by the original heat treatment).

Inadequately fine filtration would allow general abrasive wear — giving a matt appearance. Figure 26.21 indicates the scratched appearance caused by discrete particles in the oil.

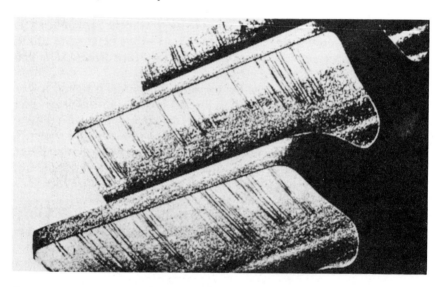

Figure 26.21 Scratching of gear teeth caused by particles in oil (GEC Traction Ltd)

Pitting (Figure 26.22) which usually manifests itself right across the tooth at about the mid-height, indicates a lubrication deficiency (perhaps associated with a slight deficiency in alignment). It is caused by local welding. It sometimes occurs on new gearing less severely than

SIGNIFICANT OPERATING PROBLEMS 527

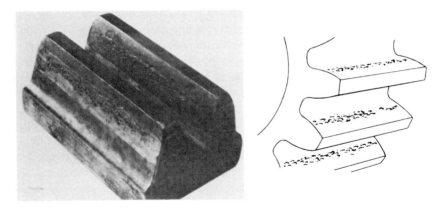

Figure 26.22(a) and (b) Pitting of gear teeth (GEC Traction Ltd)

in Figure 26.22 but heals harmlessly with further running. In general it should not be tolerated especially if it is progressive.

Spalling or flaking, that is, loss of significant areas of the case of case hardened gears, indicates deformation of the tooth, allowing the hard case to break up and become detached (Figure 26.23). It may be due to insufficient case depth, or a basic gear design fault, but it could be associated with an unforeseen overload condition, and will probably show a fatigue pattern. If crystalline it may indicate mishandling.

Figure 26.23 Spalling or flaking of case hardened gear tooth (GEC Traction Ltd)

Scuffing, galling or undercutting, is usually an indication of incorrect meshing, or of a total lubrication failure (Figure 26.24). It occurs from the pitch line inward on driven faces, and from the pitch line outward on driving faces. It stems from seizure on the pitch line followed by plastic flow.

The oil supply is therefore particularly important to the correct functioning of gearing. It is necessary to ensure that oil is flowing in the correct quantities and correct distribution, that it is adequately filtered and that the cooling is adequate. Gears within the engine depend, of course, on the total engine supply, but transmission gears can be separately monitored and any rise in the temperature difference across the gearbox oil cooler should be regarded with suspicion, as indeed any undue noise or any other change in operating behaviour.

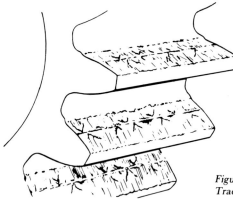

Figure 26.24 Scuffing of gear teeth (GEC Traction Ltd)

BEARINGS

While the traditional reckoning of bearing duty as total pressure divided by the projected area (length × diameter), or as load × velocity, may provide some guidance for assessing steadily loaded bearings, there are not any of these in diesel engines. (The method can be used for bearings with simple load fluctuation by using a multiplying factor.)

Sophisticated calculation methods are well developed for the major engine bearings, which plot the polar load (the total firing plus inertia load plus the direction in which it acts) throughout the firing cycle and evaluate the resulting oil film thickness subject to (at present) some simplifying assumptions. For instance, they do not yet take into account crank deflection.

The bearing lining material is chosen after assessing the maximum value of the load vector when all adjustments of the shaft system have been made, and comparing it with the safe fatigue loading for the various linings available. (See Tables 27.2 and 27.3.)

Even the angular speed of the load vector round the bearing is taken into account, since, if it slows to half speed (i.e. the point of maximum load is moving only half as fast as the journal relating to the bearing) it is theoretically unlikely to generate pressure in the oil film.

The current philosophy is that the journal and bearings are always separated by an oil film, although the thickness of this film will fluctuate during the firing cycle because of the forces involved.

In a journal bearing with a steadily loaded shaft running within it the oil film is carried round by the shaft, and this creates a hydrokinetic wedge action in the oil approaching the point of minimum clearance dictated by the applied load. The wedge action generates the pressure required to keep the journal away from the bearing.

According to the effect of the loading cycle and the magnitudes of the applied loads the thickness of the oil film at any point in the clearance in a bearing will, depending on the size of the engine, range from 200–400 micron down to as little as 2 micron (according to present methods of calculation).

The aim of oil filtration is to remove any particle which could bridge the minimum oil film thickness, but in practice 8 micron is a more realistic value for the maximum size of particle which is likely to pass the filter. Whereas this could theoretically bridge the oil film, the damage a particle of this size might do is to some extent dependent on its hardness. Provided filtration is maintained, the success achieved in practice is, in fact, very good, with negligible wear up to the major overhaul period.

Most makers line their major bearings with a thin conformable layer of a lead-based material (a lead indium mix for instance or white metal) perhaps only 20 microns thick. Although white metal is nowadays seldom used alone as a bearing lining, at least in the medium-speed and higher speed engine, because of its low strength, it is much stronger in thin layers: hence its use as a running surface.

White-metal materials have excellent anti-friction properties (provided the temperature can be otherwise controlled) and their use therefore provides some safeguard against the effects of any initial local shortage of oil when starting, or during maintenance.

It is absolutely vital that shell bearings are fitted without scraping, and clean. If alignment of the shaft requires correction, bearings are lowered by rebedding the housing to a mandrel and then restoring the nip by further machining or fitting at the joint face.

Perfect geometrical stability (or very nearly so) is essential to avoid fatigue failure of the lining.

In addition to load carrying considerations, the bearing must allow enough oil flow to cool it, as well as to lubricate it. Whereas this is often achieved by grooving the lining, it has to be recognised that an oil film in a journal bearing can only carry load where a close enough clearance exists. Moreover, at the edge of the clearance the pressure drops away to zero (Figure 26.25.)

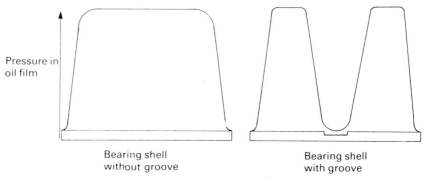

Figure 26.25 Illustrating the effect of a groove on the pressure carried in the oil film of a shell bearing

There are already losses at the edge of a bearing, so the introduction of a groove implies a disproportionate loss of bearing carrying capacity. For this reason if there is a need to carry oil through a bearing in a heavily loaded region (for instance, through a main bearing to reach a crankpin bearing, or through the crankpin bearing to reach the piston) the groove in this case is formed on the outside of the shell. Peripheral grooves also sometimes cause bearing problems due to the effect of cavitation in the oil flowing round the groove.

Cleanliness on assembly and reassembly is obviously essential, not only as regards the linings and journal themselves, but the associated oilways in addition.

The importance of correct nip and tightening has been mentioned earlier. They are also important to provide a heat path from the lining to the surface of the housing.

With bushes and ball/roller bearings, the last problem mentioned above may show as signs that the bearing has turned in its housing. If this has happened, the housing should be restored to standard but if this is impracticable it should be trued, and the outside of the bearing may be tinned to restore the fit until a proper repair is more feasible.

An old-time marine rule for running clearances in bearings allows one-thousandth of a millimetre per millimetre of shaft diameter. For diesel propelling engines this amount is normally too great; three-quarters of this amount is more representative; indeed, some engineers successfully reduce their running clearances to one-half.

Hard bearings in general require larger clearances than white-metal type bearings. This can be a problem at lower speeds if the engine relies on an engine driven luboil pump. Either a larger than necessary pump must be provided driven from the engine, or motor driven pumps must be used. If a bearing failure (or its imminence) is not found by condition monitoring or by routine inspection, the first sign it will give is a temperature rise, perhaps with a slight drop in oil pressure, followed soon by emitted noise.

Examination of a suspect bearing may indicate one or more of the following symptoms:

Wiping: If white-metal or a lead-based material, whether as the whole lining or as an overlay, is abnormally heated, it will soften or melt, and be carried round by the journal. It may indicate lubrication failure, or a fault in the lining or its adhesion (if accompanied by other faults). It may indicate overloading, or even external overheating of the housing.

Heat discoloration: is the equivalent sign in a copper based lining.

Fatigue cracking: Fine cracks leading to crazing in the lining indicate that overloading has caused the lining to backing bond to fail. (In withstanding direct pressure the lining will tend to spread.) The overload need not only be due to combustion pressure or inertia effects: an oval journal causes this kind of damage too. In traditional cast-in linings, fatigue cracking will tend to follow the dovetails on the housing (if provided). In these components faulty casting may also be a factor.

Squeezing: This is another way in which white-metal linings react to overloading whether because the load is too high, or the metal too weak, perhaps because an attempt has been made using white metal to exceed the manufacturer's permitted range of undersizes in redeeming a journal, and the resulting lining is too thick.

Scoring: Generally speaks for itself, though debris from the bearing itself may have caused the marks. The true cause will be evident on close examination.

A crazed lining will still carry some load if not disturbed, but the crazing and failure of the bond will prejudice one of the paths by which the bearing dissipates heat.

If a bearing fails the shaft journal must be examined for scoring, adhering debris and, if support has been lost, for cracks in adjacent fillets.

EXCESS OIL CONSUMPTION

Most engines nowadays need not consume lubricating oil at more than 0.7–1.3 gm/kWh at full or optimal load. Oil consumption is more dependent on speed than on load, so specific oil consumption (i.e. relating it to power generated) will rise inevitably as load is reduced. If such a rate is exceeded, it may of course be due to damaged liner or piston rings, and it may be due to leakage.

These reasons should not be too difficult to identify, and warrant no comment. Sometimes the cause is due to excess crankcase pressure, however, and this can be particularly irksome to detect, particularly as it may also provoke leakage from crankshaft seals. In such cases it is prudent to check that crankcase ventilation is functioning, including, if the engine has a separate sump tank, the breather on that.

It is, incidentally, very bad practice for any lubricating oil tank or crankcase vent to exhaust near, or for vapour to be entrained into, the air intake of the engine. Apart from fouling the turbocharger compressor and the charge cooler, it could in some circumstances entail a risk of the engine getting out of control.

LUBRICATING OIL FLOW

It is erroneous to suppose that oil pressure at any point in a forced-lubrication system is, by itself, a satisfactory criterion of oil flow. Oil flow should be interpreted in a way analogous to current flow in an electrical circuit, where pressure, resistance and flow are correlated in terms of Ohm's Law. The resistance of the whole circuit affects the flow of oil; the pressure at any one point, by itself, may be entirely misleading; it is the quantity of oil which is flowing that really matters. Lower oil pressure, by reason of, say, greater running clearances, may imply a greater oil flow, and may therefore be fully satisfactory; conversely, a higher pressure, with finer clearances, may be concomitant with a diminished and unsatisfactory oil flow. An oil flow indicator — especially at the highest point of lubricating oil circuit — is more likely to impart confidence to an engineer than a pressure gauge will do.

The pressure of lubricating oil at every point throughout a cooler, from inlet to outlet, should always be greater than the pressure of the salt water there; otherwise salt water may enter the lubricating oil, to emulsify it or to form sludge.

DEWPOINT AND CORROSION

It is a fundamental law of physical chemistry that:

1. The pressure exerted by a gas in a given space is the same whether it occupies the space alone, or with other gases (assuming they do not react);
2. The pressure exerted by a mixture of gases is the sum of the pressures which each would exert if it occupied the space alone.

If one of these gases is, in fact, a vapour, and it is saturated, that is, the amount of vapour present is the maximum possible, then the pressure and density of the vapour are uniquely related to the temperature.

In the case of water vapour these relationships are set out in Steam Tables. If these relationships do not correspond to the Steam Tables, then the vapour (steam) will either have become superheated (and behave more like a gas) or it will have started to condense to restore the Steam Tables relationship.

Taking a space of V m^3 containing saturated water vapour and, say, air at temperature t °C and pressure P bar, Steam Tables will indicate the corresponding density p_s in kg/m^3 and vapour pressure p_s in bar of saturated water vapour which could be present. Its mass will therefore be $V\rho_s$. The pressure of the air must be $(P - p_s)$ bar.

Now 1 m^3 of air at NTP has a mass of 1.291 kg. (NTP is 273 °K and 1.01 bar).

Therefore the mass of V m^3 of dry air at $(P - p_s)$ bar and t °C will be

$$ma = \frac{P - p_s}{1.01} \times \frac{273}{273+t} \times 1.291 \; V \; \text{kg} \qquad (26.10)$$

Therefore the mass of water vapour that can co-exist with 1 kg of dry air is

$$m = \frac{V\rho_s}{ma}$$

$$= \frac{1.01 \, (273 + t) \, \rho_s}{273 \times 1.291 \, (P - p_s)}$$

$$= \frac{0.00286 \, \rho_s \, (273 + t)}{P - p_s} \qquad (26.11)$$

It can be seen that increasing the temperature will increase the saturation quantity of water vapour that can be present (and therefore the density too) and increasing the pressure in the space it occupies will reduce it.

If, therefore, superheated water vapour is included in a volume of gases and the temperature is reduced (or pressure increased) it will reach the temperature (pressure) at which it is saturated, and start to condense. This is the dewpoint, that is, the temperature at which, for the pressure, the water vapour present is on the point of condensing, and satisfies Equation 26.6.

When the water vapour condenses it will form a fog of liquid particles and if it comes into contact with a cooler surface it will condense on to it.

Consider now the exhaust gases. Even if the combustion air was dry, burning a hydrocarbon fuel (see Chapter 24) will create weights of carbon dioxide and steam in the proportion roughly of 2:1. For instance, burning cetane (C_{14} H_{32}) with twice the stoichiometric quantity of dry air (30.44 times its weight) means that 200 parts of cetane burn in 1408 parts of oxygen and 4680 parts of nitrogen to produce 616 parts of CO_2 plus 288 parts of water vapour, leaving 704 parts of unused oxygen and the 4680 parts of nitrogen largely untouched.

If the vessel was operating in very hot humid conditions, and these were seen in the turbocharger intake, the water loading of the air could be of the same order as that produced by combusion, and would add directly to it.

The dewpoint of the exhaust gas could therefore be somewhere between 140 °F (at ambient pressure) and 210 °F (at 3 bar), i.e. 60 °C and 100 °C, in the absence of sulphur.

If sulphur is present in the fuel, as it usually is, it oxidises to sulphur dioxide, and some sulphur trioxide, which dissolve readily in water to create sulphurous and sulphuric acids. (Other impurities may have similar effects.) Any solution vaporises at a higher temperature than pure water would in the same conditions, so its effect is to raise the dewpoint of the vapour in the exhaust.

It has been found that the presence of 1% sulphur in the fuel will raise the dewpoint of the above conditions of 120–140 °C. 3% sulphur would raise them to 160–180 °C.

The condensation would, of course, be acidic and corrosive. It is, therefore, unwise to allow any metal exposed to exhaust gas to become as cool as this. In practice the most vulnerable areas are in the cylinder itself, or at the further end of the exhaust, including the exhaust boiler, if fitted. This is for the obvious reason that relatively cool air is blown through the cylinder once a cycle and it is almost certain that some local condensation will occur. If a cool enough surface presents itself, water, containing acidic compounds, will gather on it. This is another cogent reason for keeping coolant temperatures well up to 80 °C.

The nozzle can be particularly vulnerable if the tip is overcooled, particularly in four-strokes, because it tends to be scoured during the

Figure 26.26 Nozzle erosion due to overcooling   *(Lucas Bryce Ltd)*

scavenge period. This produces in its early stages an appearance like that in Figure 26.26, leading to more serious combustion problems if not corrected. In Figure 26.27 the nozzle is sectioned to show how it has been weakened.

Figure 26.27 Eroded nozzle sectioned
(Lucas Bryce Ltd)

COOLING SYSTEMS
(based on part of an article in *Marine Engineers Review*, May 1979, by P. S. Powell)

No engine today, apart from lifeboat engines, can be operated successfully with direct seawater cooling of the jackets. All require a sealed charge of fresh water with suitable treatment against corrosion. Corrosion prevention is frequently achieved by a combination of pH control, film-formers and anodic inhibitors. In round figures a pH value of 10 is usually recommended to assist in preservation of the metal-oxide film. In practice, this is normally achieved by the addition of caustic soda, or other alkali — or a chemical known as a pH buffer. The latter maintains a fixed pH value, resisting external influences which tend to change it.

Film-formers include the familiar 'soluble oils' which are now less popular than previously for a number of reasons. One is that in use they frequently polymerise and produce gummy deposits of insoluble material; this in turn could lead to a reduction of heat transfer with its own potential problems. In addition, it has recently been postulated that products of this type encourage growth of bacteria, leading to possible destruction of lubricating oils. On the other hand they appear to be advisable for use in the piston circuits of engines with telescopic links, although work is proceeding on alternatives. Other film-formers include sodium silicate and various forms of phosphate. The latter, however, encourage bacterial growth.

Anodic inhibitors are by far the most popular type of inhibitor currently in use for this purpose. As the name suggests, they operate by stifling the activity of the anodic areas in the metal and thus break the corrosion cycle.

Chromate in some form held sway for many years and had the advantage of revealing its presence by its yellow colour. It is now rapidly losing favour, however, owing to its considerable toxicity which makes it unacceptable for systems incorporating a fresh water generator. A further hazard is that, in the event of salt water leakage, strongly acidic conditions and rapid corrosion can result.

The most popular products at the present time are based on sodium nitrite, usually in combination with other materials. These have a number of advantages including double action, being both anodic inhibitors and capable of strengthening the oxide film by oxidisation. They provide continued protection even in the event of some salt water leakage; are not harmful to skin; and the recently introduced chemical test for their concentration has been considerably simplified.

Whatever product is selected, it is important to check that mechanical and other conditions are correct. As indicated above, some products are subject to bacterial degradation and, apart from other considerations, this will diminish their concentrations.

The effect of salt water contamination must also be checked. Excessive leakage from the system will also result in loss of concentration of the inhibitor which must be controlled by frequent testing and additions. Finally, there is the possibility of pick-up of acidic exhaust gases under certain conditions, so that the specified alkalinity or pH value must be maintained.

On the seawater side, the traditional arrangement is for the flow to pass first through the charge cooler, then the luboil cooler, gearbox cooler and jacket heat exchanger either in that order or in parallel circuits. It means, however, that the charge cooler is most vulnerable to any blockage or damage by silt, etc., since it is first in the circuit. Marine coolers usually have circular section tubes, or at least thick elliptical section, to minimise this tendency.

To obviate possible hydraulic damage in cylinders should a tube fail, designers endeavour to place the cooler below the air manifold, and to provide a small permanently open bleed so that any water leak will not enter the cylinders.

Recent developments have included centralised cooling systems where all the coolers are fed with fresh water and only one heat exchanger in the whole system comes into contact with seawater. (It is still prudent however, to retain the permanent drain on the air system.)

MICROBIAL DEGRADATION

A relatively new phenomenon to afflict diesel engines in marine service is microbial degradation of lubricating oils and coolants. It is not inevitable, but can be very damaging if conditions favour it.

The problem may involve bacteria which like warm conditions and a pH slightly on the alkaline side of 7. Or it may involve mould or yeast type organisms which prefer slightly acidic conditions (pH below 7). (pH is a shorthand for: 'The logarithm to base 10 of the reciprocal of the concentration of hydrogen ions in the solution'. pH7 represents 1 in 10 million, or 10^{-7}).

The following notes on the subject are based on an account printed in *Marine Engineers Review* for January 1978 of a paper by E. C. Hill, presented to the Institute of Marine Engineers in January 1978.

Microbes have an extraordinary ability to degrade all types of organic (and sometimes inorganic) materials. Different microbes have different preferences for 'diet' and for physical conditions (temperature, pH, light, oxygen tension and electron potential). All must have water to grow. Water is a by-product of growth, so once started they are to some extent self-sustaining. They must have a balanced 'diet' with particular respect to carbon, nitrogen and phosphorus. It is the greater availability of these last two elements, either directly as additives in modern oils, or as coolant additives when these leak into the oil, which is probably a major factor in the escalation of the lubricant problem.

There is nothing magical about the role of microbes in lubricant malfunction and corrosion. This has been described in detail in more specialised papers. We can summarise their activity in general, although individual species of microbes may accomplish only some of the processes listed:

1. Their growth products may be corrosive, e.g. organic acids, hydrogen sulphide, ammonia.
2. They reduce inter-facial tension, and hence stabilise a water-in-oil dispersion. Such a dispersion tends to be corrosive.
3. They attach some molecules of the base oil preferentially, and hence alter its viscosity.
4. They may attach the oil additives and reduce their effectiveness.
5. Local concentrations of microbes deplete that area of oxygen and hence establish electrochemical anodic pitting due to an oxygen gradient.

The time scale for a serious problem to develop can be a matter of weeks, but it is very dependent on sufficient water being available for a growth 'explosion'. Sulphide generating bacteria only appear after other organisms have flourished and have reduced the oxygen level. They prefer stagnant conditions and hence may be associated with 'lay-up'. Microbial sulphide corrosion is an intensely vicious process.

At least six factors can be recognised which may all have contributed to the escalation of marine incidents:

(a) the increased use of non-toxic coolant inhibitors to fulfil regulations for coolants also used for evaporators. Chromates, where still in use in coolants and at adequate concentration, are anti-miocrobial, but they cannot be used where the coolant is also used in drinking water evaporators. Other types of coolants may actually support microbial growth; hence a leak into the lubricant may supply both water and microbes. This appertains particularly to water cooled pistons;

(b) the increased use of sophisticated oil formulations which may constitute a complete 'diet' for micro-organisms or may become so when supplemented by nitrite from the coolant;

(c) a progressive change in base oils for lubricants. Naphthenic-base oils have partially been replaced by paraffinic-base oils which may be more prone to microbial attack;
(d) reduced use of renovating tanks. When used routinely these can function as batch sterilisers;
(e) reduced engine temperatures, due to 'slow steaming', may increase susceptibility to microbial attack.
(f) increased lay-up during which growth, particularly by sulphide-generating bacteria, may proceed unnoticed.

All of these changes have been made for very sound reasons; unfortunately when they all occurred together a potential microbiological time bomb would be created. It still, of course, requires an inoculum of those organisms which can flourish at engine temperatures in the formulations in use. It is likely that these organisms are acquired in the tropics, where the adaption for organisms growing at high ambient temperature to growth at engine temperatures is more probable; in the North Atlantic the larger temperature difference would discourage this adaption. Even when a potential time bomb has been fused (inoculated with the right organisms), oil changes and corrosion will not develop rapidly unless adequate water is available for growth proliferation. All of the evidence shows that early detection is important. In the very early stages no lubricant malfunction and corrosion are seen; at later stages progressive changes can be identified and a major casualty may occur.

All or some of the following phenomena may be observed when lubricant infection occurs. Individually each may be ascribed to some non-biological cause; they should therefore be taken as possible indications of infection which should be confirmed by microbiological tests on-board or in the laboratory:

1. Stable water content in the oil after the purifier.
2. Increased acidity of the oil.
3. Unusual smells.
4. Sliminess of the oil: this may be particularly apparent on the crankcase wall or in a glass sample bottle.
5. Honey-coloured films on the journals.
6. Corrosion pitting of the journals.
7. Rust films on machined parts.
8. Black 'graphitic' pitting in cast iron.
9. Black stains on white metal — not associated with salt water.
10. Brown or brown/black deposits.
11. Sludge accumulating in the crankcase and sump, and separating at the purifier. Filter plugging.
12. Corrosion of the purifier bowl.

Tests have been developed which can be carried out on board, though the interpretation does call for some skill.

Samples should be collected upstream of the purifier generally from a low point in the oil system, where there is most likelihood of including water if it is present.

Remedial measures for microbial attack

A whole spectrum of approaches is available depending on the extent and corrosion. In the early stage it is possible to salvage and sterilise most of the oil by renovating at 82 °C (180 °F) or more for 12–24 hours, while circulating through the purifier and its heater, and then allowing to stand at temperature for 12–24 hours. At the same time the sump and system can be cleaned out and sterilised with some of the oil containing a biocide such as MAR 71, HST2 or Vancide TH. The renovated oil can then be returned to the sump via the purifier and topped up. The renovating tank should then be cleaned out. Depending on circumstances it may be considered desirable to add a biocide to the oil in use.

No biocide should be added to oil without expert advice and the approval of the oil supplier. Almost every biocide has some adverse effect, and this should be allowed for in any anti-microbial procedures.

Precautionary measures against microbial attack

Many owners have decided to conduct routine microbiological surveys. Where substantial infection is detected it is usually advisable to open a crosshead bearing, and if possible a main bearing, to establish the extent, if any, of malfunction or corrosion.

Obviously it is sensible to be particularly careful to rectify water leakages, to renovate routinely and to check occasionally for infection. Good quality water should be used for coolant make-up. Oil and coolant temperatures should be kept well up to the maximum permissible.

EMERGENCY REPAIRS

Reference has already been made in the section headed 'Seizures' to taking a cylinder line out of service if it has been damaged, and proceeding on the remaining cylinders on the engine. (This applies to serious crankpin damage as well.) It is necessary to immobilise all the parts associated with that cylinder, and to lock them securely in place if they cannot be physically removed from the engine frame. All lubricating oil supplies, fuel supply, air and exhaust ports must, of course, be sealed off.

In addition, advice has to be taken as mentioned earlier in this chapter about the change brought about in torsional vibration conditions, and in the balance condition. The turbocharger behaviour will also be affected, and for all these reasons it will probably be necessary to adopt a different power and nominal speed to bring the vessel home.

More serious is a crankshaft fracture, not only because of the immediate disabling of the vessel, but because of the work necessary to replace it and the time needed to secure a spare one and transport it to the repair yard.

If the fracture is in a crankpin of a large built-up shaft a temporary repair may be effected by replacing the throw, that is, by making up a straight bobbin to fit between the throws and bolting it to the webs. If heavy plates are then welded across the crank throws, a stiff enough unit can be achieved, and the engine should be able to remain in service until a new shaft can be fitted. Instances have been known where such an arrangement has continued in use for six months.

If the fracture affects a journal, or is in a medium-speed engine crank, it depends how much of the shaft remains intact attached to the flywheel or output flange. It has been known for three cylinders of an engine with a broken nine cylinder crankshaft to bring the vessel to port, but the power available in such a case was so low as to make the attendance of tugs necessary. To prepare for such an emergency all camshaft followers on the redundant cylinders should be immobilised, and, if the broken section of crank could not be disengaged, the connecting rods affected should be similarly treated.

It occasionally happens that a turbocharger suffers damage and has to be taken out of service. This is achieved by locking the rotor so that it cannot turn, and once again this considerably affects the power which the engine may then develop, even if it has more than one turbocharger.

In many, if not all, cases the consequences of locking out various combinations of the turbochargers may have already been foreseen by the engine maker, and guidance may exist on board about the permissible power that may be developed. If this is absent and if no other guidance can be secured, it would be prudent to proceed within the limits of normal operating parameters such as exhaust temperature and smoke.

So far we have dwelt on the problem of getting the vessel back to port in an emergency. The repair of damaged or failed major components has then to be faced, and anything affecting the main propulsion equipment is bound to be protracted and expensive. For some components there is no option but replacement: crankshafts, and line parts. If a crankpin or journal is merely damaged on the surface it can often be ground down to a permitted undersize and specially thick (i.e. undersize) shells fitted. Sometimes the work can be done *in situ* by specialists. For structural members, however, repairs are not always impossible, even if they have to wait until a more suitable time — or until replacements can be prepared — for a permanent rehabilitation.

For steel, (and light metal, for that matter) there are specialists in complex repairs who are able to restore even quite extensive damage, and who are well versed in the precautions and procedures needed to preserve — and even to restore — alignment. Such specialists can also, in many cases, restore damaged cast iron structures. If they agree to undertake the work, they will usually guarantee it as a permanent repair, but they are liable to require extensive dismantling to give access, and may not always agree to carry out the work *in situ*.

Processes involving welding (electric or gas) are more likely to be permanent; but very effective repairs, particularly of cracks, even where sections have become detached, are possible using specialist metal stitching. This involves a minimum of dismantling, and no risk of thermal distortion.

D.A.W.

27 Engine installation

For a propulsion engine to operate successfully in a ship great care must be taken in the initial installation of the engine in the vessel. This care extends to more than just the correct alignment of the engine to the propeller shaft or gearbox and being adequately supported on firm mountings, but to the adequate sizing and mounting of suitable auxiliary machinery, the provision of auxiliary power and the measures to reduce noise and vibrations.

A topic of great importance also is the inter-action between the engine and the ship and torsional vibrations generated by the engine being transferred through the seatings to the hull. Today, engine builders are able to accurately assess the generation of unbalanced couples, moments and torsional vibrations and have adequate methods of eliminating them even with engines of very few cylinders.

ENGINE SEATINGS AND ALIGNMENT

The seating to which the main engine is bolted is of great importance because not only must it support an engine of in many cases great weight and bulk but it must also transmit the propeller thrust to the hull itself, particularly as modern engines have the thrust block built into the engine bedplate.

The normal practice is for engine seatings to be a fabricated box structure built as an integral part of the ship's double bottom and having sufficient stiffness to support the weight of the engine, transmit the thrust, withstand external couples from the engine and avoid resonance with propeller or engine excitation.

The engine bedplate is not bolted directly to the foundation but seated on cast iron or steel chocks to facilitate alignment to the propeller shaft. The tailshaft aperture is normally bored in position during construction of the vessel and thus remains in a fixed non-adjustable attitude. It is necessary, therefore, to align any intermediate shafting and the engine crankshaft with the tailshaft as accurately as possible to avoid trouble in service. While fitted chocks is the normal seating method for engines quite a number of engines today are seated on pourable resin chocks.

Special jacks are used to align the engine with the shafting and after building a dam around the vicinity of the holding down bolts along each side of the bedplate, the liquid resin is poured in and allowed to set hard and maintain the initial alignment. The engine is held down on the iron, steel or resin chocks by hydraulically tensioned bolts while side chocks are necessary to resist horizontal inertia or collision forces and side brackets usually fitted to transmit the propeller thrust from the shafting (thrust block) to the foundation. Collision chocks fitted as a stopper to prevent the engine from moving on impact of the ship with an external object, or grounding, must be carefully fitted with extreme care being taken to allow for thermal expansion of the engine structure.

Another type of engine mount is the anti-vibration type such as the Metalastik unit manufactured by Dunlop (Figure 27.1). These mounts comprise rubber elements and are so constructed to provide both support and some degree of prevention of transverse movement.

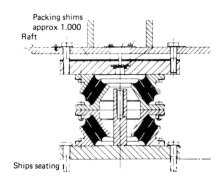

Figure 27.1 An anti-vibration engine mount (Dunlop)

An engine can be suspended on the mounts directly but a common method of installation is to have the engines rigidly bolted to a raft and this raft is in turn suspended from the foundation by anti-vibration mounts as shown in Figure 24.1. It is not uncommon for medium speed engines and reduction gears to be rigidly mounted on the raft to facilitate alignment between engine and gears with a flexible coupling between the gear output and the propeller shaft.

Flexible mounts (Figure 27.2) are most suitable for passenger vessels, research ships, and warships where it is desired to eliminate as much machinery noise and vibration as possible. On many ships, however, it is common to flexibly mount auxiliary engines as these can be the source of the most irritating noise and vibration.

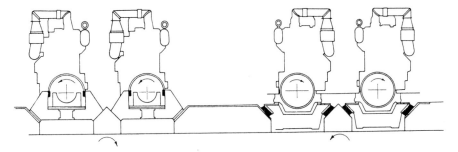

Figure 27.2 Flexible raft mounting of engines (Dunlop)

OPTICAL ALIGNMENT CHECKS

It is sometimes impossible or not accurate enough to use a piano wire to measure the relative axes of a stern tube and shafting, particularly on small bore shafting; in these instances optical methods are used. The best known methods and equipment for optical alignment checks are:
The Taylor Hobson sighting telescope;
The Legris sighting telescope;
Photo sighting;
A laser beam.

A system now in use by classification societies for checking the alignment of stern tubes, engine bearings uses a special sighting telescope which is supported on a table and a series of sighting discs, each with a centre hole, that are located in the stern tube bearing housings (Figure 27.3). The telescope incorporates a zoom lens to make it possible to focus or read the whole series of sighting marks and by micrometer adjustment misalignment can be recognised and measured.

For certain applications such as measuring variations in distortions, a laser beam can be used in addition to the telescope to more easily see the sighting axis.

REACTION MEASUREMENTS

A method of checking propeller shaft alignment is by measuring the reaction of bearings by 'weighing'. This consists of placing of dial gauges on top of the shaft and jacking up the shaft from underneath to simultaneously measure the displacement of the shafting and the load necessary to lift the shaft by the jacks. Several series of weighings are sometimes necessary to establish correct shaft alignment so a printer is

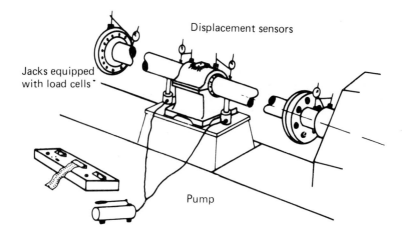

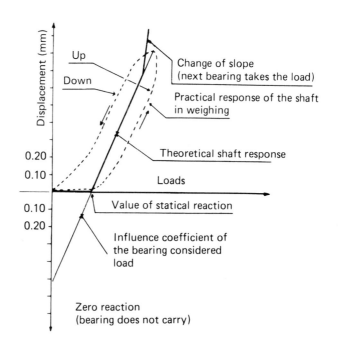

Figure 27.3 Shaft alignment by weighing and a typical recording (Bureau Veritas)

ENGINE INSTALLATION

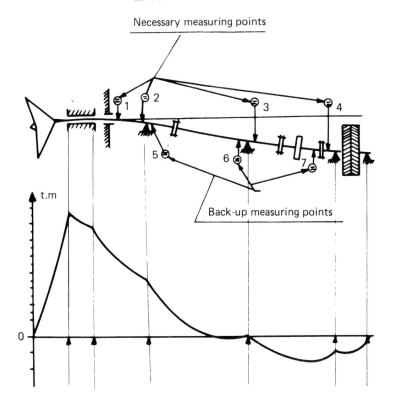

Figure 27.4 Strain gauge bridges on shafting driven from a reduction gear (Bureau Veritas)

used to record as many as five or six load and displacement signals (see Figure 27.4).

Another method is based on measuring and interpreting actual bending moments in different parts of a shafting with strain gauge bridges placed immediately next to each bearing. By using a pocket computer programme it is possible to determine *in situ* the values of static reactions from the bending moments measured. This method can also be applied to shafting in rotation.

Misalignment can cause undue wear on shaft bearings and even bending forces on the shafts. Today, shafts are aligned using laser techniques and computer calculations to determine load reactions on each bearing in the system and determine that these are within safe limits.

Alignment adjustments are usually made by working from the propeller to the main engine as once the supports are installed they cannot be altered. When the tailshaft and engine shaft have been correctly aligned by piano wire or optical methods, the object then is to align the intermediate shafting to the fixed positions of tailshaft coupling and engine thrust shaft coupling. With the current tendency towards all aft machinery on most vessels, alignment becomes far more critical because of the shorter lengths of shafting between engine and tailshaft.

This alignment, by individual adjustment of supporting bearings can be carried by measurement of the clearances between flanges before they are rigidly coupled.

In the 'gap and sag' method, the alignment starts with the propeller mounted on the shaft and measurements are taken at the top and bottom of the flanges and sides and the bearing is adjusted until all clearances are equal, before tightening of the flanges.

CRANKSHAFT ALIGNMENT

The alignment of the engine and shafting is set when a ship is built or re-engined and should not need to be adjusted in service. However, the checking of the main engine crankshaft alignment is a task that should be undertaken by the ship's engineer at least once a year or after replacing any main bearing shells or if the ship has grounded.

Crankshaft alignment is checked by measuring the deflections of the crankwebs through one revolution of the shaft (see Figure 27.5). A special dial indicator gauge is inserted between the crankweb to be measured and on rotating the shaft with the turning gear, any change in the between-web distance will be indicated by the gauge as a plus or minus reading, i.e. opening up or closing of the distance, from which it is ascertained whether the main bearings are high or low. The smaller the variation during a revolution the better aligned the shaft remains and for comparison it is important that the gauge be always mounted at the spot on each crank each time.

Readings should be taken at the TDC, BDC and horizontal crank positions and if measurements are taken with the connecting rod in position the gauge should be placed as close to the rod as possible.

The difference between the readings between BDC and TDC are recorded. Values of acceptable deflection vary according to the engine stroke but as a general guide readings of 0.05 mm to 0.02 mm for strokes from 500 to 200 mm on a linear basis are acceptable: longer variations will require investigation and remedial measures taken. It is important that the ship's state of loading and the temperature of the engine be as near as possible the same each time deflection readings are taken.

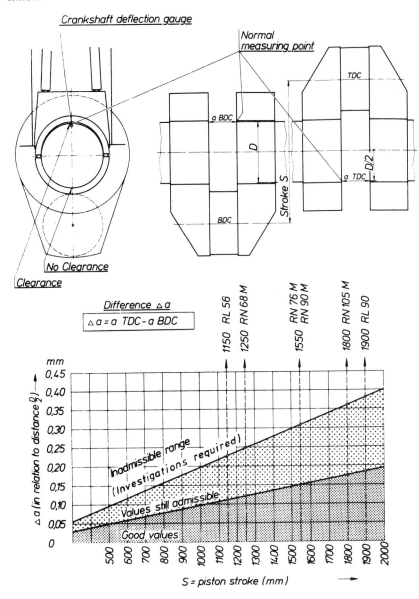

Figure 27.5 Measurement of crankshaft alignment (Sulzer)

550   ENGINE INSTALLATION

ANCILLARY SYSTEMS

With the exception of the very smallest installations no ship's propulsion engine can operate without supporting ancillary machinery systems to provide basic functions such as cooling, lubrication, exhaust arrangements, starting arrangements, fuel treatment and so on. Many types of four-stroke engines have auxiliary pumps driven directly but for larger ships, particularly those propelled by direct coupled slow speed engines, separate auxiliary machinery is normally fitted for main

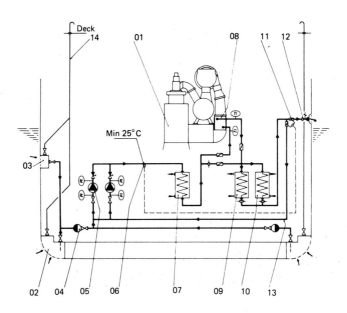

Figure 27.6 Sea water system
  1 Main engine
  2 Lower sea chest
  3 Upper sea chest
  4 Sea water filter
  5 Sea water pump
  6 Temperature feeler
  7 Lubricating oil cooler
  8 Charge air cooler
  9 Piston cooling water cooler
  10 Jacket cooling water cooler
  11 Automatic temperature control valve (butterfly type)
  12 Overboard discharge valve
  13 Warm sea water return line
  14 Air vent

engine services whilst, of course, other machinery is provided for non propulsion purposes such as power generation, domestic services, cargo handling etc. While the traditional arrangement is to have separate propulsion engines, auxiliary engines and auxiliary circuits, pumps and systems, it should be mentioned here that a new tendency is to drive auxiliary machinery such as a run alternator and major service pumps from the main engine itself as a fuel saving measure.

There have been quite a number of ships built with a so called 'standard' machinery installation but shipowners generally still prefer engine room layouts of their own choosing.

While the layout of plant and machinery is invarably different on most ships, particularly on vessels intended for different trades and

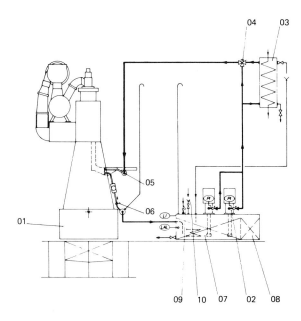

Figure 27.7 Piston cooling water system
    1 Main engine
    2 Piston cooling water pump
    3 Piston cooling water cooler
    4 Automatic temperature control valve
    5 Piston cooling water inlet pipe
    6 Piston cooling water outlet pipe
    7 Piston cooling water drain tank
    8 Piston cooling leakage water tank
    9 Filling pipe
    10 Chemical treatment inlet

ENGINE INSTALLATION

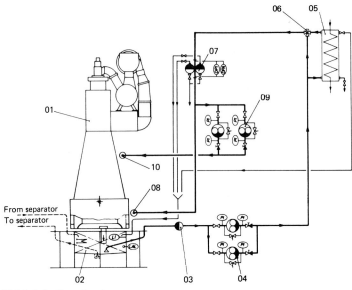

Figure 27.8 Lubricating oil system
1 Main engine
2 Oil drain tank
3 Suction oil filter
4 Lubricating oil pump
5 Lubricating oil cooler
6 Automatic temperature control valve
7 Lubricating oil filter
8 Bearing lub. oil inlet
9 Crosshead lub. oil pump
10 Crosshead lub. oil inlet

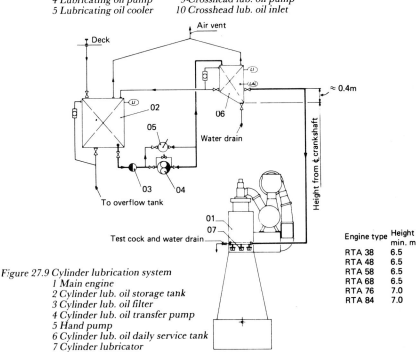

Figure 27.9 Cylinder lubrication system
1 Main engine
2 Cylinder lub. oil storage tank
3 Cylinder lub. oil filter
4 Cylinder lub. oil transfer pump
5 Hand pump
6 Cylinder lub. oil daily service tank
7 Cylinder lubricator

Engine type	Height min. m
RTA 38	6.5
RTA 48	6.5
RTA 58	6.5
RTA 68	6.5
RTA 76	7.0
RTA 84	7.0

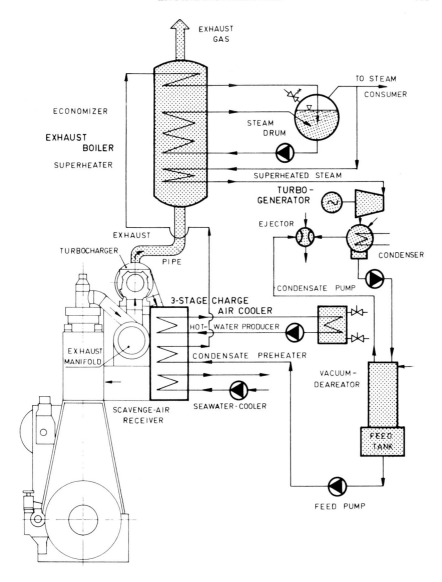

Figure 27.10 Typical waste heat recovery system

different cargoes, ancillary machinery arrangements are comparatively similar particularly for those systems allied to main propulsion.

Typical circuits for sea water circulation, fuel oil, lubricating oil, and a modern waste heat recovery plant, all as recommended by Sulzer for its RTA series engines and shown in Figures 27.6 to 27.10.

28 Materials

The materials used in marine diesel engines will reflect the duties they have to perform. Because of the paramount importance of reliability they will only venture beyond the well proven when this is shown to be absolutely essential to achieve the required duty reliably.

In a marine engine intended for commercial application, the main stress-carrying members will be ferrous. Even if weight is important, the first choice for reducing it is to go to higher speed or rating. Only the highest performance engines, intended for specialised naval applications, hydrafoils or the like, need to resort to lighter material such as aluminium, and then only for the frame.

The only exception is the piston, which for reasons of minimising inertia used often to be of aluminium; but the rating of modern engines has dictated the use of steel crowns, usually with an aluminium skirt. Heat and impact are also major design considerations in many cases.

The aim of this chapter is to provide an insight into the reasons for the choice of material. The metallurgy of steel, for instance, would easily fill this volume on its own. Rather we will set out some basic information on which the marine engineer may build, if he wishes.

Bearings are the subject most likely to tax him during his daily work, and these still display a greater variety of practice than do frames, for instance. They will therefore be discussed at greater length.

FERROUS COMPONENTS

The main frame of a marine diesel engine is invariably of cast iron, or of cast or fabricated steel (or a combination). The choice will usually be dictated more by the practicability and cost of manufacturing processes available to the manufacturer than by stress-carrying possibilities. Stress is, in fact, considerably less important than strength and rigidity. Designers strive to keep the actual stress anywhere in the frame to something like a maximum of 30 MN/m^2. Cast iron is most attractive (all other things being equal) in engines whose scantlings involve ruling sections of about 2.5 cm or less. This is partly because the physical properties of cast iron are inversely affected by size, as is shown by Table 28.1, which is actually for grey cast iron to BS 1452. Partly, the reason is that the cost advantages of cast iron tend to be greater in smaller components and those made in greater numbers.

Steel was sometimes preferred to cast iron in naval ships because of greater impact resistance. But the impact qualities of good grades of cast iron, even without the use of resilient mountings, are perfectly adequate for commercial marine duties — and in fact cast iron framed engines are now used in naval applications also.

Table 28.1 Properties of cast iron related to size of section BS1452

Main cross sectional thickness of casting (mm)	Minimum UTS (MN/m^2) Grade						
	10	12	14	17	20	23	26
Not exceeding 9	170	201	247	293	340	386	432
Over 9, less than 13	162	193	231	278	325	370	417
Over 19, less than 28	154	185	216	262	309	355	401
Over 28, less than 41	146	177	209	247	293	340	386
Over 41	139	170	201	231	278	325	370

Cast iron is distinguished from steel in that it contains more than 1.8% of carbon, often up to 3.0% or even 4.0%. In structural cast iron this carbon is partly combined chemically with iron (usually about 0.5–1.0%) and the rest is present as graphite. The graphite is usually distributed as flakes, giving the greyness of grey cast iron; but it is possible to encourage the graphite to gather into compact isolated nodules. The result, not surprisingly, is called nodular iron, or spheroidal graphite (SG) iron. The properties of each depend on the heat treatment of the casting, but in general nodular or SG iron gives better strength, ductility and impact resistance.

Annealing (heating up to about 800°C and holding the temperature long enough to encourage the atomic structure of the iron to change to the ferrite form) improves ductility, machinability and impact resistance, but reduces strength. Normalising (stress relief) is carried out usually at 500–600°C and preserves the pearlitic structure of the original casting (pearlite is a fine mixture of ferrite with cementite, or iron carbide) which preserves strength and hardness.

Both processes allow some relief (analogous to creep) of the stresses induced by uneven cooling of the casting in the mould, and which in extreme cases are high enough to cause rupture.

Of the impurities which may be present, sulphur and phosphorus increase brittleness, and for major castings are limited to 0.06% or so maximum. Silicon and manganese are used to control the embrittling effects of sulphur and phosphorous, apart from their use as alloying ingredients in their own right.

If small cast iron components are cooled quickly, the carbon is practically all taken up in chemical composition with the iron as cementite, and this gives a very hard white fine-grained cast iron with very high crushing resistance and a very hard surface. Such iron is used for piston rings and as valve seat inserts. Piston rings, incidentally, are carefully hammered in the inside to produce the free shape which will exert a uniform pressure on the walls of the liner when closed. Most engines use cast iron cylinder heads quite satisfactorily as long as the flameplate is adequately cooled and supported. It is found that if the metal temperature is kept below 300°C. the onset of creep (see Chapter 26) can be kept under control. In the size of cylinder heads, scantlings need not be as great as for frames, so that the properties of the cast iron are higher because of the scale effect.

Cast iron is also used for rubbing components because of its good oil-retaining and thus wearing properties. Examples are valve and tappet guides, and several designs of piston skirts. Cylindrical components like liners are often spun-cast to improve the properties of the casting, for example, to avoid porosity.

For steel frames, weldability is the main factor, and this is best suited by relatively low carbon content steel (0.2–0.25%). Welding, (a subject which could also fill this volume on its own) affects the mechanical properties of such a steel less than those of a higher grade, so that the quality of the resulting structure — often very large and complex and not easy to control — is more reliable. Welding is an expensive process, and depends on the quality of the joint preparation, the skill and conscientiousness of the operator and the planned sequence of operations to minimise distortion or stress. Steel castings are attractive, but not always economical or practical to use — and of course cannot eliminate welding entirely.

For crankshafts medium-speed designs have been forced by rating considerations and the inevitable high stresses to adopt very high quality steels with UTS up to $1000 MN/m^2$. To do this a number of alloying elements may be used to improve the strength of steels and also machinability and hardness. The important alloying elements are chromium, molybdenum, nickel and vanadium, which may be used singly but are more often in combinations such as chrome vanadium or nickel chrome. It is not normally considered advantageous to harden the journals of such shafts. The hardness left by heat treatment, coupled with proper control of filtration of the engine lubricant, are usually adequate to ensure a very low rate of wear in service. Correct surface finish, and the absence of stress-raising blemishes, is however, vital, and any deterioration is a serious matter.

To improve strength, medium-speed engine designers usually stipulate the continuous grain flow forging technique; that is to say, the crankthrows are formed by the forging process instead of being machined from solid billets. Forging produces a grain perpendicular to the direction of forge pressure. Steel usually has a slightly higher strength along the grain than across it. Large engines, of course, continue to use built-up shafts and have not been under the same pressure to reconcile properties with duty.

Large end bolts, on four-stroke engines in particular, usually require a very high duty steel with a high-quality, scrupulously blemish-free surface finish including rolled threads. In the case of two-stroke engines the demand is less onerous as there is not normally any stress reversal (unless a seizure occurs.)

In selecting materials for gears, it is generally good engineering practice not to run similar materials together, though they should not be too dissimilar in properties. Many marine gears use a combination of a 0.4% carbon wheel with a 0.55% carbon pinion. Originally both were usually unhardened but this is no longer possible with modern ratings, so they are carbonised and finish ground to size. Higher grade alloys are being used too, but gear designers still contrive some dissimilarity between the materials of the mating gears.

In smaller gear trains, like camshaft phasing gears, flame hardening is usual, or induction hardening, and these processes offer a very high degree of control with negligible distortion.

Valves in diesel engines have traditionally been made in a nickel-chrome-tungsten material like EN54, but the increasing use of inferior fuels has forced exhaust valves first to resort to a Stellite armoured seat face, and then to be made in solid Nimonic (or similar), a chrome-nickel alloy with additions of iron and sometimes titanium, cobalt and molybdenum. It is very expensive, but enables greater life to be obtained with much reduced vulnerability to accidental excess temperature.

Medium-speed piston crowns, despite intensive cooling, have reached very closely the limits for aluminium, which has not only a lower melting point and UTS than steel, but tends to have a lower strength when hot than when cold (depending on the alloy chosen). Higher ratings have compelled the adoption of steel crowns, and therefore of two-part pistons.

Some makers use cast iron pistons instead of aluminium, but with two-part design, aluminium is preferred for the skirt and gudgeon pin carrier, because of the influence of weight on crankshaft TV frequency.

Direct drive engine pistons have always been of two-part construction in iron/steel.

NON-FERROUS COMPONENTS

These are relatively few in most marine diesel engines, the most significant being the piston, which has been mentioned already, or nowadays perhaps only the skirt. Aluminium frames are used only on very specialised designs. Aluminium is very easy to cast and is often used for complex enclosures like valve gear covers and, sometimes, crankcase doors, or for the housings of auxiliary machines, the turbocharger compressor (housing and rator), air manifolds and the like. While aluminium may be used in the cooling water circuit it should be noted that it introduces a risk of electrolytic corrosion with iron or steel, as well as with any copper which might, however inadvertently, get into the circuit too.

In some cases glass reinforced plastic has been used for less important covers. Copper, bronze and brass are used for flanged bushes and occasionally as a mating gear in a lightly loaded auxiliary drive. The simplicity of wrapped bushes, however, is prompting designers to use these rather than purpose-made bushes.

Copper has sometimes been used for the injector tube in cylinder heads.

ELECTROLYTIC CORROSION

In the case of all metal components in contact with cooling water, or with seawater circuits, it is essential to beware of setting up an electrolytic cell.

Every metal when placed in a conducting solution will display its characteristic electrochemical potential relative to the solution. Metals (and other elements for that matter) are usually listed in terms of their potential when the other electrode is hydrogen, and for a standard strength solution. The range of possible voltages is from potassium at $+3.20$ to gold at -1.08. Where two metals in contact with an ionised solution such as cooling water have a difference in potential greater than 0.6 volts, there is a serious risk of electrolytic corrosion of the more positive one (the anode). Iron is therefore at some risk in conjunction with copper; aluminium seriously so; but only slightly against iron.

Some care is always advisable in introducing combinations of metals into contact with coolant: copper, (as brass or bronze) is almost always present. Water treatments have an influence, sometimes total control.

Sometimes the component at risk can be coated or otherwise passivated. Aluminium, for instance, in aluminium-bronze alloy is not at risk because it provides naturally a passive coating. A sacrificial zinc electrode is often included in seawater circuits. The seagoing engineer should normally only have to worry if he has to introduce a new metal into the cooling circuit, and if in doubt he should seek advice.

FINISHES

It is advisable, and virtually universal practice, to paint the inside of machinery housings with oil-resisting paint for two reasons:
1. To avoid the risk of corrosion if humidity is high before shutting down, with subsequent risk of insoluble particles finding their way into working parts.
2. To provide a clean, stable and cleanable surface.

There is not quite such need to paint external surfaces but, if they are, then oil resisting paint is needed. Once painted, it is necessary to maintain the cover since surface damage can allow corrosion in humid conditions, especially when shut down, to undermine the surrounding paint.

ORGANIC MATERIALS

Of all the various forms of organic material once used, almost all as sealings, only the various form of 'O' ring and some proprietary small shaft seals now survive. These were formerly of natural rubber but temperature, and the need to live with oil or fuel in many applications, have forced the adoption of suitably formulated nitrile rubber. Securing the correct grade is therefore crucial.

Rubber was once used as a lining on aluminium in contact with coolant water but proved problematical. Aluminium-bronze components remove the need for such extra protection.

Rubber has been an essential basis of the various proprietary lip seals on rotating and on reciprocating shafts of auxiliary machines. It has also been used in certain types of flexible coupling to transmit power either in direct compression, or in shear.

BEARINGS

Since modern designs of all sizes tend towards thin shells, which do *not* lend themselves to repairs other than by replacement, it is not proposed to discuss directly cast-in linings. Neither is it proposed to discuss ball or roller bearings beyond pointing out that they, like shell bearings, depend crucially on robust precision housings to secure them beyond any risk of movement, and to provide a heat path.

Generally the intention of shell bearing technology is that with proper lubrication and sufficient speed, the journal is lifted free of the bearing liner and floats in the bearing. The load is then transmitted via the lubricant, and the sliding properties of the bearing surface become of minor importance. However, to obtain a proper hydrodynamic lubrication, and to maintain it under heavy alternating loads, an interaction is required between the bearing surface, the lubricant and the surrounding atmosphere, to form adhesive lubrication films. Antifrictional properties is the vague term applied to this requirement.

Depending upon the size and type of load, very different bearing materials are used, including wood, plastics and metals. Among the metals there is again a wide selection mostly of alloys, based on tin, lead, copper, cadmium, silver, aluminium, zinc and iron. Tin and lead-based white metals, and copper alloys are commonly used for bearings in ships' engines, the other metals having fewer marine applications.

A good bearing metal should possess anti-frictional and running-in properties, being able to build up adhesive oil films under boundary lubrication conditions. It should not be liable to cause scoring of the steel journal, nor itself be corroded by lubricants and the deterioration products of lubricants. A certain fatigue strength is required, as well as a good capacity for carrying static loads. These requirements are to a certain extent contradictory, but they can be met by combining soft and rigid metals in composite or lined bearings, the white-metal lined bearing being a good example of this.

WHITE METAL

Generally speaking, all these alloys have good anti-frictional properties, although not entirely independent of composition. Thus it has been found that lead contributes more than does tin to the ability of white metals to react with lubricants to form the base for hydrodynamic lubrication.

The white metals as such are mechanically weak, particularly at elevated temperatures, but when thin layers are bonded to steel or bronze shells, bearings are obtained with a much increased fatigue strength and load capacity. White metals are able to accommodate a slight misalignment and to allow the embedding of small foreign particles carried in the oil, without scoring the shaft.

As with most alloys, white metals melt and solidify over a temperature interval called the melting or solidification range. At temperatures within the range the metal is in a pasty state, with some constituents molten and others solid; and this may lead to segregation of the alloying elements during the solidification of a casting. The lead-base white metals are the most prone, and the high-tin alloys the least prone, to segregation. Hence the latter are easier to cast.

For convenience, the range of white metals can be described according to their tin content. This will also usually determine their price.

One group of alloys contain about 90% tin, alloyed with antimony and about half as much copper, but no lead. Some alloys contain nickel, silver and/or cadmium. With their low alloy content, they are comparatively soft and tough metals with a high fatigue strength. This type of alloy, well used in high-speed engines, is used also in the crosshead and top-end bearings of large diesels, thus making use of their fatigue resistance.

Alloys with about 85% tin, the remainder again being antimony and copper, are harder and have a higher capacity for static loads; but their fatigue strength is somewhat lower than that of the alloys mentioned above. They have been used widely for main propulsion diesel engine bearings.

Notable for their load-carrying capacity and wear resistance is a group of alloys containing about 80% tin and 20% alloying constituents, namely antimony, copper and lead — in this order of decreasing content. Against this, they have a lower fatigue strength than any of the first two groups. Main bearings, bottom-end bearings, stern tubes and propeller shaft bearings are typical examples of application.

The presence of lead in tin alloys gives rise to the formation of a low-melting constituent, a so-called eutectic, in the metal. The tin-lead eutectic melts at 180°C, while the other constituents in white metals do not begin to melt until about 225°C. This gives rise to the phenomena of 'wiping', where just before complete failure the tin-lead will be seen to have melted. Where the heat is generated at discrete points in the bearing due to initial unlubricated movement, it is considered an advantage that such local melting occurs on a micro scale.

Lead alloys with 5–10% tin are excellent antifriction metals. They usually contain 10–15% antimony and small amounts of copper, plus additions of arsenic, cadmium or other metals. The high-density lead forms the alloy matrix; the tin, antimony and copper constituents all tend to float upwards in melting and casting operations, when passing through the melting range. This tendency to segregate renders these alloys more difficult to cast; hence, in spite of their much lower price, the use of lead-base white metals in propulsion machinery is largely restricted to bearings of secondary importance. When thoroughly cast in well-designed bearings the lead alloys do, however, compare favourably with high-tin alloys, and many small engines and auxiliary engines run entirely on lead-base bearings.

If white metal is used as the basis of main and large end bearings in high- and medium-speed engines running at more than about 400 rev/min, they are most often designed with 90% metal, for maximum fatigue strength. Increasingly the ratings of these engines have advanced to the point where a copper based lining has to be used. If used, white-metal thicknesses should be made low, certainly less than 2 mm, to avoid squeezing of the soft metal. Using very thin linings, and, with careful manufacture, lead-base metal can also be used throughout.

Reduction gears for medium- and high-speed engines need more wear resistance than fatigue strength, and thus run on bearings with 80% metal. The linings can be made thicker.

Older low-speed direct drive engines tended to use mainly 80–85% metals in loaded bearings, preferably in moderately thin (1.5 to 2 mm min) linings. Higher ratings have forced the adoption of 90% metal and thinner linings. The harder 80% metal is usually retained for main bearings and guide shoes because of its excellent wearing properties. Secondary bearings, such as those for camshafts, are regularly designed with lead-base metal; and this may also be used for the unloaded halves of primary bearings. For such applications there are no strict limitations on lining thickness.

COPPER BASE BEARING METALS

When bearing loads exceed the limits for white metals copper alloys (bronzes) can be used as bearing materials. Bronze bearings do not possess the anti-frictional properties of white metals but their fatigue strength and their load capacity are high. Owing to their hardness, they cannot adjust themselves to any misalignment; nor can they accept embedded foreign particles, as can the white metals. Moreover, harder journals are usually required to prevent scoring. Especially for crankshaft bearings, in medium- and high-speed engines it has generally been accepted that fine filtration, which inevitably means the cost of disposable filter cartridges, is worth the ability it confers on hard linings to achieve long life with extremely low rates of wear. In these engines, shaft journals tend to be heat treated for strength and mechanical properties as a whole, which includes usually some increase in hardness. Further hardening is not normally found necessary. The life of filter cartridges is in any case extended by by-pass centrifuging the oil.

The strength of very thin white metal or lead-based bearing linings is put to good use in practically all medium- and high-speed engines, to provide a running surface perhaps only 20–40 μm thick. The fatigue strength of such an overlay is virtually the same as the main lead–bronze lining. (See Tables 28.2 and 28.3). The usual overlay is lead-tin or lead-indium but the strength is virtually the same as white metal. metal.

Table 28.2 Bearing material fatigue ratings, hydraulically loaded test machine (nearly ideal conditions)

Bearing material	Load carrying capacity kN/m^2
White metal (tin based) 0.4 mm thick on steel	34 500
White metal (tin based) 0.06 mm thick on steel	48 300
Overlay plated (lead–tin 0.04 mm thick) copper–lead on steel	60 600
70/30 copper–lead on steel	76 000
Reticular 20% tin–aluminium (0.4 mm thick) on steel	>96 000

Table 28.3 Bearing material fatigue ratings (rotating out of balance load)

Bearing material	Load carrying capacity kN/m^2
White metal (tin based) 0.4 mm thick on steel	11 000
White metal (tin based) 0.06 mm thick on steel	15 200
Overlay plated (lead–tin 0.04 mm thick) copper–lead on steel	19 300
70/30 copper–lead on steel	23 500
Reticular 20% tin–aluminium (0.04 mm thick) on steel	27 600

Since the 1960s the advantages in cost, quality of manufacture and precision of fit which accrue to bearing material produced as a continuous strip (with all types of lining) have virtually swept the individually lined and machined thick shell bearing from the medium- and high-speed engine field.

The bearing lining is produced continuously, then cut, formed and the ends finished. To facilitate this, the steel shell has to be thin, usually about 2 mm, and the load bearing lining about 1 mm for (typically) a 300 mm bearing. The proportions are a compromise between the robustness needed in handling, and the ability of the composite lining to be formed by this process, and separate facings have to be provided to take thrust where this is necessary.

For bearing bushes, including those for camshafts, small ends and even valve follower rollers, the same principles have been satisfactorily applied to produce a bush by wrapping a very thin layer of lined strip to build up the wall thickness required.

These processes are very specialised and precise and the products very reliable and long lived once developed in the engine, provided that the housings are in good condition (and rigid enough) and lubrication is adequate. There is no advantage, and a high risk of failure, in replacing them with other than correct spares. The linings themselves are usually of lead-bronze which is an alloy with tin content of not more than 5%, as this material behaves better in the event of failure. Where copper–lead alone was used, the two metals tended to take the form of a matrix of copper filled with lead, rather than an alloy. If a bearing failed in this material, the shaft was liable to have to run in effect on the copper matrix alone, and this led to copper impregnation of the shaft, and its subsequent failure, even if not otherwise damaged beyond what could be reclaimed by turning down to the next undersize.

Copper–lead bearings can also fail by rapid corrosion attack, because of galvanic action between copper and lead particles, combined with some acidity of the lubricant. The use of an overlay tends to combat this.

Solid bronzes are still used in many engines for bushes, but only in engines of very old design and manufacture for crankshaft bearings. They are made either as castings or by sintering from powder. For lightly loaded ancillary applications they can be made porous and impregnated with oil so that further lubrication is unnecessary for many thousands of hours.

Most copper alloys used for bearings contain considerable amounts of tin or lead, or both, because these metals impart excellent anti-frictional properties to the alloy. Other common constituents are nickel in leaded alloys, and phosphorus in tin–bronzes. Some tin and lead-free alloys are used in bearings for special purposes.

Tin–bronzes, with 10–16% tin, are long established as bearing metals. With increasing tin content they become harder and more wear resistant, and the addition of 0.5–2.0% of phosphorus increases their durability and hardness. The 14% and 16% bronzes are rather expensive and tend to be brittle, and the 10% and 12% bronzes are usually preferred.

It has been found that lead adds extra anti-frictional properties to bronzes, as in white-metals. Roughly half the tin, but all the lead, goes to form a good bearing surface. This is because lead does not dissolve in copper but forms small globules in the structure, while tin is partly soluble and hardens the copper base. A range of leaded tin bronzes are cast into bearings, extending from 80% copper alloy with 10% each of tin and lead and similar alloys to high lead bronzes with up to 30% lead and only about 5% tin. The undissolved lead tends to segregate out of the metal during casting and the high-lead bronzes are difficult to cast.

Nickel is a useful addition, bringing about a more fine-grained solidification with improved dispersion of the lead particles.

GUNMETALS

Gunmetals are alloys of copper, tin and zinc, formerly used for casting cannon. With lead additions they become suitable as bearing materials and are called leaded gunmetals (the US term is leaded red brass); they are quite often misnamed bronzes. As with bronzes, high tin and lead improve bearing properties, while zinc is held at or below 5%. Leaded gunmetals are cheaper than bronzes, and are much easier to cast, owing to the presence of zinc. Besides being used in bronze shells for lining with white metal, they constitute the major part of all bronze bushings used, either cast to the desired shape or manufactured from centrifugally cast or continuously cast material.

Continuous casting has made gunmetals available with a very uniform and dense structure, and with properties comparable to those of some of the more costly tin bronzes. A leaded gunmetal with 85% copper and 5% of each of the three metals tin, lead and zinc, is in common use practically all over the world; it is designated '85–5–5–5', 'three fives' or 'ounce metal'. For improved wear resistance and bearing properties, other gunmetals contain 7–10% tin, up to 7% lead and down to 2% or 3% zinc. Nickel alloyed gunmetals are also available with improved grain-structure.

OTHER ANTI-FRICTION METALS

Aluminium–tin alloys with 6–7% tin have high load capacity, but their anti-frictional behaviour is inferior to white metals. The introduction of alloys with about 20% tin, rolled and heat-treated to obtain a so-called reticular tin structure, is a more recent development, and is used for crankshaft bearings in a number of medium-speed engines.

Supported on steel shells, reticular aluminium-tin shows good anti-frictional and mechanical properties.

Aluminium-zinc alloys of various compositions are sometimes called white bronze and are occasionally used for bushes in replacement of gunmetal.

Cast iron can also be used as an anti-friction metal in bearings with low speeds and low specific loads. Pearlitic iron is preferred, and the bearings must be well lubricated.

D.A.W.

Index

ABC DZ engines, 382, 383
Adiabatic compression, 1, 2
Adiabatic expansion, 1, 2
Admiralty constant, 44
Ahead operation, 216, 430
Air relief arrangement, 203
Air reservoir, 485
Air swirl, 73
Akasaka engines, 362–363
  A28(R), 362
  A31(R), 362
  A34, 362
  A37, 362
  A41, 362
Alarms, 422, 424, 425
Alco engine, 251, 405
Alignment,
  engine, 543
  propeller shaft, 545
Allen engines, 370–372
  S12, 382
  S37, 382
  type-370, 371
Alternators, 29, 31, 34
Aluminium-bronze alloy, 559
Aluminium frames, 558
Aluminium pistons, 554, 557
Aluminium-tin-alloys, 565
Aluminium-zinc alloys, 565
American high-speed engines, 403–407
Ancillary systems, 550
Anodic inhibitors, 536
Anti-vibration engine mounting, 544
Apparent propeller slip, 44
Ardrox test, 505
Ash content of fuels, 417
Astern operation, 49–51, 216, 302, 430
Automatic valve, 480–481
Automation, 304–5, 422, 425
Auxiliary blower, 134
Auxiliary engines, 29, 375–376, 486
Auxiliary machinery, 29
Auxiliary power generation, 29

Balancing, 22–24, 379, 490–491
  cylinders, 21, 22
  principle of, 22
Bearing materials, 528, 559–565
  copper base, 562–565
  fatigue ratings, 563
  gunmetals, 565
  miscellaneous, 565
  requirements, 560
  white metal, 529, 560–562
Bearings, 354
  big-end, 297, 332
  crazing, 531
  fatigue cracking, 531
  gearbox, 217
  heat discoloration, 531
  lubrication, 529–530, 560
  main, 124, 262, 283, 297, 331
  maintenance, 461–472
  operating problems, 528–531
  scoring, 531
  squeezing, 531
  thrust, 124
  wiping, 531
Bedplates, 88, 112–113, 124, 142, 167, 172, 180, 198, 217, 262, 354, 495, 543
Bergen LDM engine, 384
BICERA explosion relief valve, 494
Big-end bolt tightening equipment, 305
Bolnes engines, 372
  VDNL150/600, 401
Bolts and bolted joints, 305, 508–515
  large end, 557
  stress effects, 511–515
Boost pressure ratio, 446
Borescope, 452
Brake horsepower, 12
Brake mean effective pressure, 12, 41
Brake power, 8
Brake thermal efficiency, 7
Branched systems, 21
Brass components, 558
Bridge control systems, 97–98

567

British Ship Research Association, 496
Bronze bushes, 564
Bronze components, 558
Brown Boveri turbochargers, 63–67, 94, 103
Bryce FFOAR type flange-mounted unit pump, 82
Burmeister and Wain engines, 108–137, 372
  225/300, 386
  280/320, 386
  Alpha, 385
  auxiliary blowers, 134
  cams, 131
  camshafts, 117, 131
  chain drive, 131
  combustion chamber, 115
  connecting rods, 128
  control systems, 131
  cylinder liners, 125, 126
  cylinder lubricators, 132
  exhaust gas system, 133
  exhaust valves, 115, 126
  fuel injector, 118
  fuel oil high-pressure pipes, 131
  fuel pumps, 117, 129
  governors, 131
  in-line, 386
  K-EF, 110
  K-GF, 109, 110, 112
  K67GF, 119
  K-90GF, 111
  L-GB, 109, 110, 119–134
  L-GB/GBE, 123
  L-GF, 109, 110, 119–134
  L67GF, 119
  L-GFCA, 109
  12L90GFCA, 120, 121
  L-MC, 109, 134–137
  L-35MC, 137
  L35MC/MCE, 136, 137
  L80MC/MCE, 135
  manoeuvring system, 132
  moment compensator, 131
  pistons, 114, 128
  reversing gear, 131
  S28/V28, 387, 388
  scavenging, 133
  starting air system, 134
  T23, 387
  turbocharging, 118, 121–122, 133
  turning gear, 132
  turning wheel, 132
  valve gear, 126
  Vee, 387
  V23, 387

Calorific value, 7, 414–415
Cams, 131, 463

Camshaft, 117, 131, 188, 204–205, 245, 253, 281, 302, 324, 339, 353
Camshaft drive, 145, 462, 521
Camshaft pumps, 389
Cast iron, 499
  bearings, 565
  components, 554–557
  cracks in, 502
Caterpillar engines, 408
Cathode ray oscilloscope (CRO), 436, 506
Cetane number, 415
Chains and chain drives, 131, 521–524
  acceleration effects, 524
  polygon action of, 523–524
Charge air cooling, 56–57
Classification Societies, 21, 453, 487, 490, 494, 520
Cloud point, 418
Clutch, 29, 216–219
Combined diesel and diesel-electric plant, 33
Combustion, heat evolved by, 421
Combustion equations, 419
Combustion reactions, 419
Compression card, 437–438
Compression ratio, 2
Compression stroke, 2
Compressor, 486
Computer applications, 488
Conradson carbon value, 417
Constant-current control, 221
Constant-current loop, 221
Contract trials, 36
Control system, 32, 96, 131, 191, 213, 303–305, 357–358
  bridge, 97–98
Cooling system, 39, 208–9, 341–342, 357, 485, 536–537
  monitoring, 447–448
Copper base bearing metals, 562–565
Copper-lead bearings, 564
Corrosion,
  dewpoint effects, 533–535
  electrolytic, 558, 559
  high-temperature, 412
  low-temperature, 413
Corrosion fatigue, 503
Corrosion prevention, 536
Couples and forces, 23
Couplings, 29
Crack tests, 505
Crank angle, 3
Crankcase, 232, 275, 330
Crankcase explosions, 487, 492–497
Crankpin fracture, 540–541
Crankshaft alignment, 515–516
  measurement, 548, 549
Crankshaft defects, 517–520

INDEX

Crankshaft fillet stress levels, 500–501
Crankshaft resetting, 520
Crankweb cooling, 521
Crepelle PSN3, 384
Critical speed, 21–22
Crosshead oil pump, 146
Cummins engines, 408
Cycle efficiency, 2
Cycles,
  practical, 2
  working, 8–11
Cylinder block, 264, 275, 320
Cylinder cover, 114, 126, 169, 203–204, 230
Cylinder frame, 126
Cylinder jacket, 90, 113, 142
  maintenance, 460
Cylinder liner maintenance, 456, 460
Cylinder lubrication system, 552
Cylinder output balance, 491
Cylinder pressure, monitoring, 435–441
Cylinders, 167–168, 181–184, 229
  balancing, 21, 22
  cracking, 505
  lubrication, 93, 132, 208, 289, 327, 459

Daihatsu engines,
  DS22, 401
  DS26, 401
  DS28, 401, 402
  DS32, 367
Data logging, 422, 424, 434
Delivery valves, 80, 83
Derating, 40–41
Detroit diesel engines, 408
Deutz engines, 328–342
  bearings, 331–332
  BVM540, 328, 333, 334, 339
  BVM628, 388, 389
  camshaft, 339
  cooling system, 341–342
  cylinder head, 334–339
  cylinder liners, 339
  fuel injection pump, 339
  governor, 340
  in-line, 328, 329
  lubrication, 340
  piston rings, 333
  pistons, 333
  rocker arrangement, 338
  speed control, 340
  starter system, 342
  technical data, 328
  turbochargers, 340
  valve gear, 334–339
  Vee, 328–331, 340
Dewpoint, 533–35
Diesel cycle, 1

Diesel-electric drive, 31–35, 220–224
Diesel index number, 417
Direct drive, 28–29
Direct injection, 73, 74
Double-acting engine, 13
Doxford engines, 176–196
  camshaft, 188
  combustion chamber, 183, 195
  connecting rods, 186–187
  control system, 191
  crosshead, 186–187, 195
  cylinder liners, 181, 194
  cylinders, 181–184
  fuel injection, 188
  fuel pump, 188
  fuel system, 197
  J-type, 176–191
  76J4C, 179
  58JSC, 192–196
  lubrication, 183–186
  pistons, 177, 180, 184–186, 193
  ports, 196
  P-type, 176
  running gear, 195–196
  turbocharging, 190–191
Draw card, 3
Dribble, 75
Dynamic positioning, 33

Ear protection, 25
Efficiency, 4–6
  cycle, 2
  ideal, 2
  mechanical, 8
  propeller, 44
  propulsive, 29
  thermal, 6–7
'Elastic curve', 16
Electrical loading, 33, 375
Electrical machines, 376
Electrochemical potential, 558
Emergency repairs, 540–542
Endoscope, 452
Engine alignment, 543
Engine bedplate, 543
Engine costs, 29, 32, 374, 409
Engine frame; see Frame construction
Engine installation, 543
Engine location, 32
Engine performance, 36–51
Engine rating, 36; see also Derating
Engine reversal, 51, 216, 339, 430
Engine seatings, 543
Engine selection, 27–35
  factors influencing, 27–28
  size effects, 28
  uniformity, 31

Engine, slowing down, 433
Engine speed, 220
Engine stopping, 49, 429
Engine trials, 47
Entablatures, 497
Enterprise RV series engines, 373
Exhaust gas, 133
  dewpoint, 534
  visible, 432
'Exhaust lead', 11
Exhaust manifold system, 54
Exhaust stroke, 2
Exhaust system, 133, 235
Exhaust temperature, 39–40
  monitoring, 445–447
Exhaust valve, 115, 126, 231, 268, 299, 321
Expansion stroke, 2

Fairbanks-Morse 38TD8 engine, 408
Fatigue, 498–506
Fatigue cracks, 502, 531
Fatigue failure, 487, 503
Fatigue limit, 499
Fatigue ratings, bearing materials, 563
Fatigue stress, 499
Ferrous components, 554–557
Fire precautions, 497–498
Firing problems, 432
Fixtures, 454
Flame-plate distortion/support, 389
Flash point, 418
Flexible raft mounting, engine, 545
Forces and couples, 23
Forcing frequencies, 16, 19
Four-stroke engines, 2, 8–10, 53–54, 222, 239–258, 377, 397, 442–443, 457, 557
Franco Tosi QT320 SSM engine, 390
'Free-wheeling', 48
Freshwater circuits, 306
Fretting, 506–507
Friction torque, 8
Fuel/air mixture, 73
Fuel circuit, 307–308
Fuel coefficient, 43
Fuel consumption, 15, 28, 32, 38, 39, 40, 54, 134
Fuel economy, 53, 109
Fuel injection, 39, 72–85, 170–171, 188, 233, 253–254, 269, 280, 302, 357
  adjustment, 85
  and combustion, 72–75
  electronically controlled, 150–152
  equipment, 223
  problems of, 432
  timing, 40–41, 81–83, 106
  uniformity, 84–85

Fuel injection pump, 79–85, 96, 149, 170, 281, 339
  calibration tolerances, 85
  maintenance, 475–479
Fuel injection valve, 280
Fuel injector, 75–77, 118, 303, 468–469
Fuel line pressure diagram, 77
Fuel lines, jacketed, 497
Fuel oil high-pressure pipes, 131
Fuel pumps, 103–107, 117, 129, 171, 188, 205–206, 233, 281, 303
Fuel surcharging pump, 480
Fuel system, 197, 209–210
Fuel valve lift, 467–468
Fuel valve priming, 464
Fuel valve testing, 472–473
Fuel valves, 206, 210, 465–467
Fuels, 35, 409–421
  abrasive impurities, 413
  ash content, 417
  blended, 29
  burnability, 411–412
  calorific value, 414–415
  cetane number, 415
  corrosion, 412–413
  harmful components, 409
  heavy, 31, 410–413
  ignition quality, 415
  properties of, 413–421
  quality of, 413
  storage problems, 410–411
  sulphur in, 417, 534
  temperature/viscosity chart, 416
  water content, 418
  water effects, 411
Fuji diesel engine, 367

Gallery brackets, 132–133
Gearbox, 29
  lubrication, 217
  non-reversing, 218
  plain reduction, 217
  reverse/reduction, 215, 216
Geared propulsion, 216–220
Gears, 217, 525–528
  abrasive wear, 526
  case hardening, 526
  design aspects, 525
  lubrication, 528
  materials, 557
  mating, 526
  pitting, 526
  scuffing, galling or undercutting, 528
  spalling or flaking, 527
General Electric 7FDL engine, 407
General Motors engine, 404

INDEX

Generators, 6, 29, 376
  stand-by, 376
GMT2-stroke engines, 164–175
  'B' series, 164, 167–175
  B600, 165, 166
  'C' series, 164, 167
  CC600, 171–175
  combustion chamber, 168, 173
  connecting rods, 169, 173
  cylinder head, 172
  cylinders, 167–168
  fuel injection, 170–171
  fuel pumps, 171
  performance curves, 174
  piston rings, 170
  pistons, 170
  running gear, 169
GMT4-stroke engines, 292–309
  A420, 292
  A420L, 292, 293
  automation, 304–305
  B230, 388, 389
  B420, 292
  B420V, 293, 294
  B550, 292, 298, 300
  B550V, 294, 295
  B550 8L, 296
  B550.10, 292
  B550 16V, 296
  bearings, 297
  bedplate, 295
  camshaft, 302
  connecting rods, 298
  control system, 303–305
  cylinder heads, 299–301
  cylinder liners, 298–299
  exhaust valves, 299
  freshwater circuits, 306
  fuel circuit, 307–308
  fuel injection, 302
  fuel injector, 303
  fuel pumps, 303
  general characteristics, 292
  governor, 304
  lubrication, 307
  piston rings, 298
  pistons, 298, 307
  running gear, 298
  safety devices, 303–304
  seawater circuit, 306
  starting air circuits, 308
  valves, 299–301
Goodman diagram, 515
Götaverken engines, 198–213
  850/1700 VGA-U, 198
  850/1700 VGS-U, 199
  camshafts, 204–205
  control system, 213
  cooling water systems, 208–209
  crosshead, 201
  cylinder cover, 203–204
  cylinder liners, 202–203
  entablatures, 199–200
  exhaust gas manifold, 210
  fuel pumps, 205–206
  fuel system, 209–210
  fuel valves, 206, 210
  lubrication, 208
  manoeuvring system, 211–213
  pistons, 201–202
  starting air system, 207–208
  turbochargers, 210–211
  VGS-U, 198
Governors, 131, 304, 340, 463
Graviner oil-mist detector, 496
Gunmetals, 565
  continuous casting, 565

Hanshin engines,
  6LUS40, 360
  6LUS54, 360
Hearing loss, 25
Heat cycle, 1
Heat evolved by combustion, 421
Hedemora Verstäder, 400
High-speed engines, 225, 374–408, 438
  European designs, 377
  principles of, 374–375
  representative selection, 377
Hooke's Law, 508
Horsepower, 11–13, 32, 41

Ignition problems, 432
Indicated mean effective pressure, 12
Indicated power, 8
Indicated thermal efficiency, 7
Indicator cards,
  2-stroke engines, 441–442
  4-stroke engines, 442–443
Indicator diagram, 3, 4, 12
Indicator valve, 127
Indirect drive, 29–35
Inspections, 453
Installation, engine, 543
Irregularity of running, 433

Jacks, 544
Japanese engines, 359–367, 401–403

Knocking, 432–433

Lanchester balancing system, 24, 379

Lead alloys, 561, 562
Lead-bronze bearings, 565
Liquid nitrogen, 520
'Long stroke' engine, 28
Lower calorific value, 7
Lubricating oil, 35, 485–486
Lubricating oil consumption, 31
Lubricating oil flow, 532
Lubricating oil system, 552
Lubrication, 183–184, 307, 340, 356–357
  bearings, 529–530, 560
  cylinder, 93, 132, 208, 289, 327, 459
  gearbox, 217
  gears, 528
  hydrostatic, 144
  monitoring, 448–449

Maintenance,
  bearings, 461–472
  cylinder jackets, 460
  cylinder liners, 456, 460
  engine, 30, 31, 32, 35, 223–224, 236–238, 258, 270–272, 290–291, 358, 422, 424, 452–486
  fuel injection pumps, 475–479
  nozzles, 470–472
  piston rings, 456–458
  pistons, 455
  slow-speed engines, 454
  starting gear, 480–482
  turbochargers, 480
MaK engines, 310–327
  320 mm bore, 311
  450 mm bore, 316
  580 mm bore, 322–325
  camshaft, 325
  combustion chamber, 324
  connecting rods, 314, 320
  crankshaft, 320, 323
  cylinder block, 320
  cylinder head, 314, 318, 324
  cylinder liners, 311, 323
  exhaust valves, 321
  in-line, 311, 316, 318, 326
  lubrication, 327
  M35, 310, 325–327
  M332, 390
  M332 AK, 390
  M451, 311
  M453, 311, 312, 313, 325
  M551, 316, 317, 321, 325
  M552, 318–321
  M601, 322, 323
  pistons, 314, 318, 321, 324
  technical data, 310
  valve cages, 324
  valve seat, 314–315

valves, 315, 324
Vee, 311, 317–329
MAN 2-stroke engines, 138–152
  camshaft drive, 145
  combustion chamber, 145
  cylinder liners, 146
  fuel injection, electronically controlled, 150–152
  K3EZ 50/105C/CL, 152
  K8SZ 90/160, 141
  K8SZ90/160B/BL, 140, 142
  KSZ90/160, 139
  KSZ-A, 138, 146
  KSZ-B, 138, 140, 142
  KSZ-C, 138, 142, 145
  KSZ C/CL, 140
  performance curves, 141
  piston and piston rod, 147
  piston cooling, 147
  running gear, 144
  turbocharging, 146
MAN 4-stroke engines, 239–258
  32/36, 240–254
  ASV 25/30, 392
  camshaft, 245, 253
  connecting rods, 242, 248, 256
  cylinder head, 244, 250
  cylinder liners, 241, 243, 257
  fuel injection, 253–254
  L20/27, 391
  L40/45, 245
  L52/52, 254–258
  maintenance, 258
  pistons, 243, 249
  turbocharging, 245, 252
  valves, 251–252
  Vee, 239
  V52/52, 254–258
  vibration damper, 247
  VV40/54, 239
  VV52/55, 239
Manoeuvring devices, 211
Manoeuvring gear, 482–483
Manoeuvring system, 132
Materials, 543–554
Maximum rating, 37–39
Mean effective pressure, 12, 41
Mean piston speed, 14
Measurement, crankshaft alignment, 548, 549
Measurements, reaction, 545
Mechanical efficiency, 8
Medium-speed engines, 29–31, 34, 35, 214–224, 259, 273, 292, 343, 359, 422, 438, 439, 455, 468–469, 482, 566
Microbial attack,
  precautionary measures against, 540
  remedial measures for, 540

INDEX

Microprocessor, fuel injection system, 150
Mirrlees Blackstone engines, 343–358
  bearings, 354
  camshaft, 353
  connecting rods, 355
  control system, 357–358
  cooling system, 357
  crankshaft, 354
  cylinder heads, 351
  cylinder liners, 348
  E, 377
  E Mark 2, 379–381
  ESL Mark 2, 381
  fuel injection, 357
  in-line, 343, 347
  K Major, 343
  K Major Mark 2, 343
  K Major Mark 3, 347
  lubrication, 356–357
  maintenance, 358
  MB190, 400
  MB275, 393
  pistons, 349–351
  technical data, 343
  turbocharging, 357
  valve gear, 353
  valves, 351–353
  Vee, 343, 347
Mitsubishi engines, 153–163
  combustion chamber, 159
  connecting rod, 158
  crosshead, 158
  exhaust and scavenging, 158
  frame construction, 157
  turbochargers, 69
  two-stage turbocharged, 162–163
  UEC-E, 153, 155, 156, 162, 163
  UEC85/180E, 161
  UEC-H, 153, 155, 156
  UEC-HA, 155
  6UEC45/115H, 160
  6UEC52/125H, 153
  UE-H, 163
  UEV42/56C, 363
Mitsui engines,
  42M, 364–367
  60M, 364–367
Moment compensator, 131
Monitoring, 422, 424, 433–452
  condition, 435
  continuous, 449
  cooling system, 447–448
  cylinder pressure, 435–441
  data, 433–435
  exhaust temperature, 445–447
  lubrication, 448–449
  techniques, 223
  turboblowers, 447
  vibration, 451
Mountings, anti-vibration, 26
MTU V652TB, 399
MWM 440 and 441, 393
MWM TBD510, 368
MWM TBD511, 368

Napier turbochargers, 67–69
Niigata engines, 367
  28BX, 402, 404
Nohab F20 engine, 394
Noise, 24–26, 35
Noise reduction, 25
Noise regulations, 223
Noise screening, 26
Non-ferrous components, 558
Normo LDM engine, 384
Nozzles,
  dewpoint effects, 534
  maintenance, 470–472

Oil consumption, 31, 532
Oil coolers, 484
Oil filters, 483
Operating problems, 487–542
Operating procedures, 424–433
Opposed-piston engine, 13
Optical aids, 452
Optional alignment checks, engine, 545
Organic materials, 559
Overhaul, 30, 452–486

Paxman RPH engine, 398
Performance aspects, 36–51
Performance curves, 37, 141
pH effects, 536, 537
Pipe, high-pressure, 479
Piston cooling water system, 551
Piston rings, 170, 228, 298, 333
  maintenance, 456–458
Piston speed, 39
Pistons, 94, 114, 128, 170, 177, 180, 184–186, 193, 201–202, 228, 243, 249, 264–267, 285–287, 298, 307, 314, 318, 321, 324, 333, 349–351
  aluminium, 543
  cast iron, 557
  maintenance, 455
  steel crowns, 557
Plastics, glass reinforced, 558
Ports, 196
Pour point, 418
Power, 12, 13, 33
Power build-up, 45
Power distribution, 32
Pressure charging, 52–69

574 INDEX

Pressure charging (cont.)
  2-stroke engines, 55–56
  4-stroke engines, 53–54
Pressure ratio, 447
Pressure sensors, 450
Propeller blade resistance, 19
Propeller design, 44
Propeller efficiency, 6, 44
Propeller law, 42–43
Propeller performance, 44–45
Propeller shaft alignment, 545
Propeller size, 215
Propeller slip, 41–42
Propeller speed, 214, 215, 220
Propellers,
  fixed pitch, 219, 220
  locking, 48
  total shaft horsepower at, 45
  trailing, 48
  variable pitch, 6, 218, 219
Propulsion data, 46, 47
Protection, 422
Proximity sensors, 449
PV diagram, 41

Quadruple-screw ship, 49
Quasi-propulsive coefficient, 45

Reaction measurements, 545
Reciprocating forces, 24
Reciprocating masses, 23
Refrigerated vessels, 375, 376
Reliability, 31, 424
Repair, 30
Resonance frequency, 21
Reversible engines; see Engine reversal
Reversing gear, 131
Ricardo pre-chamber, 73
Rocker arrangement, 338
Rotocap, 335
Rubber seals, 559
Running gear, 169, 195–196, 298
Running in, 458–459
Ruston engines,
  AP230, 379
  6AP230, 380
  AT, 372
  AT350, 372
  RK, 377
  RKC, 378

SACM engines, 394
  MGO, 398
Safety devices, 269–270, 303–304
Safety legislation, 223
Safety valve, 127
Salt water contamination, 537

Sankey diagram, 5, 6
Scavenge belts, 464, 497
Scavenging, 39, 54, 57–59, 108, 109, 133, 155, 158, 164, 179
Schwermaschinenbau VD26/20 engine, 384
Sea trials, 36, 45, 49, 51
Seals, 548
Seatings, engine, 543
Seawater circuit, 306
Seawater corrosion, 536-537
Seawater system, 550
Seizures, 507–508
SEMT Pielstick engines, 225–238
  combustion chamber, 74
  connecting rod, 228
  crankcase, 232
  crankshaft, 227–228
  cylinder cover, 230
  cylinders, 229
  exhaust system, 235
  exhaust valves, 231
  fuel injection, 233
  fuel pump, 233
  general characteristics, 225
  maintenance, 236–238
  PA, 225
  PA4-185, 398, 399
  PA6, 395
  PC1, 225
  PC2, 225, 227, 228, 230
  PC2-5, 225, 227–30, 232, 233, 234, 236
  PC2-6, 225, 236
  PC3, 225, 232
  PC4, 225, 233, 234, 236, 237
  PC4-2, 227
  piston rings, 228
  pistons, 228
Servicing, 424
Shallow water, 428
Ship stopping data, 51
Ship stopping trials, 50
Shock absorbers, maintenance, 479
Slow-speed engines, 86, 134, 138, 153, 164, 176, 438, 454
Smit-Bolnes engines, 372
S-N curve, 498–500, 502
Sound absorbing materials, 25
Sound pressure level, 24
Specific gravity, 418
Spectroscopic analysis, 450–451
Speed control, 340
Speed/power curves, 48, 49
Spill port closure, 83
Starter system, 342
Starting air system, 134, 189, 207–208, 308
Starting gear, maintenance, 480–482
Starting preparations, 425–426

INDEX 575

Starting problems, 431
Starting procedure, 426–427
Starting valve, 105, 127, 128, 189
Steel components, 543–546
Steel frames, 556
Stoichiometric mixture, 492
Stopping procedure, 49, 429
Stork Werkspoor engines, 259–272
   connecting rods, 264, 265–267
   crankshafts, 263, 264
   cylinder block, 264
   cylinder head, 267–268
   cylinder liners, 267
   exhaust valves, 268
   fuel injection, 269
   main bearings, 262
   maintenance, 270–272
   pistons, 264–267
   safety devices, 269–270
   TM410, 259, 260, 262, 265–268, 270
   TM620, 259, 261, 262, 265, 266, 268, 271
   valve gear, 230, 237, 268, 269
   valves, 267–268
Strain gauge bridges, 547
Stuffing box, 126
Suction stroke, 2
Sulzer 2-stroke engines, 86–107
   combustion chamber, 90–93
   controls, 96–98
   crankshafts, 89, 100
   cylinder lubrication, 93
   fuel pumps, 103–107
   piston assembly, 94
   RD, 86
   RL, 86–98
   RLA, 86
   RLA56, 86, 89 90
   RLA90, 91
   RLB, 86
   RLB90, 87
   RND, 86
   RND-M, 86
   RTA, 86, 98–103
   RTA84, 101
   turbocharging, 94–95
   under-piston supercharging, 95
   variable injection timing, 106–107
Sulzer 4-stroke engines, 273–291
   A25, 273
   AL20/24, 396, 397
   AS25/30, 392, 396, 397
   auxiliary pumps, 289
   camshafts, 281
   connecting rods, 283–284
   crankcases, 275
   crankshafts, 282–283
   cylinder block, 275

   cylinder head, 277
   cylinder liners, 277
   cylinder ratings, 273
   fuel injection, 280–281
   fuel-injection pump, 281
   fuel-injection valve, 280
   fuel pumps, 281
   lubrication, 289
   main bearings, 283
   maintenance, 290–291
   pistons, 285–287
   turbocharging, 287
   valves and valve gear, 279
   Z40, 273, 289
   Z40, 277, 279, 285
   6Z40, 283
   ZA, 279, 288
   ZA40, 273, 278, 281, 282, 283, 285, 287, 289
   ZA40, 285
   ZAL40, 276
   ZAL40, 274
   ZAV, 288
   ZL40, 276
   ZL40, 274
   8ZL40, 275
   ZV40, 287
   ZV40, 274, 276
Supercharged engines, 486
Supply vessels, 33
SWD DRO210K engine, 398
SWD F240 engine, 383
Synchronous speeds, 376

Temperature sensors, 450
Theoretical heat cycle, 1
Thermal efficiency, 6–7
Thrusters, 33, 34
Time between overhauls (TBO), 424
Timing of cylinder lubricators, 459
Tin alloys, 561
Tin-bronze bearings, 564
Tin-lead alloys, 561
Tools, 454
Top dead centre, 2, 3
Torque, 13–14
Torsional critical speed, 21
Torsional vibration; see Vibration
Torsionmeter, 6, 41
Turboblowers, 52–69
   characteristics, 60
   matching, 60
   monitoring, 447
   surge, 61
Turbocharged engines, 52–69, 446
Turbochargers, 52, 103, 133, 210–211, 340, 376
   Brown Boveri, 63–67, 325
   maintenance, 480

Turbochargers (cont.)
  Mitsubishi, 69
  Napier, 67–69
  repair, 541
  Super MET, 155, 158, 163
  types, 61–71
Turbocharging, 39, 52–69, 118, 311, 357
  constant pressure, 55, 62, 94, 109, 121–122, 140, 149, 155, 190–191, 195, 245, 252
  cyclic variations, 56
  impulse, 62, 109
  injector drive, 146
  multi-pulse, 63
  pulse, 53, 55, 62, 287
  two-stage, 162–163
Turning gear, 132
Turning wheel, 132
Twin-screw vessel, 48, 49
Two-stroke engines, 10–11, 16, 31, 37, 55–56, 86, 98, 108, 110, 138, 153, 176, 198, 222, 273, 368, 372, 400, 441–442, 454, 457, 466–468

'Ultra long stroke' engine, 134
Unbalanced secondary force, 24
United States engines, 403–407
Unmanned operation, 424

Valve cage, 251–252, 324, 335–337
Valve guides, 252
Valve seat, 314–315
Valves and valve gear, 126, 127, 230, 237, 251–252, 267–269, 279, 299–301, 315, 324, 334–339, 351–353, 557
Vibration, 15–22, 35
  axial, 22
  control of, 22, 23

Vibration (cont.)
  forced, 15
  forms of, 491–492
  linear, 22
  most significant masses, 21
  resonant, 15, 492
  sympathetic, 25
  three-node, 489
  torsional, 15, 16, 22, 487, 489–492
  two-node, 489
  see also Noise
Vibration dampers, 247, 490
Vibration monitoring, 451
Viscosity, 415

Ward/Leonard system, 221
"Wartsila" engines, 397
Waste heat recovery, 29
  system, 553
Watch keeping, 427–429
Water content of fuels, 418
Water vapour effects, 533
Waukesha VHP engine, 408
Wear sensors, 450
Weldability, 566
Welding repairs, 542
White metal, 529, 560, 561
Wichmann engines, 368–370
  AXA, 370
  VX, 400
Windage, 4
Woodward PGA58 pneumatic governor, 96

Yammar engines,
  GL, 402
  T220KL, 402
  ZL, 402